On Dreams and the East

A list of Jung's works appears at the back of the volume.

Figure 1. Carl Gustav Jung and Heinrich Zimmer at Eranos, Ascona (Switzerland) in August 1938. Ph. Margharethe Fellerer. © Eranos Foundation, Ascona.

On Dreams and the East

Notes of the 1933 Berlin Seminar

C. G. JUNG AND HEINRICH ZIMMER

EDITED BY GIOVANNI V. R. SORGE

Translated by Mark Kyburz and John Peck

PHILEMON SERIES

Published with the support of the Philemon Foundation
This book is part of the Philemon Series of the Philemon Foundation

PRINCETON UNIVERSITY PRESS
PRINCETON AND OXFORD

Requests for permission to reproduce material from this work should be sent to permissions@press.princeton.edu

Published by Princeton University Press

41 William Street, Princeton, New Jersey 08540

99 Banbury Road, Oxford OX2 6JX

press.princeton.edu

Figures 3 and 4 are reproduced by permission of the Jung Foundation.

All Rights Reserved

ISBN 9780691250557

ISBN (e-book) 9780691250540

British Library Cataloging-in-Publication Data is available

Editorial: Fred Appel and James Collier

Production Editorial: Karen Carter

Text and Jacket Design: Wanda España

Production: Erin Suydam

Copyeditor: Francis Eaves

Jacket Image: CPA Media Pte Ltd / Alamy Stock Photo

This book has been composed in Sabon LT Std

Printed in the United States of America

10 9 8 7 6 5 4 3 2 1

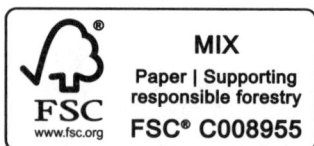

Contents

Acknowledgments   vii

Editorial Note   xi

Abbreviations   xiii

Introduction, by G.V.R. Sorge   1

  1. Special Features of the Seminar   1

  2. Germany, 1933: A Bird's-Eye View   6

  3. "Where danger is . . .": The (Obscure) Background
     of the Seminar   12

  4. The Seminar, the Radio Berlin Interview, and the
     Society for Psychotherapy   16

  5. The Participants and the Analysand   30

  6. Fear, Amazement, and Trembling   33

  7. Alchemical Germination, and Nostalgia for the Center   37

  8. West versus East, or the Berlin Seminar as a Phase in the
     Making of a Complex Psychology   40

  9. Heinrich R. Zimmer: A Biographical Profile   47

10. Amazed and Amused: Impressions from the
     Jung–Zimmer Encounter   53

11. Yoga, Individuation, and Western Psychotherapy
     According to Zimmer   57

12. Mandala, Yantra, and the "Machine": Zimmer's Influence
     upon Jung   62

13. Concluding Reflections   72

HEINRICH ZIMMER: ON THE PSYCHOLOGY OF YOGA

Editor's Note to Zimmer's Lecture   79

On the Psychology of Yoga   81

THE BERLIN SEMINAR OF C. G. JUNG

Transcriber's Note [original manuscript]   125

Transcriber's Note, by Ernst Falzeder   125

Lecture 1   Monday, 26 June 1933   127

Lecture 2   Tuesday, 27 June 1933   157

Lecture 3   Wednesday, 28 June 1933   188

Lecture 4   Thursday, 29 June 1933   212

Lecture 5   Friday, 30 June 1933   245

Lecture 6   Saturday, 1 July 1933   279

Index   309

Acknowledgments

FIRSTLY, I wish to thank the Foundation of the Works of C. G. Jung and the Board of the Philemon Foundation for making this edition possible.

I also wish to express my gratitude to Sonu Shamdasani for his invaluable help and suggestions in the intriguing and sometimes challenging task of editing the manuscripts of Jung's Seminar and Heinrich Zimmer's lecture. During the process of translating the editorial apparatus, Mark Kyburz and John Peck provided several suggestions and insights for which I am grateful; and a particular acknowledgment goes to John Peck for his assistance (supported by Professors Brian Swannand and Paul G. Zolbrod, and Dr. Jerome Bernstein) in my research into a Navajo Indian primary source quoted by Jung, which unfortunately has remained unidentified. I am indebted too to Ernst Falzeder and Martin Liebscher for their unfailing friendly support in answering my questions.

I wish to thank both Ulrich Hoerni and Andreas Jung for their willingness and cooperation in answering specific questions related to primary and secondary sources at the JA and JFA. The Stiftung der Werke von C. G. Jung authorized access to unpublished sources from the JA (our thanks to Thomas Fischer and Carl Jung), and Andreas Jung kindly allowed me to consult a Chronicle of the C. G. Jung Society of Berlin, Jung's personal copy of the Berlin Seminar, and other books from Jung's library in Küsnacht, including his copy of Zimmer's *Kunstform und Yoga im indischen Kultbild*. On behalf of the Philemon Foundation I would also like to thank the Eranos Foundation for permission to reproduce Margarethe Feller's photo of Jung and Zimmer at Eranos in August 1938.

I wish to express my appreciation to the staff of the Special Collections Department for their professional and considerate assistance during my archival researches at the JA. My particular thanks go to the director of the ETH Zurich University Archives, Christian J. Huber, and to the former JA archivist Yvonne Voegeli, as well as to the librarian Marion Wullschleger.

I also am indebted to the DLA (Deutsches Literaturarchiv-Marbach) for permission to reproduce a number of unpublished documents from the literary estate of Zimmer's long-term companion Mila Esslinger-Rauch. A sincere acknowledgment is due to the Indologist Markus Schuepbach, of the Indo-Germanic Seminar, University of Zurich, for his constant willingness to discuss philological and philosophical issues in Zimmer's lecture. Further consultation on the same was provided by Professor Karin Stüber, also of the Indo-Germanic Seminar at the University of Zurich; and in a more recent phase by Elisa Ganser, currently a lecturer at the Institut für Indologie und Tibetologie, LMU Munich.

Others who provided information on archival sources are referred to in the footnotes. Here I wish to thank the archivists at the Wheelwright Museum of the American Indian, Santa Fe, and the the American Museum of Natural History in New York; John Pack, assistant librarian and archivist at the Center of Southwest Studies, Durango, Colorado; and his colleague, Associate Professor Elayne Silversmith of the Delaney Southwest Research Library at the Center.

I am grateful to Julie Sgarzi, Elise Collins Shields, Safron Rossi, and Bob Walter, for information on Zimmer and the Joseph Campbell Estate at the Pacifica Graduate Institute, as well as to Richard Buchen and Lorna Paechin for providing a copy of "Some Biographical Remarks about Henry R. Zimmer." Special thanks go also to Luca Baschera for his valuable insights into some specific theological questions treated in the Seminar. Claudio Bonvecchio, Giovanni Casadio, Attilio Mastrocinque, Bernardo Nante, Robert Wimmer, Lance Owens, and Jörg Rasche congenially supplied timely hints for parts of my research.

My gratitude goes also to the late Lukas Rauch who, during an insightful conversation in Winterthur, kindly shared memories and impressions of his father Heinrich Zimmer, as well as comments around the scholarly reception of his work. To him goes our gratitude for permission to include Zimmer's lecture in this edition. The late Ximena de Angulo Roelli, with her warm hospitality and inimitable style, shared during several conversations in Cavigliano (Ticino) further impressions of Zimmer. Indirectly, she led me to a distant time: one that she and her mother, Cary de Angulo Baynes, had known at first hand.

The numerous published works, both by scholars and other authors, which proved indispensable in editing both Jung's Seminar and Zimmer's lecture are of course acknowledged in the apparatus. Nonetheless, I wish to express here too my sincere gratitude for their work.

My thanks go also to Russell P. Holmes for assistance with passages in ecclesiastical Latin, and to Thomas Fischer (Stiftung der Werke von C. G. Jung) for providing additional information, as well as to the late Donatella Buonassisi for rereading some of the final passages. Last but not least, I'm grateful to Francis Eaves from Princeton University Press for his very careful and patient assistance in the revision of the final text.

Editorial Note

TERMS IN THE LECTURE TEXTS which had originally been incorrectly transcribed have been silently corrected, to an extent by reference to the several but not exhaustive corrections (in particular of Latin terms and expressions) that Jung made to his personal copy of the Seminar. Unless otherwise indicated, non-English sources for which no authoritative translations are available have been translated by Mark Kyburz and John Peck, in collaboration with the editor. For economy, clarity, and consistency, the simplest possible English transcriptions of Sanskrit terms (i.e., without diacritics) have been adopted (other than in quotations and titles of cited works). All evidence within quotations (or quoted works) are by their author.

Abbreviations

PRINT SOURCES

CW
The Collected Works of C. G. Jung, edited by Sir Herbert Read, Michael Fordham, and Gerhard Adler, translated by R.F.C. Hull, executive director William McGuire, 20 vols and 2 supplementary vols (Bollingen Series XX) (New York: Pantheon; 1953–66; Princeton, NJ: Princeton University Press, 1967–1979

DAS (footnotes); *Dream Analysis* (text)
C. G. Jung, *Dream Analysis: Notes of the Seminar Given in 1928–30*, edited by William McGuire (Bollingen Series XCIX) (Princeton, NJ: Princeton University Press, 1984)

MDR
C. G. Jung and Aniela Jaffé, *Memories, Dreams, Reflections*, recorded and edited by Aniela Jaffé, translated by Richard and Clara Winston, revised edn (New York: Vintage Books 1989 [1963]; orig. Ger. edn: *Erinnerungen, Träume, Gedanken* [Zurich and Stuttgart: Rascher, 1962])

PL
Patrologia Latina, edited by Jacques-Paul Migne, 221 vols. (Paris, 1844–64)

ARCHIVAL SOURCES

APC
Archive of the Psychologischer Club (Psychological Club), Zurich

AZ
Archive of the Zentralbibliothek (Central Library), Zurich

DLA
Deutsches Literaturarchiv (German Literature Archive), Marbach (Sammlung Mila Esslinger Rauch)

HHA
Archive Harnack Haus, Berlin

JA   C. G. Jung-Arbeitsarchiv (C. G. Jung Papers Collection), Swiss Federal Institute of Technology Zurich, University Archives (Hochschularchiv der ETH or Eidgenössiche Technische Hochschule), Zurich; by courtesy of Ulrich Hoerni and Thomas Fischer, and subsequently Carl Jung

JFA   C. G. Jung-Familienarchiv (C. G. Jung Family Archive); by courtesy of Andreas Jung, and subsequently Susanne Eggenberger-Jung, Küsnacht-Zurich.

SECTIONS IN THE PRESENT VOLUME

[I]   Introduction

[MT]   Main Text (Jung's Seminar)

[ZL]   Zimmer's Lecture

On Dreams and the East

IN BERLIN TRAF EIN:

Dr. C. G. Jung (Zürich),

der bedeutendste Schweizer Psychiater und Psychologe, Begründer der Internationalen psychoanalytischen Gesellschaft, der im Vorjahr den Literaturpreis der Stadt Zürich für seine tiefschürfenden Analysen dichterischer Kunstwerke (besonders der Werke von Spitteler und James Joyce) erhielt. Jung ist nicht nur Theoretiker, er ist Arzt und gründlicher Kliniker. Als Mitarbeiter des bedeutendsten Schweizer Psychiaters seiner Zeit, E. Bleuler, machte er in den Jahren 1906—1910 das Assoziationsexperiment der Affekt- und Neurosenforschung dienstbar. Von der Freudschen Lehre trennt ihn die Auffassung, dass jeder Mensch sozusagen ein Museum ist, in dem Erinnerungen aus grauer Vorzeit aufbewahrt sind. Was er die unpersönliche Seele nennt, ist Niederschlag der unzähligen Existenzen unserer Ahnen. Jung begnügt sich nicht mit dem Unbewussten, das mit dem Individuum geboren wird und auch vergeht. Er nimmt ein überindividuelles, ein Kollektiv-Unbewusstes an, ein Menschheitsgedächtnis von mehreren Millionen Jahren, das säkuläre Träume gebiert.

Der Gelehrte ist Gast der Gesellschaft für analytische Psychologie.

Bohrturm in Flammen

Explosion auf dem Erdölgelände bei Boberg

PEINE, 26. Juni. (T. U.)

Am Montag vormittag explodierte auf dem Erdölgelände bei Boberg ein Bohrturm der Erdölbohrgesellschaft. Die Detonation war in einem Umkreis von 10 Kilometern vernehmbar. Es entwickelte sich eine Stichflamme, die bis zu einer Höhe von fünfzig Metern emporschlug, so dass der Bohrturm in Nu in Flammen stand, denen eine sehr starke Rauchentwicklung folgte. Die Feuerwehr stand dem Brande machtlos gegenüber. Der Bohrturm wurde in kurzer Zeit ein Raub der Flammen. Einige Arbeiter erlitten leichte Verletzungen. Man führt den Brand auf die Entzündung von Gasen zurück, die sich im Bohrturm gesammelt hatten.

Figure 2. Berliner Tageblatt, 27 June 1933. Staatsbibliothek Berlin, Preussicher Kulturbesitz.

Introduction

G.V.R. SORGE

1. SPECIAL FEATURES OF THE SEMINAR

When Jung opened the Berlin Seminar in June 1933, one month before his fifty-eighth birthday, he had probably not expected an audience of over one hundred and fifty. His fame by then was considerable, however, and the number of honors went on growing, not only in his own country—in 1932 the city of Zurich had awarded him its Literature Prize—but also abroad, particularly in Germany, Great Britain, the United States, and India, which a few years later would award him with three honorary degrees. The *annus horribilis* which saw, in January, the Nazi rise to power, was a crucial one in Jung's own life. In particular, a few days before arriving in the German capital with his wife Emma and some of his students,[1] he had officially assumed the presidency of the Allgemeine Ärztliche Gesellschaft für Psychotherapie, or General Medical Society for Psychotherapy, a burden whose professional and political implications he was still far from foreseeing. Nineteen thirty-three was marked by his intense activity as a lecturer abroad—and travels. In March he went with his friend Hans Eduard Fierz on a cruise ship to the eastern Mediterranean, visiting Italy, Malta, Turkey, Palestine, Egypt, Greece, and Rhodes.[2] In August, he lectured at the first Eranos conference in Ascona, Canton Ticino—the legendary cross-cultural project regarded by its founder, the Dutch artist and spiritualist Olga Fröbe-Kapteyn (1881–1962), as an "army of constructive forces [. . .] against the destructive ones which seem

---

[1] See Barbara Hannah, *Jung: His Life and Work: A Biographical Memoir* (New York: Putnam's Sons, 1976), 209ff.

[2] See Andreas Jung, "Carl Jung and Hans Fierz in Palestine and Egypt: Journey from March 13th to April 6th, 1933"; Thomas Fischer, "1933: The Year of Jung's Journey to Palestine/Israel and Several Beginnings," in Erel Shalit and Murray Stein, eds, *Turbulent Times, Creative Minds: Erich Neumann and C. G. Jung in Relationship (1933–1960)* (Asheville, NC: Chiron, 2016), 131–34 and 135–49.

to be ruling the world."[3] So from its very beginning Jung joined, with his fourteen lectures up to summer 1951, an international community of scholars and researchers unified by the wish to create a dialogue both among different disciplines and between East and West—a dialogue that matched profoundly Jung's deepest epistemological concerns—and that also informs the structure of the present seminar. Then in the winter semester of 1933 Jung began a successful cycle of lectures at the Eidgenössische Technische Hochschule (Swiss Federal Institute of Technology) in Zurich, which continued through 1941 and was dedicated to discussing his comparative psychology through a variety of topics including ecclesiastical history, Gnosticism, alchemy, comparative symbolism, and Eastern spirituality.[4] In so doing, he was resuming teaching after twenty years: he had resigned his academic post as a lecturer in the Faculty of Medicine at the University of Zurich in 1914, after his split with Freud, his concomitant distancing from psychoanalysis, and his descent into a pioneering "confrontation with the unconscious" which would give birth to the *Liber Novus (The Red Book)*.[5]

The year of the Berlin Seminar should be linked back to 1928, when Jung's own mythopoetic experiment reached a turning point thanks to the sinologist and missionary Richard Wilhelm's transmission to him of his translation of *The Secret of the Golden Flower*, a Taoist meditation manual on inner alchemy which to the two of them would publish jointly the following year.[6] As explained in the 1959 epilogue to *Liber Novus*, it was

---

[3] William McGuire, *Bollingen: An Adventure in Collecting the Past* (Bollingen Series CVIII) (Princeton, NJ: Princeton University Press, 1982), 24. On Eranos, in addition to McGuire's book, see particularly: Hans Thomas Hakl, *Eranos: An Alternative Intellectual History of the Twentieth Century*, trans. by Christopher McIntosh with the collaboration of Hereward Tilton (Bristol, CT: Equinox, 2013); Riccardo Bernardini, *Jung a Eranos: Il progetto della psicologia complessa* (Milan: Franco Angeli, 2011).

[4] C. G. Jung, *The History of Modern Psychology: Lectures Delivered at ETH Zurich, Volume 1: 1933–1934*, ed. and trans. by Ernst Falzeder (Philemon Series) (Princeton, NJ: 2019).

[5] C. G. Jung, *The Red Book (Liber Novus)*, ed. by Sonu Shamdasani, trans. by Mark Kyburz, John Peck, and Sonu Shamdasani (New York: W. W. Norton, 2009). See also its actual basis, C. G. Jung, *The Black Books, 1913–1932*, trans. by Martin Liebscher, John Peck, and Sonu Shamdasani, 7 vols (Philemon Series) (New York: W. W. Norton, 2020); and C. G. Jung and Aniela Jaffé, *Memories, Dreams, Reflections*, recorded and edited by Aniela Jaffé, trans. by Richard and Clara Winston, revised edn (New York: Vintage, 1989 [1963]; henceforth *MDR*), ch. 6; Henri F. Ellenberger, *The Discovery of the Unconscious: The History and Evolution of Dynamic Psychiatry* (New York: Basic Books, 1970), 447–48, 672–73, 889–91.

[6] C. G. Jung and Richard Wilhelm, *The Secret of the Golden Flower: A Chinese Book of Life* (London: Routledge & Kegan Paul [1929], 1931 [2nd edn 1962]), CW 13.

this treatise that thirty years earlier had eventually lead him to turn to the study of alchemy, in order to confirm his intuitions about the existence of a universal, symbolic psychic structure witnessing a fundamental development process toward unity within the human soul. According to Sonu Shamdasani, "for nigh on forty-five years, in ways that are only now beginning to come clear, he tried to envisage a transformation of psychotherapy and a science of complex psychology that in critical respects were tributaries to his work on the parallel opus of *Liber Novus*."[7] So, after his "confrontation with the unconscious," Jung commenced a "confrontation with the world,"[8] through growing commitments to sharing his thoughts, lecturing, and publication. In 1928 the seminal *The Relations between the I and the Unconscious* was published, a "little book" meant to summon "twenty-eight years of psychological and psychiatric experience," which Jung would later regard as closely tied to his "Commentary on the *Secret of the Golden Flower*" (1929).[9] Equally, in 1928 Jung attended for the first time the (third) Psychotherapy Congress in Baden-Baden; and last but not least, in the fall of that year he started, at the Zurich Psychological Club, a Dream Analysis Seminar (published as *Dream Analysis: Notes of the Seminar Given in 1928–30* [1984], henceforth referred to as *Dream Analysis*; in notes, *DAS*) which lasted through spring 1930, devoted to amplifications and explorations around a long oneiric series of a patient—more precisely, an analysand—of his: the one who also features in the Berlin seminar.

This leads us to indicate the two special features of the Berlin Seminar: first, that it focuses on five dreams which Jung had dealt with in the Dream Analysis Seminar;[10] and second, that it is introduced by the

---

[7] Sonu Shamdasani, "After *Liber Novus*," *Journal of Analytical Psychology* 57 (2012): 364–77 at 365.

[8] Sonu Shamdasani, "Introduction: *Liber Novus* or 'The Red Book' of C. G. Jung," in C. G. Jung, *The Red Book (Liber Novus)*, ed. by Sonu Shamdasani, trans. by Mark Kyburz, John Peck, and Sonu Shamdasani (New York: Norton, 2009), 218.

[9] As briefly but conspicuously noted in his 1935 preface to the second edition: "The reader will find a development of the last chapter in my commentary to the *Secret of the Golden Flower*" ("Preface to the Second Edition" [of *The Relations between the I and the Unconscious*], CW 7, 123–25 at 124). Jung states a continuity between the two publications: more precisely, between the "last chapter" of the 1928 book—notably dedicated to the "mana personality," and including a discussion on the Self only in its very last pages—and his "commentary" to the Taoist manual, where the Self instead rises to crucial prominence. See also below [ZL], n. 304.

[10] That seminar's final session on June 25, 1930 preceded the beginning of the Berlin Seminar by three years and one day. The above-mentioned five dreams were treated in twelve sessions (from November 7, 1928 to February 28, 1929).

Sanskritist Heinrich Zimmer's lecture "On the Psychology of Yoga." Hence one may wonder,

1. Why take up again the initial dreams from a series already analyzed in a former seminar?
2. Why decide to inaugurate the Berlin Seminar with Zimmer's lecture without supplying any systematic commentary on it?

With regard to the first question, Jung declares that the dreams considered aim at showing "what the path that leads to inner self-development looks like at its inception."[11] In fact the oneiric imagery, once activated by the therapeutic relationship, seems to anticipate the development of a psychic process in an entelechial manner (incidentally, this might be considered consistent with Jung's conviction of the prospective tendency of children's dreams with respect to the unfolding of their subsequent life, a perspective manifestly indebted to the recapitulation theory according to which ontogeny, or the development of an organism, recapitulates phylogeny, or its evolutionary story).[12] Indeed in Berlin, Jung chose to focus on a handful of significant dreams well suited to illustration of the transpersonal aspects of the challenging, labyrinthine journey through the individuation process. Hence the seminar topic allowed him to reconsider a specific oneiric segment in light of his new explorations—including his encounter with Zimmer's work—and give us the chance to take stock of the ongoing development of Jung's hermeneutics, and involvement with Oriental thought.

As for the link between a seminar on dreams and a lecture on the psychology of yoga, Jung likewise gives us an indication at the outset, when he invites the audience to bear in mind the "highest perfections"[13] in Zimmer's description of millennial Hindu efforts to relink the *individuum* to the divine. In contrast to the Eastern perspective outlined by his Sanskritist friend, he expressly adopts an earthier—more chthonic—approach, considering Western psychology still to be occupied with the "beginnings":[14] a perspective that was already familiar, for instance, to the participants in the seminar on the psychology of the Kundalini yoga held together with

---

[11] See below [MT], 129.

[12] See C. G. Jung, *Children's Dreams: Notes from the Seminar Given in 1936–1940 by C. G. Jung*, ed. by Lorenz Jung and Maria Meyer-Grass, trans. by Ernst Falzeder with Tony Woolfson (Philemon Series) (Princeton, NJ: Princeton University Press, 2008), 1; cf. Sonu Shamdasani, *Jung and the Making of Modern Psychology: The Dream of a Science* (Cambridge: Cambridge University Press, 2003), 157–58 and 300.

[13] Below [MT], 128.

[14] Ibid.

Jakob Wilhelm Hauer the previous winter.[15] So Jung sets up his argument in integrative counterpoint to Zimmer's exposition, pointing to the "inexperienced" youth of medical psychology (however impressive its epistemology), which had barely worked itself free of its bonds to philosophy and theology—having only begun, we might say, to penetrate the Baudelarian "symbolic forest" of the realms of the unconscious.[16] And the *via regia* to access it, no differently than for Freud, was the exploration of dreams.

This double specificity (a unicum with respect to Jung's other seminars) which inserts, so to speak, an old *fabula* in a new structure, has been taken into account in the layout of the present edition. Compared with the seminar on the psychology of Kundalini yoga, in which Jung provided a thorough commentary on the exposition of Jakob Hauer, Zimmer's overture to the Berlin seminar relates to it by way of a lighter, more suggestive connotation. His lecture, already notable for a good deal of psychological sensitivity, draws on the *Yoga-sutras*, the *Bhagavadgita*, and Tantric texts, thus providing an inspiring background scenario for the Jung's subsequent analysis.

On the top of that, Zimmer's contribution is germane to Jung's epistemological understanding of, as he termed it in the 1930s, a "complex psychology," hence making of this seminar a significant step in the development of a metapsychology grounded upon a comparative, interdisciplinary, cross-cultural base. By expanding its clinical and epistemological tenets, Jung was looking to de-provincialize Western psychotherapy, by means of a dialogic approach toward a variety of disciplines (from among both the natural and the human sciences), worldviews, and different cultural traditions—as Eastern spirituality was at that time considered. In this dialogue with the widest range of differing declensions of "otherness," he

---

[15] See C. G. Jung, *The Psychology of Kundalini Yoga: Notes of the Seminar Given in 1932*, ed. by Sonu Shamdasani (Bollingen Series XCIX) (Princeton, NJ: Princeton University Press, 1996): e.g., 65–67, and Hauer's English lecture, October 8, 1932, in ibid., Appendix 3, 108–10.

[16] One can even think of the unusual relation between micro- and macrocosms underpinning a text such as the *Brihad-Aranyaka-Upanishad* (5.5.3–4), one of the earliest Upanishads about the Atman, in which the realms of "Earth," "Mid-heaven," and "Heaven" (*bhuh, bhuvas, svah*) referred to in the hymnody of the Gayatri Mantra (see below [ZL], nn. 324 and 325) are correlated respectively with the head, the arms, and the feet of the person. It has been noted that "[t]his hints at an archaic teaching about the human being springing from Heaven (involution) rather than from an earthly womb (evolution). The work of Yoga consists in finding our feet, or roots, in Heaven" (Georg Feuerstein, *The Deeper Dimension of Yoga: Theory and Practice* [Boston, MA: Shambala, 2003], 301). Cf. "The Philosophical Tree" [(1945) 1954], *CW* 13, esp. chs 11, 12, and 19.

nonetheless did not seek a generalizing syncretism, but rather maintained a rootedness within the historical (and psychohistorical) tenets of the West. Such was, with all its strengths and weakenesses, lights and shadows, Jung's specific, differentiated approach to "the East."

With this seminar in the German capital Jung was furthermore giving a signal to the country he considered crucial to the development of psychoanalysis and psychotherapy. In this he found himself in line with Freud, who in 1920 had chosen to found the first Psychoanalytic Institute, which would serve as model for the subsequent training institutes (also those for Jungians), not in his native Vienna, but in Berlin. Germany indeed was the land which a few years before had seen the founding of the first official association intended to represent all the various strands of psychotherapy. It was also, however, the land which was about to give birth to an unparalleled totalitarian regime.

2. GERMANY, 1933: A BIRD'S-EYE VIEW

The year 1933 falls in the midst of a period of dramatic sociopolitical turmoil. Among the consequences of the Great Depression that stemmed from the financial crash of 1929, ending the so-called Roaring Twenties, was an exacerbation of the nationalistic, militaristic, and authoritarian tendencies which, in the wake of Italian fascism, were emerging across a swathe of European countries. Between the two world wars, in Hungary, Poland, the Baltic republics, Yugoslavia, Greece, Albania, Spain, Romania, and Portugal, liberal and democratic regimes were supplanted by authoritarian ones. To these should be added, of course, the case of communist revolutionary Russia, which fascist systems aimed to counteract. On January 30, 1933 Adolf Hitler, riding the wave of the resentment that had hamstrung the fragile Weimar republic (fueled by a sense of injustice at the Versailles treaty and concomitant discredit for the "Allied conspiracy"), took power in Germany. After his advent as chancellor, propaganda blaming the communists for the Reichstag fire on February 27 and a vote giving him full powers on March 23 provided the context to foster an increasing dehumanization of all "racial" and political opponents. The insertion of the "Aryan paragraph" in the Law for the Restoration of the Professional Civil Service enacted that April finalized the expulsion of "non-Aryan" citizens, especially Jews, as well as political enemies such as social democrats and communists, from any positions of public and state authority. Subsequent laws, among which was one against the reconstitution of political parties, enacted on July 14, completed the Nazi Party's totalitarian conquest of power: by then the regime corresponded to the state. Its campaign of race propaganda, intensified by the promulgation of the Nuremberg Laws on September 15, 1935, would eventually culminate in the extermination camps. In addition to the extermination of approximately six million Jews, around five million non-Jewish civilians were murdered, among them Roma, Poles, Jehovah's Witnesses, homosexuals, other supposed "anti-socials," and a host of intellectuals and other opponents of various kinds.

The scale of the enormity that was to be perpetrated, first in Germany itself and subsequently across its occupied territories, goes beyond human understanding and imagination. The Holocaust would become an unprecedented emblem of evil, and a watershed in the history of Western civilization. "To write poetry after Auschwitz," Theodor Adorno declared in his *Negative Dialectics* (1966), "is barbaric"; while the very notion of theodicy was challenged by intellectuals and survivors such as Primo Levi—see his *The Drowned and the Saved* (1986)—or philosophers such as Hans Jonas, who in *The Concept of God after Auschwitz* (1984) affirmed that the Holocaust had irremediably challenged the omnipotence of God.[17]

In 1933, of the approximately sixty-seven million inhabitants of the German Reich (including the Saar region still under the administration of the League of Nations), the Jewish communities constituted less than 1 percent (mostly settled in urban centers, primarily in Berlin, where they comprised at most 160,000 people). Such relatively small numbers did not prevent the Nazis from regarding them as the ideal scapegoats for all that was considered to afflict the *Herrenrasse*, or "master race."[18] All evils were subsumed, as Nicholas Goodrick-Clarke aptly observes, into the notion of

a monstrous Jewish plot to destroy Germany. Only the total destruction of the Jews could thus save the Germans and enable them to

---

[17] For an analysis of the scholarly discussion around the (controversial) unicity of the Nazi phenomenon, see Ian Kershaw, *The Nazi Dictatorship: Problems and Perspectives of Interpretation* (London: Bloomsbury Academic, 2015 [1985]), esp. chs 1, 2, and 5. Hannah Arendt's classic *The Origins of Totalitarianism* (London: Penguin, 2017 [1951]) also retains considerable value in regard.

[18] For more on this, see George L. Mosse, *The Crisis of German Ideology: Intellectual Origins of the Third Reich*, with a critical introduction by Steven E. Aschheim (Madison, WI: The University of Wisconsin Press, 2021 [1964]), esp. ch. 7. "Antisemitism could conveniently encompass the charges of racial adulteration, economic sabotage, and absolute enmity to the German *Volk*. It could also, just as conveniently, picture the Jews as the incarnation of the inferior race, as capitalism, or as Bolshevism" (288).

enter the promised land. The chiliastic promise of a Third Reich echoed medieval Joachite prophecy and remained a potent metaphor in the fantasy-world of so many in Germany who bemoaned the lost war, the harsh terms of the peace settlement, and the misery and chaos of the early Weimar Republic.[19]

The "German revolution" was imbued with cleverly channeled salvific expectations. Philosophy graduate Joseph Goebbels in his capacity as minister of propaganda substantially contributed to the creation of the myth of the Führer as a new "Aryan" messiah by means of a demonically orchestrated propaganda machine and impressive mass demonstrations, exploiting the mass media and new technologies—and finding an ideal aid in the radio, which he termed "the eighth power."

Moreover, in the general context of the eugenics movements that had been flourishing in most of western Europe and the United States, as well as in the Soviet Union, a peak was reached in Nazi "racial hygiene" (*Rassenhygiene*) policies, resulting in the infamous euthanasia program Aktion T4, involving the deliberate, planned suppression of so-called "unworthwhile lives": a label applied not only to those afflicted by congenital "feeblemindedness," but also sufferers from "illnesses" that included mental defects and antisocial behavior. The allegedly scientific body of thought behind this— foreshadowed, by way of example, in the influential tract by psychiatrists Karl Binding and Alfred Hoche, *Permission for the Destruction of Life Unworthy of Life* (1920)—envisaged the need to "enhance" the process of natural selection and to separate the "unfit," or *Minderwertigen* (literally, "less valuable") from the rest of society, for the *Gesundheit* (health) of the latter: an "ethical" task, indeed, in line with Nazi biological populism and the motto "Gemeinnutz geht vor Eigennutz" (approximately "the common good before self-interest") which, sadly, frequently met with sympathy and support from a broad range of intellectuals, philosophers, literati, and even clerics, convinced that the authoritarian regimes could serve as a bulwark against the socialist or communist threat.[20]

Such socio-cultural upheaval, magisterially interpreted by Nietzsche, dated back at least to the previous century, and was intensified by the

---

[19] Nicholas Goodrick-Clarke, *The Occult Roots of Nazism* (London: Tauris Parke, 2005), 203.

[20] See Léon Poliakov, *The Aryan Myth: A History of Racist and Nationalist Ideas in Europe*, trans. by Edmund Howard (London: Chatto & Heinemann for Sussex University Press, 1974), 290–304; Michael Burleigh, *The Third Reich: A New History* (London: Macmillan 2000), 166–67 and passim.

Great War. Along with widespread cultural pessimism in terms of the failure of the myth of the progress, and antimodernist critique of the West as famously explicated in Oswald Spengler's successful (though barely readable) *Decline of the West* (1918) and Hermann Graf Keyserling's equally successful *Travel Diary of a Philosopher* (1919), which conterpoused declining Europe to the ancient, supreme wisdom of China and India, the notion of decadence came to be invoked not only within an intellectual, literary, and artistic elite, but also amongst "ordinary people" frustrated by the implosion of meaning, the *Entzauberung* (disenchantment) explored by Max Weber, and feelings of a *Fragmentierung* (fragmentation) and *Zerrissenheit* (approximately, being "torn" or "disjointed"): sick components, that is, of the nexus of forces generally ascribed to "modernization." There was, as J. J. Clarke puts it,

> a fascination with ideas of degeneration and decadence, and a willingness to explore new seas of thought. The very speed of progress, the rapid transformation from traditional to modern social and economic formations, the growth of science-inspired materialist philosophies, and the ever-slackening hold of ancient religious beliefs and rituals, all of these combined to create a mood of discontent with the comforts and promises of western civilization, and to encourage a search for more satisfying and meaningful alternatives.[21]

The crisis which was spreading through interwar Europe catalyzed expectations for a renewal of the West, considered to be suffering from a progressive, fatalistic "decay of values," to which a broad array of "diseases of civilization" (a notion enhanced by a social-Darwinistic scientific patina that gained wide currency outside medical circles) bore witness: the products of an urban lifestyle that came to be counterposed to the natural, rural, primeval life.[22] Meanwhile, the trend of extending medical (especially psychiatric and later also psychoanalytic) categories to the "diagnosis" of the condition of society as a whole developed to a hitherto unmatched extent. In this cultural humus various forms of palingenetic myth were nurtured by authoritarian political movements seeking, and presenting themselves as, a third way between capitalism and

---

[21] J. J. Clarke, *Oriental Enlightenment: The Encounter between Oriental and Western Thought* (London: Routledge, 1997), 96.

[22] See Volker Roelcke, *Krankheit und Kulturkritik: Psychiatrische Gesellschaftsdeutungen im bürgerlichen Zeitalter (1790–1914)* (Frankfurt am Main: Campus, 1999).

communism—and often displaying, in the wake of Italian fascism, a totalitarian tendency with religious or para-religious significance.[23]

Roger Griffin in his seminal *The Nature of Fascism* observes,

> Whether we approach the question from the point of view of comparative religion, the history of ideas, sociology, political science or psychoanalysis, there is no shortage of theories which can be adduced to suggest that the convinced fascist is to be seen as someone whose flight from inner chaos has found expression in a powerful elective affinity with an intensely mythic ideology of national regeneration. This ideology, though objectively representing just one partial system of values among so many others he (and even on occasion she) experiences affectively as an absolute vehicle of personal salvation. By promising to redeem the nation from the quagmire "demo-liberalism," fascism offers its converts the prospect of a renewed sense of meaning, of transcendence, of ritual time, of imminent rebirth in a world otherwise threatened with inexorable decadence, a decadence which, however real the objective factors of social disfunction at the time, at bottom was no more than their inner world of anxiety and chaos projected on to contemporary history.[24]

Influential intellectuals and proponents of social Darwinism such as—to cite just two names from two centuries—French aristocrat and anthropologist Arthur de Gobineau, author of the (in)famous *Essay on the Inequality of the Human Races* (1853–1855) and German zoologist and philosopher Ernst Haeckel, popularizer of the recapitulation theory but also the founder in 1906 of the (racially oriented) Monist League—had propagated the conviction that different human beings represent different

---

[23] Nazi ideologist Alfred Rosenberg proclaimed in his *Myth of the Twentieth Century*, "Today a new faith is awakening: the myth of blood, the faith that the divine essence of mankind is to be defended through blood; the faith embodied by the fullest realization that Nordic blood represents the mystery which has supplanted and surmounted the old sacraments" (Alfred Rosenberg, *Der Mythus des 20. Jahrhunderts: Eine Wertung der seelisch-geistigen Geszaltenkämpfe unserer Zeit* [Munich: Hoheneichen, 1930], cited by Richard Steigmann-Gall, "Nazism and the Revival of Political Religion Theory," in Constantin Iordachi, ed., *Comparative Fascist Studies: New Perspectives* [London: Routledge, 2010], 297–315 at 305). See on this subject Karla Poewe, *New Religions and the Nazis* (New York: Routledge, 2006). It might be recalled here that in 1945, the year of his death, Ernst Cassirer argued, "It's probable that the most important and unnerving aspect of modern political thought is the apparition of a new power: the power of mythical thinking" (Ernst Cassirer, *The Myth of the State* [New Haven, CT: Yale University Press, (1946) 1966], 3).

[24] Roger Griffin, *The Nature of Fascism* (New York: Routledge, 1993), 199.

levels of evolution, thus paving the way for the idea of the supremacy of the "Nordic-Aryan" race. Nazism responded to pessimistic views of degeneracy and the inevitability of racial contamination by investing the chosen *Volk* with a vision and a fated task.[25] The *völkisch* ideology channeled gloomy social-darwinistic conceptions toward the bright lure of a mission imbued by the mystical longing for the salvation of a *Gemeinschaft*, or community, emphasizing the virtues of a (rural) "rootedness" as the regenerative natural state of creative (German) man, in contrast to the materialistic, individualistic worldview and "uprootedness" that characterized the Jew.

The "Jewish doctrine" of psychoanalysis, regarded as the wellspring of a materialistic and individualistic mentality, with the aggravating factor of its alleged Bolshevik tendencies, became one of the favorite targets of Nazi anti-Semitism, branded with the stereotype of the *wurzellose jüdische Seele*, or "rootless Jewish soul," and relentlessly attacked as antithetical to the *deutscher Geist* or "German spirit" especially because of its "pansexuality." As each scientific organization was eventually forced into "Aryanization" or "self-Aryanization" (the so-called *Gleichschaltung* or *Selbstgleichschaltung*, as Regine Lockot thoroughly explains),[26] which imposed conformity at least upon its board, in November 1933 the three Jewish board members of the German Psychoanalytic Society—Max Eitingon, Ernst Simmel, and Otto Fenichel—were replaced by the "Aryan" psychoanalysts Felix Boehm and Carl Müller-Braunschweig. This decision, approved by Freud himself, was needed to prevent the dissolution of the Society and to allow the practice of psychoanalysis to survive.[27] In 1935 the non-Jewish members were invited to "voluntarily" resign. The Society was eventually suppressed in 1938 (when Freud, after the German annexation of Austria, fled to London, where he was to spend his last year of life). Perhaps more emblematic of the development of psychoanalysis in

---

[25] See Mosse, *Crisis of German Ideology*, cit.: "Idealized and transcendent, the *Volk* symbolized the desired unity beyond contemporary reality. It was lifted out of the actual conditions in Europe onto a level where both individuality and the larger unity of belonging were given scope. The *Volk* provided a more tangible vessel for the life force that flowed from the cosmos; it furnished a more satisfying unity to which man could relate functionally while being in tune with the universe. *Völkisch* thought made the *Volk* the intermediary between man and the 'higher reality'" (17; see also passim).

[26] Regine Lockot, *Erinnern und Durcharbeiten: Zur Geschichte der Psychoanalyse und Psychotherapie im Nationalsozialismus* (Frankfurt am Main: Fischer, 2002 [1985]), ch. 5.

[27] Karen Brecht, Friedrich Volker, Ludger H. Hermanns, Isidor J.Kaminer, and Dierk H. Juelich, eds, *"Hier geht das Leben auf eine sehr merkwürdige Weise weiter . . .": Zur Geschichte der Psychoanalyse in Deutschland* (Giessen: Psychosozial-Verlag, 2009 [exh. cat. 1985]), 94–95. See also Lockot, *Erinnern und Durcharbeiten*, cit., 110ff.

Nazi Germany is what happened to the Berliner Psychoanalytisches Institut, founded in 1920 by Karl Abraham and Max Eitingon; in 1936 it was closed and incorporated in the newly founded Institut für psychologische Forschung und Psychotherapie, the so-called "Göring Institute": an institution run by Matthias Heinrich Göring, a neuropathologist with Adlerian training and a cousin of the notorious Feldmarschall, who became overnight the ideal man to ensure for psychotherapy a significant degree of independence and security whilst remaining in accordance the regime's directives. This institution, along with the protection of the Society for Psychotherapy, was to some extent instrumental in a continuation of the teaching and practice of the discipline (including, notwithstanding all the difficulties and dangers, psychoanalysis), in line with Nazis' refusal to "apply a racial-biological standard to Aryans who exhibited lesser, and more common, neurotic conflict," for such "mental disorder within the master race could not be genetic or essentially organic," as assumed by psychiatric paradigms, but should be correctable.[28]

3. "Where danger is . . .": The (Obscure) Background of the Seminar

A few days before the seminar, Jung wrote to the student of Jewish origin to whom he was closest, the Hungarian-Swiss Jolande Jacobi,

> I shall be traveling to Berlin next week, to conduct a week-long seminar on the interpretation of dreams. After feeling the full pressure of the political suspense during the winter, this will allow me to take in the new German atmosphere. I ask myself, more and more often, what else is going to be distilled from Europe's smoldering cauldron. In Germany, we are currently experiencing an eruption of the *puer aeternus* that is truly unforeseeable. The new regime's foreign policy antics would be pretty amusing if they weren't fueled by quite so much dangerous enthusiasm. Sometimes it almost feels as if preparations were underway for a new Blood Wedding [*Bluthochzeit*].[29]

---

[28] Geoffrey Cocks, *Psychotherapy in the Third Reich*, 2nd revised and expanded edn (New Brunswick, NJ: Transaction, 1997 [1985]), 12–13. On the Göring Institute and its leader, see respectively ch. 6 and 110–21; Lockot, *Erinnern und Durcharbeiten*, cit., 79–87 and ch. 6.1. See also Geoffrey Cocks, "Repressing, Remembering, Working Through. German Psychiatry, Psychotherapy, Psychoanalysis, and the 'Missed Resistance' in the Third Reich," *The Journal of Modern History* 64 (1992) (Supplement: "Resistance against the Third Reich"): 204–16.

[29] Jung to Jacobi, June 19, 1933, JA (Hs 1056:2142).

Despite his worries, hinted at by a thoughtful nod toward the Saint Bartholomew's Day massacre,[30] Jung's gaze at the situation sounded somehow remote—almost *per speculum alchemiae*, one might venture to say, considering his references to the cauldron and distillation—and featured a sense of expectation toward a vital "eruption of the *puer aeternus*" in contrast to the political stalemate of the hobbled Weimar Republic.[31] In a series of lectures held in February 1933 in Köln and Essen, the Swiss psychologist had diagnosed the current age as "a time of dissociation and sickness,"[32] and compared it to the twilight phase of the Roman Empire. Although the contemporary condition of collective suffering and turmoil was affecting every social, political, religious, and philosophical aspect of life, he was sensing—or hoping for—a rejuvenating potential in the crisis, for "the sickness of dissociation in our world is at the same time a process of recovery, or rather, the climax of a period of pregnancy which heralds the throes of birth."[33] He went on to recall that when "les extrêmes se touchent," that is, when a given tendency climaxes, its opposite germinates, in accordance with the law of the Tao.

Jung's (meta)psychological conception envisaged the faculty of the destructive forces of unconscious, constellated in times of crisis, of being able to reverse into something new. After all, he maintained, a spiritual rebirth followed Rome's breakdown. Maybe the new "barbarians" could usher in a renovation: indeed, at the end of the First World War Jung had already sought a reappraisal of the "genuine barbarian in us."[34] He valorized a confrontation with "otherness" to compensate the narrowness of the

---

[30] The massacre on Saint Bartholomew's Eve was carried out in Paris between 23 and 24 August 1572 by the Catholic faction against Parisian Huguenots in revenge for defeat at the Battle of Lepanto. The slaughter occurred six days after the marriage of Margherita, the king's sister, to the Protestant Henry III of Navarre (later Henry IV of France).

[31] In a later interview Jacobi referred to Jung's response to a letter of hers addressing her concerns about Nazism: "'Keep your eyes open. You can't reject the evil because the evil is the bringer of the light.' Lucifer means light-bringer. [. . .] For him this [the Nazi movement] was an inner happening which had to be accepted as a psychological pre-condition for rebirth" (Jolande Jacobi interview, December 23, 1969, C. G. Jung Biographical Archives, Rare Books Department, Francis A. Countway Library of Medicine, Harvard Medical School, Boston, MA, p. 24).

[32] "The Meaning of Psychology for Modern Man" [1933/1934], CW 10, § 290. See also Fischer, "The Year of Jung's Journey," cit., 136, and n. 2.

[33] Ibid., § 293.

[34] "The Role of the Unconscious" [1918], CW 10, § 19. Jung also addressed "the chthonic quality [which] is found in dangerous concentration in the Germanic peoples" whereby "the Jew lives in amicable relationship with the earth, but without feeling the power of the chthonic. [. . .] This may explain the specific need of the Jew to reduce everything to its material beginnings" (§ 18). Then, referring to "Freud's and Adler's reduction of

Western civilization, and possibly for the sake of its spiritual renewal: just as, in his estimation, the introverted Eastern basic attitude could help the hyper-extroverted attitude of the West, so the enhancement (and reanimation) of the instinctive, primordial dynamism of the collective psyche could counterbalance the rationalistic one-sidedness he ascribed to the Western mentality. And addressing the psychohistorical split in the German psyche, he had argued in a letter to Oscar A. H. Schmitz (author of *Psychoanalyse und Yoga*) in 1923 that "every step beyond the existing situation has to begin down among the truncated nature-demons. In other words, there is a whole lot of primitivity in us to be made good." He maintained that cultural development relies on going "back behind our cultural level, thus giving the suppressed primitive man in ourselves a chance to develop [. . .] for only out of the conflict between civilized man and the Germanic barbarian will there come what we need: a new experience of God."[35] Much later, in his essay "Wotan," tinged with gloomy forebodings, Jung would interpret the phenomenon of Nazism as a revival of a specifically German-pagan archetype, and characterize the condition of the Germans as one of possession—or being fatally gripped or seized (*ergriffen*)—by such collective unconscious forces; however, as has been noted, Wotan notoriously "did not cure himself."[36]

Although admitting that it was "exceedingly difficult to judge events that are happening right under our noses," Jung blatantly underestimated the unprecedented gravity of certain events, while assessing them as pathological "symptoms." Among these were the bonfires of "forbidden" books conducted by the German student union under the auspices of the "Campaign against the Un-German Spirit," with the aim of purging the *Volk* of Jewish, 'decadent' literature—including works of Freud and Adler.[37]

---

everything psychic to primitive sexual wishes and power-drives," he maintained that "these specifically Jewish doctrines are thoroughly unsatisfying to the Germanic mentality" (§ 19).

[35] C. G. Jung, *C. G. Jung: Letters*, ed. by Gerhard Adler with Aniela Jaffé, trans. by R.F.C. Hull, 2 vols (Bollingen Series XCV) (Princeton, NJ: Princeton University Press, 1973), 1:40 (May 26, 1923).

[36] Giuseppe Maffei, "Jung e il nazismo," in Patrizia Puccioni Marasco, ed., *Jung e l'ebraismo* (Florence: Giuntina, 2002), 41–71 at 63. See "Wotan" [1936], *CW* 10.

[37] In the session of May 10, 1933 during his Visions Seminar, Jung commented contemptuously on the burnings of the works of psychoanalyst, activist, and homosexual Magnus Hirschfeld "because I happen to know something of that stuff"; he then turned to focus on the fact that "it is not very far from the burning of books to the burning of witches. If that should follow after a while it would also be interesting, not for the people at the stake, but as a good specimen for any museum. You see that it is obviously a symptom" (C. G. Jung,

This can suffice, perhaps, for us to infer that Jung's initial view derives from an unfortunate extension of his hermeneutics of individual psychic development (together with the persuasion of the need of a rediscovery of the "primitive" springs of the collective psyche) to the contemporary socio-political arena, against the background of cyclical, mythic, palingenetic view. This was supported by a Heraclitean enantiodromic principle, in line with Hölderlin's famous lines, so dear to him: "Where danger is, there / arises salvation also";[38] which may indeed meet the case in terms of individual compensatory psychic dynamics, as long as the process is supported by and directed toward a responsible enlargement of consciousness. But what were the chances of historical renovation stemming from a truly chaotic situation exacerbated by authoritarian tendencies? And of the

---

Visions: Notes of the Seminar Given in 1930–1934, ed. by Claire Douglas, 2 vols [Bollingen Series XCIX] [Princeton, NJ: Princeton University Press, 1997], 1:971. The same day about twenty-five thousand books were burnt in Berlin under the slogan "Against the overvaluation of instinctual urges that destroy the soul, for the nobility of the soul" and to celebrate, as proclaimed by the pamphlet "Wider den undeutschen Geist!" (Against the un-German spirit), a "symbol of the revolution" and a "sign of victory of a new value belief" (Volker Dahm, "Die nationalsozialistische Schrifftumspolitik nach dem 10. Mai 1933," in Ulrich Walberer, ed., *10. Mai 1933: Bücherverbrennung in Deutschland und die Folgen* [Frankfurt am Main: Fischer, 1983], 36–83 at 36). Yet May 1933 was still regarded as a joyous time by many, including clerics prone to regard the new outburst of official anti-Semitism with favor and to believe that the Nazi "Holy Crusade could be accommodated within the Christian dogma" (Mosse, *Crisis of German Ideology*, cit., 308).

[38] Jung often invoked this passage from Hölderlin's poem "Patmos" (1802–3): in *Symbols of Transformation* [1912/1952], CW 5, § 630, for example, and, notably, in "The Spiritual Problem of Modern Man" ([1928/1931], CW 10), an essay about the transformations of his time. "Instinctively," he wrote, "modern man leaves the trodden paths to explore the by-ways and lanes, just as the man of the Greco-Roman world cast off his defunct Olympian gods and turned to the mystery cults of Asia. Our instinct turns outward, and appropriates Eastern theosophy and magic; but it also turns inward, and leads us to contemplate the dark background of the psyche" (§ 192). Contextually he affirmed: "If we are pessimists, we shall call it a sign of decadence; if we are optimistically inclined, we shall see in it the promise of a far-reaching spiritual change in the Western world" (§ 191). Jung evidently tended toward the second option, by exhibiting enthusiasm, perhaps with a kind of vitalistic, Nietzschean tone, for the regenerating qualities of the unconscious. He regarded the growing interest in psychic research and psychoanalysis as symptoms of a healthy counter-movement to the overextroverted attitude of the West, and welcomed the increasing attention to the body, sport, and modern dance, but also jazz and cinema: in sum the tuning up of the West for "a more rapid tempo—the American tempo—the exact opposite of quietism and world-negating resignation" (§ 195). Then, after quoting the Hölderlin verses (ibid.), he concluded the essay by proclaiming, "An unprecedented tension arises between outside and inside, between objective and subjective reality. Perhaps it is a final race [*ein letzter Wettlauf*] between aging Europe and young America; perhaps it is a healthier or a last desperate effort to escape the dark sway of natural law, and to wrest a yet greater and more heroic victory of waking consciousness over the sleep of the nations. This is a question only history can answer" (§ 196).

development of a collective consciousness or individuality, which in a letter to Erich Neumann loaded with psychological and theological implications Jung would define as "a *contradictio in adjecto*"?[39]

4. THE SEMINAR, THE RADIO BERLIN INTERVIEW, AND THE SOCIETY FOR PSYCHOTHERAPY

One must bear in mind that the Berlin Seminar coincided with Jung's decision to take on the presidency of the Society for Psychotherapy, as we shall see in greater detail. In relation to Jung's other seminars in those years, this one took place during an interlude in the Visions Seminar held between 1930 and 1934 at the Psychological Club.[40] It also followed the so-called German Seminars of 1930 and 1931, held respectively at his home and then at the Hotel Sonne in Küsnacht, and which included two closely related lectures on "Indian Parallels," devoted to Tantric yoga and the symbolism of the chakras, whose main sources were most likely the work of two brilliant popularizers of the treasure of Tantra, Arthur Avalon (pseudonym of the British orientalist—and High Court judge at Calcutta—Sir John Woodroffe) and Henrich Zimmer.[41] In the winter of 1932, along with Sanskritist and former missionary Jakob Wilhelm Hauer,[42] Jung gave, at the Psychological Club, the aforementioned seminar

---

[39] Jung to Neumann, April 27, 1935, in Martin Liebscher, ed., *Analytical Psychology in Exile: The Correspondence of C. G. Jung and Erich Neumann* (Princeton, NJ: Princeton University Press, 2015), 98–109 at 108. Cf. Liard's reflections on Jung's tendency to interpret the regressive character of the socio-cultural events of those years as possible "seeds of a higher culture" (Veronique Liard, *Carl Gustav Jung, "Kulturphilosoph"* [Paris: Presses de l'Université Paris-Sorbonne, 2007], 107 and passim).

[40] See Jung, *Visions*, cit.

[41] See Jung, *Psychology of Kundalini Yoga*, cit., 71 n. 1 and Sonu Shamdasani, "Introduction: Jung's Journey to the East," in Jung, *Psychology of Kundalini Yoga*, cit., xvii–xlvi at xxxiv, and n. 65. The 1930 text "Indian Parallels" is in Appendix 1 of *Psychology of Kundalini Yoga*, 71–87. The *Bericht über das Deutsche Seminar von Dr. C. G. Jung 5–10 Oktober 1931 in Küsnacht-Zürich*, compiled by Olga von Koenig-Fachsenfeld (Stuttgart, 1932) is forthcoming with the title *On Active Imagination: Jung's 1931 German Seminar*, ed. by Ernst Falzeder, trans. by Ernst Falzeder with Tony Woolfson (Philemon Series) (Princeton, NJ: Princeton University Press). On the seminars of 1930 and 1931, see Hannah, *Jung: His Life and Work*, cit., 202–3.

[42] Jakob W. Hauer (1881–1962) would found in the following year the "German Faith Movement" (Deutsche Glaubensbewegung) intended to embody the deep meaning of National Socialism and seeking a Germanic—that is, "Aryan"—religious renaissance. Jung would interpret Hauer's movement in his essay "Wotan" (1936) as symptomatic of the contemporary psychic condition of the German *Volk* (CW 10, § 397ff.). According to the sources, Jung and Hauer kept in contact until 1938. For more on this, see Poewe, *New Religions and the Nazis*, cit., and Petteri Pietikainen, "The *Volk* and its Unconscious: Jung,

"The Psychology of Kundalini Yoga," at which he addressed "Western Parallels to Tantric Symbols."[43] Furthermore, in 1934 Jung was to inaugurate the Zarathustra Seminar, dedicated to a thorough analysis of Nietzsche's powerful, visionary book, his relationship to the Persian prophet, and his concept of the Superman or Overman (*Übermensch*).[44]

The transcript of the Berlin Seminar was printed in cyclostyle for private distribution (as was customary for Jung's seminars), in a sole edition in 1933. Besides the transcript of the six daily sessions (Monday, 26 June to Saturday, 1 July), the *Report* included the "Lecture by Heinrich Zimmer 'On the Psychology of Yoga' to the C. G. Jung Society on June 25, 1933" and a transcription of the "Colloquy between Dr. C. G. Jung and Dr. A. Weizsäcker from the Berlin Radio broadcast of June 26, 1933."[45] This interview is omitted here as it has already been published and is readily available.[46]

Hauer and the 'German Revolution'," *Journal of Contemporary History* 35, no. 4 (2000): 523–39.

[43] Shamdasani, "Introduction: Jung's Journey," cit., xxxviii n. 83.

[44] C. G. Jung, *Nietzsche's Zarathustra: Notes of the Seminar Given in 1934–1939*, ed. by James L. Jarrett, 2 vols (Princeton, NJ: Princeton University Press, 1988). On Jung and Nietzsche, see Paul Bishop, *The Dionysian Self: C. G. Jung's Reception of Friedrich Nietzsche* (Berlin: de Gruyter, 1995); Martin Liebscher, *Libido und Wille zur Macht: C. G. Jungs Auseinandersetzung mit Nietzsche* (Basel: Schwabe, 2011); Gaia Domenici, *Jung's Nietzsche: Zarathustra, The Red Book, and "Visionary" Works* (London: Palgrave, 2019).

[45] *Bericht über das Berliner Seminar von Dr. C. G. Jung vom 26. Juni bis 1. Juli 1933. Anhang: Zwiegespräch Dr. C. G. Jung und Dr. A. Weizsäcker in der Funkstunde Berlin am 26. Juni 1933. Vortrag Prof. H. Zimmer: 'Zur Psychologie des Yoga' in der C. G. Jung-Gesellschaft am 25. Juni 1933, Berlin, 1933*. There also exists a version, likewise printed in 1933, including only Jung's seminar. No preparatory notes for the seminar, nor related photographic material, are preserved in either JFA or JA (with thanks to Andreas Jung and Ulrich Hoerni). Jung's personal copy of the seminar protocol deposited at the JFA bears in its first half a few comments along with a fair number of marks, underlinings, and minor corrections, primarily of foreign terms, in Jung's hand (which have been reported or else silently corrected in the present edition). Thus one may surmise a plan for a second, corrected edition of the seminar, which was never produced.

[46] The broadcast's English translation—with the title "An Interview on Radio Berlin"—is in William McGuire and R.F.C. Hull, eds, *C. G. Jung Speaking: Interviews and Encounters* (Bollingen Series XCVII) (Princeton, NJ: Princeton University Press, 1977), 59–66. The interview (entitled "Ziegespräch, wiedergegeben auf Schallplatte in der Berliner Funkstunde am 26. Juni 1933," *Bericht über das Berliner Seminar*, cit., 166) appears to have been broadcast in fact on July [*sic*] 26, 1933, with the title "Aufbauende Seelenlehre: Ein Gespräch über die neuen Aufgaben der Psychologie [Constructive psychology (lit., "soul doctrine" or "soul teaching"): a conversation about the new tasks of psychology]. C. G. Jung/A. Weizsäcker." Unfortunately, no tape recording or transcription of it is preserved in the archives of the *Deutscher Rundfunk* (my thanks to the archivist, Jörg Wyrschowy).

The volume does not identify the transcriber in its front matter: this is another peculiarity of this seminar. Barbara Hannah merely stated that the transcription was made "in an unusually good stenogram and multigraphed for the use of the class in almost verbatim form."[47] The transcription may plausibly be ascribed to Olga von Koenig-Fachsenfeld, a German psychologist and teacher who produced other transcripts.[48] Omitted from the *Report* were also the names of participants—with the exceptions of Heinrich Zimmer and Gustav Schmaltz[49]—who interacted during the sessions; an additional unusual fact that can be hypothetically attributed to the intention of preserving the participants' anonymity in consideration of the political situation.

The Berlin Seminar was organized by the Gesellschaft für Analytische Psychologie, or Society for Analytical Psychology, founded on September 24, 1931 in Berlin by the German analysts Eva Moritz, Wolfgang Müller Kranefeldt, and Hanna Kummerlé, with the aim, according to an early manuscript chronicle, of encouraging "interest in C. G. Jung's psychology in a convivial social setting."[50] For ten years, until its suppression by the Nazis in 1941, the Society promoted lectures, courses, and seminars on various subjects related to Jung's depth psychology.[51] This

---

[47] Hannah, *Jung: His Life and Work*, cit., 215. Included in Jung's personal copy is a typescript index written by Max Silber, but their correspondence (in the JA) makes no reference either to this or to Silber's possible involvement in the transcription.

[48] Ernst Falzeder has observed that the typewriter used for this seminar's transcript matches the one for the German Seminar, and that "the style of the title page, the presentation as a whole, as well as the wording of the prefatory note are nearly identical." See Falzeder's "Introduction" to Jung, *On Active Imagination*, cit. [forthcoming]. He also informs me that the Koenig-Fachsenfeld family archive retains no documentation from Koenig-Fachsenfeld concerning the Berlin Seminar. Olga von Koenig-Fachsenfeld began analysis with Jung in 1929. Subsequently, she worked as a member of the "Göring Institute," directing its Educational Counseling Service department from 1939 to 1942. See Cocks, *Psychotherapy in the Third Reich*, cit., 185 and passim; Falzeder, "Introduction," cit.

[49] Gustav Schmaltz (1884–1959), German engineer and physician, owner of a machine-tool factory, and director of a woodworkers' association. He was interested in astrology and mandala symbolism, and underwent analysis with Emma Jung. Jung wrote a foreword to his *Komplexe Psychologie und körperliches Symptom* (Stuttgart: Hippokrates, 1955): see *CW* 18, § 839–40.

[50] "Chronik der Gesellschaft für Analytische Psychologie," unnumbered, JFA. This is a manuscript volume, a gift from the Society to Jung, recounting the Society's activities—lectures and seminars—up until 1934, and including a list of thirty-one members (for the years 1931–33).

[51] Among the speakers during the first few years were W. M. Kranefeldt, H. Zimmer, J. W. Hauer, G. R. Heyer, J. Kirsch, G. Schmaltz, H. Schmid-Guisan, K. Bügler, A. Weizsäcker, E. Rousselle, and H. Keyserling. For more on the Berlin Society, see Lockot, *Erinnern und Durcharbeiten*, cit., 50–52.

being the first of its seminars to be held by Jung himself, the occasion was extremely well attended, and was celebrated on June 25 by the conferring upon Jung of the title *Ehrenvorsitzender* (honorary president) of the Society.

The event was arranged by Eva Moritz, in collaboration with Wolfgang Kranefeldt and Toni Wolff. Jung initially relied on Moritz to propose the choice of topic, reportedly in deference to her knowledge of the audience and the general situation.[52] On account of his heavy workload, the seminar was almost canceled in April, but a few weeks later was reinstated.[53] In the (in places incomplete) set of organizational correspondence Jung appeared somewhat dubious of its feasibility and even reluctant to proceed. His main reservation concerned the risks he feared for the Berlin Society as a consequence of an equation of his psychological system with that of psychoanalysis. In a letter to Moritz he lamented the boycott of his theory "for decades" by Freud's and Adler's followers, and hoped that the seminar would not perpetuate such "confusion" [*Verwechslung*] and "deceit" [*Betrug*], so as to "protect the Berlin Club from the possibility that, as a result of the abolition of Freud's ideology, it would disappear under the table."[54] In the same vein he wrote to Otto Curtius, a psychotherapist close to him, that his "main concern" about running the seminar was that "this would give the Berlin Club a dangerous opportunity for publicity, because I'm a little worried that our cause [*Sache*: literally, "thing"], which is as anti-Jewish as possible [*sic*], might end up being lumped in [*in einen Topf geworfen (werden könnte)*: literally, "thrown into a pot"] with Freudian psychoanalysis, as frequently happens."[55] This

---

52 Toni Wolff wrote to Kranefeldt: "Ms. Moritz asked me something concerning the seminar. I wrote to her at the request of Dr. Jung that she should choose the subject. She ought to do it, Jung decidedly prefers it that way, as he does not know the people. From that one should spare him [. . .]. Dr. Jung will surely take into account the situation there" (Wolff to Kranefeldt, March 25, 1933, AZ, correspondence W. Kranefeldt, Sign. Ms Z VII 395).

53 As in the letter of Jung to J. Kirsch of May 6, 1933, in Ann Conrad Lammers, ed., *The Jung–Kirsch Letters: The Correspondence of C. G. Jung and James Kirsch*, trans. by Ursula Egli and Ann Conrad Lammers (London: Routledge, 2011), 37.

54 Jung to Moritz, May 6, 1933, JA (Hs 1056:2527), quoted in Giovanni Sorge, *Psicologia analitica e anni Trenta: Il ruolo di C. G. Jung nella "Internationale Allgemeine Ärztliche Gesellschaft für Psychotherapie" (1933–1939/40)* (dissertation, University of Zurich, 2010/2018), 138 n. 601.

55 Jung to Curtius, May 6, 1933, JA (Hs 1056:2331), quoted in Giovanni Sorge, *Bestandbeschrieb der Akten zur Geschichte der Präsidentschaft von C. G. Jung in der Internationalen Ärztlichen Gesellschaft für Psychotherapie, 1933–1940 im Nachlass von C. A. Meier* (ETH Zürich Research Collection, 2016 [online]), 74 and n. 147.

phrase gives an idea of the extent of Jung's propensity to "take into account the situation there" (to repeat the expression used by Wolff in her letter to Kranefeldt quoted previously): that is, to adjust (at least) his semantics to the anti-Semitic propaganda for reasons surely ascribable, though not limitable, to a strategic agenda. Curtius, who had good political connections,[56] is credited with having arranged Jung's mysterious, reportedly short encounter with Joseph Goebbels during his stay in Berlin.[57] This encounter, as the correspondence permits us to infer, was the direct result of Curtius's unrealized plan to invite Goebbels himself to the seminar, which Jung reportedly agreed to.[58] What today cannot but appear as at least grotesque—the participation of an infamous Nazi criminal in a Jung seminar—was evidently considered by those involved an attempt to seek some form of protection from a high-ranking government official.

Jung's interview with Radio Berlin on the eve of the seminar must likewise be considered in the light of his strategy to distinguish his own psychological system—and perspective—from that of psychoanalysis *also* for the sake of his colleagues in Germany, although it ended up sounding, to an extent at least, like a species of endorsement of the new political course. Jung was introduced by his student, the neurologist, psychiatrist (and Nazi Party member) Adolf Weizsäcker as "the most progressive psychologist of modern times."[59] Jung portrayed a sort of divergence between the old guard, trapped in an intellectualistic mentality and unable

---

[56] Curtius's brother Julius had been former chancellor and foreign minister for the Deutsche Volkspartei, a conservative-liberal political party during the Weimar Republic (Cocks, *Psychotherapy in the Third Reich*, cit.,129; Jay Sherry, *Carl Gustav Jung: Avant-Garde Conservative* [New York: Palgrave Macmillan, 2010], 104). From 1935, Nazi Party member Otto Curtius would serve as a German editor of the *Zentralblatt*, the official journal of the International General Medical Society for Psychotherapy.

[57] See Hannah, *Jung: His Life and Work*, cit., 211; Cocks, *Psychotherapy in the Third Reich*, cit., 129; Deirdre Bair, *Jung: A Biography* (Boston, MA: Little, Brown and Co., 2003), 424–25; and Sherry, *Carl Gustav Jung*, cit., 104–5.

[58] As reported in a letter of Jung's secretary to Curtius of May 29, 1933, JA (Hs 1056:2332) (quoted in Sorge, *Bestandbeschrieb der Akten*, cit., 73–74).

[59] "Zwiegespräch Dr. C. G. Jung und Dr. A. Weizsäcker," in *Bericht über das Berliner Seminar*, cit., 166–73 (henceforth "Zwiegespräch"), 166; Tilman Evers, *Mythos und Emanzipation: Eine kritische Annäherung an C. G. Jung* (Hamburg: Junius, 1987), Anhang 1, 241–77 at 241; "An Interview on Radio Berlin," in McGuire and Hull, *C. G. Jung Speaking*, cit., 59–66 at 60. This followed a short announcer presentation (not included in the English version in *Jung Speaking*, cit.), which described Jung as the "psychologist who countered Sigmund Freud's corrosive [*zersetzende(n)*] psychoanalysis with his constructive theory of the soul" ("Zwiegespräch," cit., 166; Evers, *Mythos und Emanzipation*, cit., 241). For more details (also about Weizsäcker's background), see Sherry, *Carl Gustav Jung*, cit., 104–8.

to comprehend the "integral man," and the new movement, regarded as an eruption of collective psychic forces constellated in a particular historical *kairos*; for Germany, as a young nation still constructing its unity and, as he was accustomed to say, suffering from a long-lasting inferiority complex, was catching up with the rest of Europe. Contextually he affirmed that

[t]imes of mass movement are always times of leadership [*Führertum*]. Every movement culminates organically in a leader [*gipfelt organisch im Führer*], who embodies in his whole being the meaning and purpose of the popular movement. [. . .] The need of the whole always calls forth a leader regardless of the form a state may take.[60]

This attestation of unquestionable, almost naturally justified leadership, was followed by one of Jung's most controversial declarations:

Only in times of aimless quiescence does the aimless conversation of parliamentary deliberations [*ziellose Konversation parlamentarischer Beratungen*] drone on, which always demonstrates the absence of a stirring in the depths or of a definite emergency.[61]

At the same time, he called for a "self-development [*Selbstentwicklung*] of the individual": that is,

the supreme goal of all psychological endeavor, [that] can produce consciously responsible spokesmen and leaders of the collective

---

[60] "Zwiegespräch," cit., 173; Evers, *Mythos und Emanzipation*, cit., 246; "An Interview on Radio Berlin," cit., 65. Jung here echoed, and went beyond, the core thesis of Gustav Le Bon's influential *The Crowd: A Study of the Popular Mind* (1895) with regard to the emotional, rather than rationally based, depersonalizing power of mass psychology. See Shamdasani, *Jung and the Making of Modern Psychology*, cit., 286–87.

[61] "Zwiegespräch," cit., 173; Evers, *Mythos und Emanzipation*, cit., 246; "An Interview on Radio Berlin," cit., 65–66; Sherry, *Carl Gustav Jung*, cit., 104ff. Jung continued, "Even the most peaceable government in Europe, the Swiss Bundesrat, is in times of emergency invested with extraordinary powers, democracy or no democracy. It is perfectly natural that a leader should stand at the head of an elite, which in earlier centuries was formed by the feudal [*feudale*; omitted in the English version] nobility. The nobility believes by the law of the nature in the blood and exclusiveness of the race [an das Blut und an Rassenausschliesslichkeit]. Western Europe doesn't understand the special psychic emergency of the young German nation because it does not find itself in the same situation either historically or psychologically." This could resonate easily with Jung's audience since, as observed by Mosse, the German "revolution" was expected to be "democratic but not parliamentary. *Völkisch* thought had always been concerned with having all the people participate in the *völkisch* destiny and at the same time preventing the *Volk* from being atomized into political parties" (Mosse, *Crisis of German Ideology*, cit., 285).

movement. As Hitler said recently, the leader must be able to be alone and must have the courage to go his own way. But if he doesn't know himself, how is he to lead others? That is why the true leader is always one who has the courage to be himself, and can look not only others in the eye but above all himself [an expression we'll find repeated in the seminar, in relation to the Shadow].[62]

At this point it is worth observing that the biennium 1933–34 presents us with Jung's most problematic and controversial remarks with regard to politics and collective (especially national) psychologies—including such considerations about alleged differences between Jewish and German psychologies as, by way of example, "the Jew, who is something of a nomad, has never created a cultural form of his own," counterposed to evocation of the youthfulness of the "Aryan" unconscious, which contains creative tension and "seeds of a future yet to be born."[63] Such differentiations belonged to a larger epistemological plexus inherent to Jung's notion of a phylogenetically stratified unconscious, including levels such as nation and race.[64] Yet although he had already previously voiced the distinction of his system from the Freudian,[65] grounded in his broader opposition to nineteenth materialism and positivism, he now seems to have been insufficiently aware that his call—notably in the *Zentralblatt*, official organ of the Psychotherapeutic Society he had just begun to direct—for recognition of the "scientific validity" of such differences, far from being neutral, in fact had

---

[62] "Zwiegespräch," cit., 171; Evers, *Mythos und Emanzipation*, cit., 245; "An Interview on Radio Berlin," cit., 64. See below [MT], 156 and 174. One year later Jung wrote to Kirsch that his interview didn't "express anything concerning politics" (Jung to Kirsch, May 29, 1934, in Lammers, *The Jung–Kirsch Letters*, cit., 44–47 at 45). In a later statement he would affirm, "Hitler has never gained a healthy relationship to this female figure, which I call the anima. The result is that he is possessed by it. Instead of being truly creative he is consequently destructive. This is one reason why Hitler is dangerous, he does not possess within himself the seeds of true harmony [. . .]. I [do not] think he will turn into a normal human being. He will probably die in his job" ("Jung Diagnoses the Dictators" [*The Psychologist*, London, May 1939], in McGuire and Hull, *C. G. Jung Speaking*, cit., 136–40 at 140).

[63] See "The State of Psychotherapy Today" [1934], CW 10, § 353. See also Jung's "Editorial" [1933], CW 10.

[64] For the broad intellectual, anthropological, ethnological, and philosophical context of such notions, see Shamdasani, *Jung and the Making of Modern Psychology*, cit., 158–59, 182–89, 213–27, and 232–43. See also Nicholas Adam Lewin, *Jung on War, Politics and Nazi Germany: Exploring the Theory of Archetypes and the Collective Unconscious* (London: Karnac Books, 2009), ch. 5; and Michael Vannoy Adams, *The Multicultural Imagination: "Race," Color, and the Unconscious* (New York: Routledge, 1996).

[65] In, e.g., "The Role of the Unconscious" [1918], CW 10, and his lecture at the fourth General Medical Congress for Psychotherapy, "The Aims of Psychotherapy" [(1929) 1931], CW 16. Cf. Poliakov, *The Aryan Myth*, cit., 304–25.

an inevitable political echo.[66] On top of this, despite his explicit statement that he did not intend any devaluation of Jewish psychology, his argument concerning a different Jewish relationship to the earth in the end could not but reflect the *völkish* topos that attributed to the Jews a lack of rootedness, in contrast to the purely instinctual nature of the Germans.

We can surmise that Jung must have had such (mis)judgments in mind when, in 1947, he would admit to the educator and community leader Rabbi Leo Baeck that he had "slipped up"—as reported in a letter of Gershom Scholem to Aniela Jaffé,[67] Jung's loyal secretary who in turn judged his choice to distinguish, at that historical moment, between Jewish and non-Jewish psychology as "a grave human error."[68]

Jung himself admitted in 1935 that he like others was "affected" by the events which were upfolding in Germany when he was there.[69] And after the war, he avowed that

[a]t that time, in Germany as well as in Italy, there were not a few things that appeared plausible and seemed to speak in favor of the regime. An undeniable piece of evidence in this respect was the disappearance of the unemployed, who used to tramp the German highroads in their hundreds of thousands. And after the stagnation and decay of the post-war years, the refreshing wind that blew through the two countries was a tempting sign of hope.[70]

---

[66] According to Cocks, such statements "fit both his own anthropological tendency toward national and racial characterization as well as the practical demands of the moment" (*Psychotherapy in the Third Reich*, cit., 131). They also "reveal a destructive ambivalence and prejudice that may have served Nazi persecution of the Jews. But Jung," Cocks concludes, "conceded much more to the Nazis by his words than by his actions" (134). Cf. Sherry, *Carl Gustav Jung*, cit., ch. 4.

[67] Jung's words were, according to Scholem, "Jawohl, ich bin ausgerutscht" (Indeed, I stumbled); in Aniela Jaffé, *Parapsychologie, Individuation, Nationalsozialismus: Themen bei C. G. Jung* (Einsiedeln: Daimon, 1985), 163–64. Scholem's letter was published in full in Gershom Scholem, *Briefe II: 1948–1970*, ed. by Thomas Sparr (Munich: Beck, 1995), 94–95, and is translated into English in Aniela Jaffé, "C. G. Jung and National Socialism," in Jaffé, *From the Life and Work of C. G. Jung*, trans. by R.F.C. Hull and Murray Stein, expanded edn (Einsiedeln: Daimon, 1989), 78–102 at 100.

[68] Ibid., 86. Conversely, Jaffé praised Jung's decision to collaborate with an "incriminated nation" in the interests of a "co-existence" (ibid., 82).

[69] "I saw it coming, and I can understand it because I know the power of the collective unconscious. But on the surface it looks simply incredible. Even my personal friends are under that fascination, and when I am in Germany, I believe it myself, I understand it all, I know it has to be as it is. One cannot resist it. It gets you below the belt and not in your mind, your brain just counts for nothing, your sympathetic system is gripped" (*The Tavistock Lectures* [1935], CW 18, § 164).

[70] "After the Catastrophe" [1945], CW 10, § 420.

In retrospect he stated that since Hitler's seizure of power he had realized that "a mass psychosis was boiling up in Germany"; nonetheless he felt confidence in "one of the most differentiated and highly civilized countries on earth, besides being, for us Swiss, a spiritual background to which we were bound by ties of blood, language, and friendship."[71] He added that during his Berlin stay he had

> received an extremely unfavorable impression both of the behaviour of the Party and of the person of Goebbels. But I did not wish to assume from the start that these symptoms were decisive, for I knew other people of unquestionable idealism who sought to prove to me that these things were unavoidable abuses such as are customary in any great revolution. It was indeed not at all easy for a foreigner to form a clear judgment at that time.[72]

Jung's optimism is hardly explicable without considering his medical-clinical attitude for, as he himself stated,

> [i]t is part of the doctor's professional equipment to be able to summon up a certain amount of optimism even in the most unlikely circumstances, with a view to saving everything that it is still possible to save. He cannot afford to let himself be too much impressed by the real or apparent hopelessness of a situation, even if this means exposing himself to danger.[73]

Let us recall, in terms of context, that less than a week before the seminar, on June 21, the summer solstice, Jung had officially assumed the presidency of the General Medical Society for Psychotherapy, which at that time represented psychotherapy's medical avant-garde. Founded in Germany in 1927, the society was characterized by a predominantly conservative- and nationalist-oriented membership with a substantial interest in religious questions (for example, how psychotherapy related to the *Seelsorge*, or cure of souls). Psychotherapists aspired to a proper, autonomous, epistemological status and adequate recognition from the medical establishment, in particular the much longer-established field of academic psychiatry, which was predominantly concerned with hereditarian and organicist factors in the etiology of psychic disorders, whereas psychotherapy primarily engaged with unconscious, socio-cultural, interrelational, and

---

[71] Epilogue to "Essays on Contemporary Events" [1946], CW 10, § 472 and § 474.
[72] Ibid., § 472.
[73] Ibid., § 474.

also spiritual factors. Conversely, a somewhat holistic approach of this type could end up being consistent with, and manipulable in terms of, the *völkisch* tenets underlying Nazi ideology, imbued as these were with post-romantic aspirations towards a unity of body and soul for the sake of *Gemeinschaftsgefühl* (a sense of community), in the context of so-called national "psychic hygiene." The main obstacle here was the need to "free" psychotherapy from its alleged "Jewish roots," implicit in the indisputable leadership assumed by psychoanalysis from the early years of the century, by establishing a nationalistic, Germanic version—hence the "Aryan" Jung, with his emphasis upon the unfathomable depths of the psyche, was deemed fit to play a leading role. Jung accepted the presidency (having served as a vice-president since 1930) following the leading psychiatrist Ernst Kretschmer's resignation in response to "harassing" pressure from German colleagues concerned by the risk of an imminent dissolution of the Society or its incorporation into the much more powerful German Psychiatric Society.[74] The latter in fact would continue in the years that followed to plan such an amalgamation, and this would be in large measure prevented by Jung (in this finding himself at one with M. H. Göring). Jung's decision was not in the event immediate and unreserved. He accepted the office "provisionally" and with it the "special [that is, Aryan] commission" which had been meanwhile been formed under the guidance of psychiatrist Johannes Heinrich Schultz in order to ensure the continuance of the Society.[75]. In so doing, he displayed an

---

[74] On the top of such pressures there came a letter, dated April 21, 1933, JA (Hs 1056:1997), from the neuropsychologist Walter Cimbal, a co-founder of the Society and an active supporter of the psychotherapeutic cause. Cimbal, who that year joined the Nazi party, although without fully sharing its commitment to eugenic and sterilization policies, skillfully pushed on all the buttons in an effort to convince the Jung: the fatally uncertain moment; Jung's responsibility before psychotherapy itself, its future, and its new—Germanic—course; the risk of the dissolution of the Society and the guarantee of the full inviolability of Jung's "geistige Führung," or spiritual leadership; the intrinsically Germanic connotations of Jung's psychology; the importance of the German revolution in putting a stop to the narcissism and the degradation embodied in Jewish influence and in turn propagated by psychoanalysis. (See Sorge, *Psicologia analitica e anni Trenta*, cit., ch. 2, 2.3; and Ann Conrad Lammers, "Professional Relationships in Dangerous Times: C. G. Jung and the Society for Psychotherapy," *Journal of Analytical Psychology* 57, no. 1 [2012]: 99–119).

[75] Following Jung's decision communicated on June 8 (Jung to Cimbal, June 8, 1933, JA [Akz. 2022-18, C. G. Jung–W. Cimbal letters, as yet uncatalogued]) the official announcement of his formal assumption of the presidency, on June 21, was then disclosed by Walter Cimbal, as the newly re-named general secretary of the Society, in a circular letter to the members of the above-mentioned commission on June 27 (Cimbal to Herrn Vorsitzenden und den Mitgliedern der geschäftsführenden Vorstandskommission der AAGP, June 27,

attitude marked by a kind of *Realpolitik*, as echoed in his appeal to the biblical injunction "Give to Caesar what belongs to Caesar, and to God what belongs to God,"[76] which we find, for instance, in his letter written to approve the subsequent foundation on September 15 of the national German Society for Psychotherapy, under the presidency of Matthias Heinrich Göring, which was subject both to the regime's rules and to the General (soon to be International) Society headed by Jung himself. Here Jung once again—or still—remained sanguine with regard, notably, to the capacity of the psychotherapeutic profession to acclimatize itself to every political wind:

> For me it is wholly self-evident that a dominating political move-ment pulls everything into its sphere. It is completely senseless for individuals to resist, I wish to say, this metereological condition. Just as psychotherapy must prove itself worthy of all different kinds of patients, so must it also rise to all outward realities.[77]

Returning to the progress of actual events: once assured of the con-tinuation of the German Society as a section of the General Medical Society, in May 1934 Jung refounded the latter as the Internationale Allgemeine Ärztliche Gesellschaft für Psychotherapie, or International General Medi-cal Society for Psychotherapy (IGMSfP); that is, as a supranational, apo-litical, non-confessional federation. He oversaw the byelaws of the statutes with the assistance of the famous Jewish lawyer Wladimir Rosenbaum, establishing a limitation upon the voting power of the larger sections (first and foremost the German one) and the authority to enroll members inde-pendently of their own national groups (the so-called rule of the *Einzel-mitglieder*, or individual members, a measure specifically to benefit Ger-man Jews). Together with his student the Swiss psychiatrist C. A. Meier (1905–1995)—later to be Jung's successor as professor of psychology at the ETH Zurich—in his role as general secretary of the Society, in the years to follow Jung fostered the foundation of national psychotherapeutic Societies in Switzerland, Holland, Sweden, Denmark, Austria, and Great Britain, promoted three international congresses (in Bad Nauheim in 1935, Copenhagen in 1937, and Oxford in 1938), and co-organized an

---

1933, JA [Akz. 2022-18, Jung–Cimbal letters, as above]), the second day of the seminar here presented. On Schultz, see Cocks, *Psychotherapy in the Third Reich*, cit., 72–76.

[76] Mark 12:17.

[77] Jung to Cimbal, September 3, 1933, quoted in Giovanni Sorge, "Jung's Presidency of the International General Medical Society of Psychotherapy: New Insights," *Jung Journal: Culture & Psyche* 6, no. 4, (2012): 31–53 at 35.

international symposium in Basel (1936). He also supervised publication in the official organ of the Society, the *Zentralblatt für Psychotherapie und ihre Grenzgebiete* together with Meier as managing editor (and with the German editors, first Walter Cimbal and then Otto Curtius; in 1936 Göring would become co-director of the journal).

Jung embarked on this endeavor both to promote his own system and to support psychotherapy at a very critical moment. Demonstrating an increased awareness of his role *super partes*, he counterbalanced the predominance of the German Society branch through a remarkable enlargement of the federation, and managed to retain its presidency until 1939, subsequently remaining in office as honorary president until 1940, when Göring, whose power steadily grew also thanks to the ministerial support he obtained for his Institute, managed to take over the headquarters and run the Society from Berlin.

Overall, Jung's attitude confronted by an objectively complex situation showed a continuous balancing between resistance and conformity, along with considerable diplomatic and twisting and turning, and no little Machiavellianism in his efforts to maintain a "neutral" position, seeking at least to "save what could be saved." The profile of events briefly sketched here presents lines of both continuity and discontinuity. For instance, while Jung was little concerned that his differentiations between Jewish and German psychologies in 1933–34 might be adding fuel, indirectly, to the *völkisch* and even anti-Semitic fire, at the 1938 Oxford conference he presented a list of "fourteen points of common understanding" arising from a combined effort in the "democratic spirit of Switzerland," and aimed at promoting the principles shared by the different psychotherapeutic schools.[78] Despite his unceasingly sympathetic yet watchful concern for Germany, from time to time he firmly rejected requests from his German colleagues: such as, for example, to dedicate an international conference to the topic of race, or to host a review of Alfred Rosenberg's infamous book *The Myth of the Twentieth Century* in the pages of the *Zentralblatt*. Thus the man who seemed to serve as the ideal figurehead for the *neue deutsche Seelenheilkunde* ("new German soul treatment" [psychotherapy]) ended up being reproached by Matthias Heinrich Göring as "too

---

[78] "Vierzehn Punkte gemeinsamen Einverständnisses." See "Presidential Address to the 10th International Medical Congress for Psychotherapy, Oxford, 1938" [1938], CW 10, § 1072–73; Sorge, *Psicologia analitica e anni Trenta*, cit., ch. 6, 3.1. Subsequently the list, which echoed the list of principles for peace famously presented by President Woodrow Wilson in 1918 to the United States Congress, did not attract much attention.

old" to understand the German revolution (in a 1940 letter which condemned Jung's newly published *Terry Lectures*).[79]

Any measured assessment of the vexed question of Jung's alleged philo-Nazism and anti-Semitism must take into account the intersections between the institutional and the theoretical planes. As for the latter, the principal landmarks surely include, in addition to the Radio Berlin interview and his writings from 1933–34 quoted above, the essays "Wotan" (1936) and "Psychology and National Problems" (1936), the *Terry Lectures* ([1937] 1938/1940), his interview "Diagnosing the Dictators" ([1938] 1939), and "After the Catastrophe" (1946).[80] In these a gradually increasing level of criticism is perceptible, directed against mass movements, dictatorships, and "-isms," although in tandem with a persistent tendency to gloss over the concrete role played by a cunningly

---

[79] Göring to C. A. Meier, October 9, 1940, JA (Hs 1069:1424), cited in Sorge, "Jung's Presidency," cit., 44 and in Sorge, *Psicologia analitica e anni trenta*, cit., 294. In the *Terry Lectures* Jung referred provocatively to "the adventurious Germanic tribes with their characteristic curiosity, acquisitiveness, and recklessness," and addressed "the amazing spectacle of states taking over the age-old totalitarian claims of theocracy, which are inevitably accompanied by suppression of free opinion. Once more we see people cutting each other's throats in support of childish theories of how to create paradise on earth. It is not very difficult to see that the powers of the underworld—not to say of hell—which in former times were more or less successfully chained up in a gigantic spiritual edifice where they could be of some use, are now creating, or trying to create, a State slavery and a State prison devoid of any mental or spiritual charm" ("Psychology and Religion" [(1937) 1938/1940], CW 11, § 83). Lockot notes that although the *Lectures* included "only an indirect criticism, Jung was placed on the Nazi 'blacklist'" (*Erinnern und Durcharbeiten*, cit., 108). Jung's stance here appears to be consistent with that of Eric Voegelin, who is credited with being among the first (with Raymond Aron) to introduce the concept of "political religion" into sociopolitical analysis, referring to authoritarian political systems seeking and promising an imminent, "Gnostic" incarnation into a transcendent world ("When God is invisible behind the word, the contents of the word will become new gods," Voegelin proclaimed in *The Political Religions*: see *The Collected Works of Eric Voegelin*, 34 vols, vol. 5: *Modernity without Restraint: The Political Religions; The New Science of Politics; and Science, Politics, and Gnosticism*, ed. by Manfred Henningsen [Columbia, MO: University of Missouri, 1999], 60). On this subject, see Emilio Gentile, "The Sacralization of Politics," in Iordachi, *Comparative Fascist Studies*, cit., 257–89; as well as Gentile's key monograph *Politics as Religion* (Princeton, NJ: Princeton University Press, 2006). For more on the scholarly debate on the nature of the Nazis' "theology of race," see Steigmann-Gall, "Nazism and the Revival," cit.

[80] The essays are in CW 10. See moreover "Diagnosing the Dictators," *Hearst's International Cosmopolitan*, January 1939, in McGuire and Hull, *C. G. Jung Speaking*, cit., 115–35. It may be added here, quoting Stanley Grossmann, that 1937 represented "a clear shift in [Jung's] opinions" (Grossman, "C. G. Jung and National Socialism," in Paul Bishop, ed., *Jung in Context: A Reader* [London: Routledge, 2003 (1999)], 9–21 at 113–14), probably linked to his September sojourn in Berlin to deliver (still unpublished) lectures on archetypes, and coincidental with Mussolini's visit to Germany, which gave him the opportunity

orchestrated propaganda machinery capable of manipulating symbols and mythic narratives to sustain a political agenda. That was the other "side of the picture" that George Mosse, quoting Jung's essay on Wotan, highlighted as crucial factor, besides the "preoccupation with the Ur forces, the primeval instincts, combined with a mystically oriented dynamic" investigated by Jung; for "[t]he ideology was formalized. The archetypes were not allowed free play. And as the ideology was tamed, it came to express itself through an internal logic of its own which took on concrete, outward forms."[81]

Finally, this picture would be not be complete without also taking into consideration Jung's subsequent collaboration with the controversial Bern-based director of the US Office of Strategic Services, Allen Welsh Dulles.[82]

---

to observe the two dictators closely (as reported in his later Knickerbroker interview). So, while in 1936 Jung had not yet excluded the possibility that National Socialism could represent a sort of "reculer pour mieux sauter" ("Wotan" [1936], CW 10, § 399), and that "an era of dictators, Caesars, and incarnated states" could have "accomplished a cycle of two thousand years" such that "the serpent" could meet "with its own tail" again ("Psychology and National Problems," [1936], CW 18, § 1342), the following year he saw nothing but an "interregnum" full of dangers in which the hubris of the Western world, various -isms, and "superhuman" personalities were successfully taking over ("Psychology and Religion" [(1937) 1938/1940], CW 11, § 144).

[81] Mosse, Crisis of German Ideology, cit., 317.

[82] On this, see Allen Dulles, Germany's Underground: The Anti-Nazi Resistance (New York: Macmillan, 1947), the book by his wartime colleague (and mistress) Mary Bancroft, Autobiography of a Spy (New York: William Morrow, 1983), and Deirdre Bair, Jung: A Biography, cit., ch. 31 ("Agent 488"), 481–95. For a closer look at the whole matter, see two pivotal, groundbreaking works—both deeply informed by archival documentation—already cited above: Cocks, Psychotherapy in the Third Reich, and Lockot, Erinnern und durcharbeiten. Among the large body of available literature I select for mention: Aryeh Maidenbaum and Stephen A. Martin, eds, Lingering Shadows: Jungians, Freudians, and Anti-Semitism (Boston, MA: Shambhala, 1991), and Grossman, "C. G. Jung and National Socialism," cit. Regarding Jung and politics, see also Walter Odajnyk, Jung and Politics: The Political and Social Ideas of C. G. Jung (New York: New York University Press, 1976), Lewin, Jung on War, cit., and the accurate analysis of Jung's intellectual, social, and editorial network by Jay Sherry (Sherry, Carl Gustav Jung, cit.), which argues for Jung's proximity to the line of thought of the "conservative revolution" (as it was termed by political theorist Armin Mohler in his The Conservative Revolution in Germany 1918–1932 [1989]). See also the essays included in the monographic issue Jung Journal: Culture & Psyche 6, no. 4 (2012); William Schoenl and Linda Schoenl, Jung's Evolving Views of Nazi Germany: From the Nazi Takeover to the End of World War II (Asheville, NC: Chiron, 2016); my inventory of (and introduction to) the records of Jung's and Meier's activity in the International General Medical Society for Psychotherapy preserved in the JA: Sorge, Bestandbeschrieb der Akten, cit.; and Sorge, Psicologia analitica e anni Trenta, cit. For an insightfully critical analysis, see Carrie B. Dohe, The Wandering Archetype: C. G. Jung's "Wotan" and Germanic-Aryan Myth and Ideology (London: Routledge, 2016).

5. THE PARTICIPANTS AND THE ANALYSAND

We can now turn to the Berlin Seminar in more detail. The six daily sessions following Zimmer's lecture took place regularly from 5 to 7 p.m. at Harnack-Haus, a prestigious venue built in 1929 in the Dahlem district.[83] The manuscript chronicle of the Gesellschaft contains a list of 153 signatures of seminar participants.[84] They came mainly from Germany, but also from Austria, Switzerland, Sweden, England, and America; most were physicians and professionals in psychotherapy, including members of the Society for Psychotherapy such as Otto Körner, Hans von Hattingberg, and Ernst Speer, as well as the aforementioned Johannes Heinrich Schultz, whose successful book on autogenic training had been published the previous year.[85] Harnack Haus's guest record bears a list of eighteen guests registered for the so-called "Tagung der Gesellschaft für analytische Psychologie," including a "Dr. Read" from London (on June 26–27) who can probably be identified as Sir Herbert Edward Read, a British scholar of art history and poet, later co-editor of Jung's *Collected Works*.[86] Other participants included clerics, teachers, and counselors (for instance, Pastor Ivan Alm, Assilie Kollwitz, Hildegard Stackmann, Felicia Froboese-Thiele, and Charlotte Geitel). Some participants had already attended the Dream Analysis Seminar (including Toni Wolff, Barbara Hannah, Gustav Schmaltz, and James and Hilde Kirsch). Importantly, practically all Jung's Jewish German students were present. All of these with the exception of Käthe Bügler (who was half-Jewish, and remained in Germany under the protective wing of G. R. Heyer, whom we'll encounter again below), were soon to leave the country: Erich Neumann together with his wife Julie (later emigrants to Palestine), James Kirsch and his wife Hilde Silber (later to leave for Palestine, London,

---

[83] S. Eckart Henning, *Das Harnack-Haus in Berlin-Dahlem: "Institut für ausländische Gäste", Clubhaus und Vortragszentrum der Kaiser-Wilhelm-/Max-Planck-Gesellschaft* (Munich: Max-Planck-Gesellschaft, 1996). As attested by circular letter by Eva Moritz of May 1, 1933, the first choice had been the (smaller) Haus der Presse's conference room. The advertised price was thirty-two German marks, and there was no mention of Zimmer's lecture (Bundesarchiv Koblenz, Kleine Erwerbungen, dossier 173, 275). For more on the context, see Jörg Rasche, "C. G. Jung in the 1930s: Not to Idealize, Neither to Diminish," *Jung Journal: Culture & Psyche* 6, no. 4 (2012): 54–73.

[84] "Chronik der Gesellschaft für Analytische Psychologie," unnumbered, JFA (see above, n. 50).

[85] J. H. Schultz, *Das Autogene Training (konzentrative Selbstentspannung): Versuch einer klinisch-praktischen Darstellung* (Leipzig: Thieme, 1932).

[86] Harnack-Haus guest registry for June 1933, signature 1, Rep. 1A, No. 2513/2, 51–52, HHA. My thanks to the archivist, Marion Kazemi.

and California), Max Zeller (later an emigrant to Los Angeles with his wife Lore), Gerhard Adler and his wife Hella (to leave for London in 1936), Toni Sussmann (who also emigrated to London in 1937), Ernst Bernhard (who fled to Italy together with his second wife Dora Friedländer in 1936), and Heinz Westmann and Werner Engel (later emigrants to New York).[87] In 1963 Käthe Bügler remembered that the number of attendees, "greater than Jung had expected," gave rise to "an exceptionally lively seminar situation, with queries and responses," since "Jung allowed participants to put written questions to him, to let him gauge their outlook and position." She added that "we were still drawing on memories of this seminar years later."[88]

Let us now consider Jung's analysand, also on the basis of *Dream Analysis*. He was a man in his mid-forties, a well-known businessman accustomed to international travel, including in the East. Behind the screen of a slight neurosis, he was suffering an existential crisis: he felt impeded in his own development, an indication of a balked spiritual quest. This man had been born in Africa, and like Jung himself was the son of a pastor, who had died long before the analysis.[89] He had "a good intellect," and was "cultivated, prosperous, very polite and social, married, with three or four children [. . .]; his main trouble is that he is irritable and particularly anxious to avoid situations where someone might reproach him or hurt him."[90] His typology presents some degree of contradiction: conventional and very correct by temperament, "he is a sensation type"[91] (a few pages later "a thinking type," eight months earlier "an intellectual type," and perhaps most helpfully, still later "a thinking-sensation type"); and, "looked at theoretically, the dreamer is a thinking type with sensation as his secondary function, a man of reality. He has intuition to a certain extent. Obviously his blind spot is feeling":[92] so feeling is definitely his inferior function. He is introduced as having "a very complex nature. [. . .] He has no neurosis, but great intellectual

---

[87] It is worth recalling that Jung provided several certificates aimed at facilitating the professional settlement of his emigré pupils and acquaintances in their destination countries (see Sorge, *Bestandbeschrieb der Akten*, cit., 96–99).

[88] Käthe Bügler, "Die Entwicklung der analytischen Psychologie in Deutschland," in Michael Fordham, ed., *Contact with Jung: Essays on the Influence of His Work and Personality* (London: Tavistock Publications, 1963), 29.

[89] *DAS*, 158.

[90] Ibid., 6.

[91] Ibid., 302.

[92] Ibid., respectively 319, 109, 314, and 316.

interest,"[93] as well as a "hygienic streak."[94] However, "he has trouble that disturbs his sexuality."[95] We are dealing with a quite well-educated person who "speaks Italian and [. . .] also knows Greek and Latin,"[96] and a voracious reader who approached in a somewhat confused manner esoteric, occult, and theosophic literature.[97] He had also read "a good deal of psychology and books about sex," is "quite informed about symbolism"[98]—and had attended a lecture by Jung on psychological types. The subject had undergone analysis with Jung for about two years before and also during the Dream Analysis Seminar,[99] which, quite notably, he even attended with—as Jung reported—beneficial impact on his feeling.[100]

When he turned himself over to Jung, this businessman had unsuccessfully explored two avenues. First came unsatisfying extramarital "adventures with demimondaines," which led him to feel increasingly inauthentic, since his expectations revolved around "a higher type of love and real devotion which is not in such women—really a most decent

---

[93] Ibid., 111.

[94] "So he takes exercises in the morning beginning with the bath, probably singing in the tub—that is exceedingly healthy—and then he would drink a non-alcoholic coffee and eat a particular kind of bread" (ibid., 533).

[95] Ibid., 261.

[96] Ibid., 133.

[97] It should be noted that theosophy, with its syncretistic faith based on a mixture of Eastern spirituality, modern science, and the Darwinian theory of evolution, enjoyed in the early twentieth century a considerable vogue in Germany, Switzerland, and Austria and became intertwined with other movements seeking alternative lifestyles, such as the *Lebensreform* (life reform) which, besides finding overlaps with the *völkisch* movement, had discovered at the turn of the century, at so-called Monte Verità in Ascona in Canton Ticino, a haven for free spirits, experimenters, vegetarians, freemasons, intellectuals, and dancers. For more on this, see Harald Szeemann, *Monte Verità: Berg der Wahrheit; Lokale Anthropologie als Beitrag zur Wiederentdeckung einer neuzeitlichen sakralen Topographie* (Milan: Electa, 1978); Martin Green, *Mountain of Truth: The Counterculture Begins; Ascona, 1900–1920* (Lebanon, NH: University Press of New England, 1986).

[98] *DAS*, 7 and 500.

[99] "The dreamer," Jung said on November 28, 1928, "is a man whom I occasionally still see—that means analysis has not killed him yet!" (*DAS*, 43). Analysis continued until at least July 1929, Jung reported on October 9, 1929, yet giving to understand that besides a "more positive relation to the work" he was still on a "long way (. . .) full of risks and dangers," including a possible "latent trouble in his unconscious" which could even "lead to a local schizophrenia" (301–2)–which suggests that the treatment was far from being concluded. One can also infer that some discrepancies germane to the patient's diagnosis during the seminar may be related to the simultaneity of his treatment. No indications have been found with regard to the analysand's possible participation in the Berlin Seminar as well.

[100] *DAS*, 431.

tendency [. . .]. So he turned away from that and landed in theosophy, he dreamed himself into an artificial world of images, a stupid [*sic*] place where one can lose oneself in all sorts of heavens, and since it was a substitute, it was like flirting with a sort of spiritual cocotte, and he got very tired of it."[101] Thus, worn out between the amusements of the *bon viveur* and a fascination with the occult, this restless individual developed a mild neurosis and turned to Jung "not really for treatment. He had come to psychoanalysis"—Jung went on wryly—"through his studies, and thought it might be better 'dope' than theosophy, for he had heard of the Freudian sublimation idea by which nature is transformed by magic into playing the piano or living a saintly life; one thinks in a marvelous way and sex is wiped out."[102] Thus Jung proposed that he should to focus on seeing "what your nature, physical as well as spiritual, will produce. You have to be patient, as I have to be. There is no prescription."[103] Development does not lie merely in the analyst's indications, but rather in listening to "the Old Man who is a million years old"—a metaphor which we will encounter, variously declined, throughout the seminar.

6. FEAR, AMAZEMENT, AND TREMBLING

Dream life is an eminent channel used by the million-year-old "Old Man" to make himself perceptible. In 1932, Jung wrote to Gustav Schmaltz, concerning an "extremely remarkable" dream, of how

> it always seems to me quite extraordinary with what precision the unconscious anticipates events and you really have to ask yourself what degree of consciousness should be attributed to anticipations of this nature. Sometimes one cannot avoid the impression that a superior agency [überlegene Regie] is at work. Our immanent causality then seems to me like a tissue of deception, a reckoning we make without our host.[104]

The presentiment of a numinous, inconceivable Otherness, the sense of *Ehrfurcht* (close to "reverence and awe") toward the lively presence of an *unus mundus* behind the limits of a Schopenhauerian veil of Maya, runs through Jung's entire body of work, alongside his need to look at

---

[101] Ibid., 300.
[102] Ibid., 301.
[103] Ibid., 317.
[104] Jung, C. G. *Jung: Letters*, cit., 1:92 (April 9, 1932).

this from a psychological, empirical perspective, without resort to meta-physical assumptions.[105] This nexus is particularly evident in his notion of synchronicity, matured over decades through his involvement with Taoist philosophy, research on parapsychic phenomena, and dialogue with his patient—and winner of the Nobel Prize in physics—Wolfgang Pauli. Jung conceived synchronicity as "an intellectually necessary principle which could be added as a fourth to the recognized triad of space, time, and cau-sality" and which, notably, he placed somewhere between the cause–effect principle and the Leibnizian thesis of a pre-established harmony.[106] So, though maintaining the numinous character of archetypes and their onto-logically transcendent (and transcendental, in Kant's sense) nature, Jung abstained from theorizing a sphere above the (collective) unconscious or the possibility of an *Überbewusste*, or superior consciousness. The crucial act remains, in his judgment, inquiry into the world of images and sym-bols—in other words, the *mundus imaginalis*, that intermediate realm be-tween matter and the spirit which was thoroughly investigated by the leading Islamist Henry Corbin.[107] In distinction from Freud's theory of dream, whose ascendancy, as observed by Shamdasani, "had the effect of privatizing the dream, which was seen to be solely concerned with the in-timate sphere of subjectivity, and its all too human concerns," Jung's stance was closely linked with his interest for religion, spirituality, and his quest for the common roots of transgeographical symbolism. Hence

[Jung's] notion that some dreams had a suprapersonal source in the collective unconscious, together with his validation of the view that

---

[105] This may resonate with "the belief that there is an unseen order, and that our su-preme good lies in harmoniously adjusting ourselves thereto. This belief and this adjust-ment are the religious attitude in the soul," as it was put by William James in *The Varieties of Religious Experience* (Cambridge, MA: Harvard University Press, 1985), 51. Such an attitude underpins the entire work of one of the fathers of pragmatism, who sought a sci-ence of spontaneous religious manifestations free from theological dogmatism, which was deeply inspiring for Jung. For more on this, see Eugene Taylor, *William James on Con-sciousness beyond the Margin* (Princeton, NJ: Princeton University Press, 2011 [1996]).

[106] "Synchronicity: An Acausal Connecting Principle" [(1951) 1952], *CW* 8, § 958 and passim. See also C. G. Jung, *Dream Symbols of The Individuation Process: Notes of C. G. Jung's Seminars on Wolfgang Pauli's Dreams*, ed. by Suzanne Gieser (Princeton, NJ: Prince-ton University Press, 2019).

[107] Thus "spiritual Imagination," according to Corbin, is both "a cognitive power" and "an organ of true knowledge" which reconnects with the "*mundus imaginalis*"—his trans-lation of Arabic *ālam al-mithāl*. (Henry Corbin, *Swedenborg and Esoteric Islam*, trans. by Leonard Fox [West Chester, PA: Swedenborg Foundation, 1995], 16). See in particular Corbin's *Mundus Imaginalis, or The Imaginary and the Imaginal*, trans. by Ruth Horine (Ipswich: Golgonooza Press, 1976).

they could be a source of guidance, wisdom, and ultimately religious experience, recovered the religious and metaphysical significance that had traditionally been assigned to the dream.[108]

In this vein, the seminar presented here exemplifies Jung's phenomenological approach toward the transpersonal, prefigurative skills of the oneiric realm. In dealing with dreams, he recommends maintaining a special receptiveness coupled with a sort of naivity, for they are "simply nature—undistorted, impartial [*unparteiisch*] nature."[109] The *Bildungskraft*, or power of imagination, can disclose, through the prospective, teleological capacity of symbols, a *tertium quid* pregnant with meaning which impacts consciousness itself; yet "the synthesis of the attitude cannot be produced by consciousness, but only by the unconscious."[110] Consciousness then is asked to re-enhance its archaic, primeval springs, so that it may travel—*per modum imaginandi*—along the path of its enlargement, towards an increased *Bewusstwerden*, or "becoming (more) conscious." At the same time, fantasy can also be depersonalizing, toxic, addictive. Jung characterizes it as "the greatest danger of all" when there is a lack of rootedness,[111] perhaps also alluding to contemporary collective expectations for a salvific "new order."

To counteract such risks, Jung took care to increase his audience's awareness of the potential danger of the psyche. To Barbara Hannah, who reports the anxiety-filled atmosphere she sensed in driving to Berlin and during the Berlin Seminar, Jung would say, "These people are all in a panic, they are scared stiff and have no idea where all this is leading. I am afraid nothing can save them and that they are heading for the inevitable disaster, but at least we will earn the merit of trying to help them as long

---

[108] Shamdasani, *Jung and the Making of Modern Psychology*, cit., 156.

[109] Below [MT], 136. See also "The Meaning of Psychology for Modern Man" [1933/1934], CW 10, § 137ff.

[110] Below [MT], 195. This did not imply full reliance upon it. As Jung wrote in 1918, "nature is not, in herself, a guide, for she is not there for man's sake. Ships are not guided by the phenomenon of magnetism. We have to make the compass guide." Unconscious emanations such as dreams must be therefore considered "with the necessary conscious correction that has to be applied to every natural phenomenon in order to make it serve to our purpose" ("The Role of the Unconscious" [1918], CW 10, § 34). See also "The Practical Use of Dream Analysis" [1934], CW 16, esp. § 294ff.

[111] Below [MT], 215. See also *DAS*: "Certain forms of fantasy may be the worst poison for the person who is not reasonably adapted. But when you find germs of imagination in a man who is firmly rooted, perhaps imprisoned, in his environment, they should be treated as the most valuable material, for out of this material he can win his freedom" (85). Cf. Jung's considerations on the importance of having a "point of support in 'this world" while he himself was engaging with his fantasies in *MDR*, 189 and his *Aion* [1951], CW 9.2, § 46.

as we can."[112] Hannah also writes that during the seminar Jung "did not [. . .] refer to the political situation *at all*."[113] (He did only allude, dispiritedly, to the unsuccessful Geneva disarmament conference which had begun in 1932).[114] "But he did do his utmost to open their eyes to the reality of the psyche and the inner life. He also spoke a good deal of the danger of being unconscious and of getting caught in *participation mystique* and mass emotion, but he always spoke in general terms and left the individuals in the audience to apply it to their own present situation. He certainly managed to calm his audience."[115]

Looking now at the dream interpretations in the Berlin Seminar, they do not diverge substantially from the corresponding ones of 1928–29 in *Dream Analysis*. In the five dreams under review, the Anima, the Shadow, and the inferior function emerge along with images evoking the archetypes of the *puer* and the savior (or, for certain aspects, the figure of the mana personality). Finally, a presentiment of the Self becomes perceptible. Nonetheless the oneiric sequence (re)considered in Berlin acquires a somewhat more unified, and in a way paradigmatic character, such that it rises to become a sort of explanatory picture of Jung's process of individuation. With regard to Jung's diagnostic evaluation, patently less scrutiny is given to the sexual component in the etiology of the analysand's neurosis. On a semantic level, terms used in *Dream Analysis* and liable to recall more or less directly Freud's theory, such as "sublimation" and "totemism," disappear (albeit the latter had famously been discussed by social anthropologist James George Frazer and was regarded by him as the primeval form of religious conceptions). Instead, the symbolism of the center runs throughout the seminar as a kind of magnetic catalyzation point around which revolve Jung's cross-cultural inquiries into contemporary scholarly anthropological, ethnological, and theological literature. Contextually, the center of the mandala (and of the yantra), regarded as the quintessential locus of the meditation work and the culmination point of the ritual

---

[112] Hannah, *Jung: His Life and Work*, cit., 211.

[113] Ibid., 215.

[114] Below [MT], 197. It may be recalled that to September 1932 dated Freud's "Why War?" ([1932] 1933) (collected in *The Standard Edition of the Complete Psychological Works of Sigmund Freud*, ed. and trans. by James Strachey with Anna Freud [London: Hogarth Press, 1953–74], vol. 22: *1932–36* [1964], 197–215), based on his correspondence with Alfred Einstein, a copy of which he notoriously donated to Mussolini.

[115] Hannah, *Jung: His Life and Work*, cit., 215. Jörg Rasche notes that "Jung worked similarly to a group therapist, not talking about the political situation, but beginning with the case of a completely 'normal man' finding his way to self-development" ("C. G. Jung in the 1930s," cit., 57).

practice, constitutes a subterranean leitmotif able to shape the labyrinthine wanderings that weave a *circumambulatio* both around and toward the Self as the archetype of wholeness.

7. ALCHEMICAL GERMINATION, AND NOSTALGIA FOR THE CENTER

The Berlin Seminar makes extensive use of the method of amplification, proceeding with cross-cultural incursions into primitive psychology, Greco-Roman Antiquity, early Christianity and Gnosticism, medieval ecclesiastical dances, Hindu and Tantric ritual practices, and Native American healing ceremonies. Supported by an impressive erudition, Jung makes it apparent that current life goes on fermenting with occurrences—"living fossils," to invoke an Eliadean speleological metaphor—from the deepest levels of the collective psyche. Jung's reflection of the similarity between mythological contents and unconscious symbolism dates back at least to his first edition of *Symbols of Transformation* (1912) and runs throughout his entire research.

As he would say years later, Jung regarded the mandala as a suitable representation of the central symbol of the collective unconscious, for

> [t]he mythological motifs whose presence has been demonstrated by the exploration of the unconscious form in themselves a multiplicity, but this culminates in a concentric or radial order which constitutes the true centre or essence of the collective unconscious. On account of the remarkable agreement between the insights of yoga and the results of psychological research, I have chosen the Sanskrit term *mandala* for this central symbol.[116]

Around this topic, representing a kind of universally symbolic *Urform*, revolves a whole panoply of cross-cultural explorations of this Seminar which would provide inspiration to scholars such as Karl [Károly]

---

[116] "The Psychology of Eastern Meditation" [1948], CW 11, § 945. Elsewhere, by considering the spontaneous appearance of the mandala symbolism in "[Western] people who cannot project the divine image any longer," Jung inferred that "the mandala denotes and assists exclusive concentration on the centre, the self. This is anything but egocentricity. On the contrary, it is a much needed self-control for the purpose of avoiding inflation and dissociation" ("Psychology and Religion" [(1937) 1938/1940], CW 11, § 156). For more on this, see in particular "Concerning Mandala Symbolism" [1950], and "Mandalas" [1955], CW 9.1. See also J. J. Clarke, *Jung and Eastern Thought: A Dialogue with the Orient* (London: Routledge, 1994), 134–40, and below [ZL], n. 358.

Kerényi[117] and Mircea Eliade.[118] Already in *Dream Analysis*, in referring to the (notably Taoist) mandala, Jung was wondering,

> When a Taoist priest meditates on a mandala and gradually concentrates his libido on the center, what is the meaning of the center? The center of consciousness is the ego, but the center represented in the mandala is not identical with the ego. It is outside of consciousness, it is another center. The naive man projects it into space; he would say it is outside somewhere in the world. The aim of the exercise is to shift the guiding factor away from the ego to a non-ego center in the unconscious, and this is also the general aim of analytical procedure.[119]

---

[117] In *Essays on a Science Mythology*, co-authored with Jung in 1941, and in his lecture of the same year at Eranos, the Hungarian classicist and historian of religion explored the meaning of the "origin" (*Ursprung*) in relation to the foundation ritual for ancient cities, particularly in Rome, as a symbolic system that organizes space in line with the mythologem of the mandala as a centering archetype. He observed that "the fourfold division of the Roman city-plan would seem [. . .] to be the result of [. . .] a natural north–south and east–west orientation, and the quadripartite *heaven* [of the Buddhist mandala] to be the common basis of all mandala-plans" (C. [*sic*] Kerényi, "Prolegomena," in C. G. Jung and C. Kerényi, *Science of Mythology: Essays on the Myth of the Divine Child and the Mysteries of Eleusis*, London: Routledge & Kegan Paul, 1985 [1949], 15; first published as *Einführung in das Wesen der Mythologie: Der Mythos vom göttlichen Kind und Eleusinische Mysterien*, Amsterdam: Akademische Verlagsanstalt Pantheon, 1941 [new edn: Olten: Walter, 1999]; see also Karl Kerényi, "Mythologie und Gnosis," *Eranos-Jahrbuch VIII/1940–1941* [Zurich: Rhein, 1942], 157–229). Kerényi went on to argue that the mandalas discovered and theorized by Jung "might as well be the reflection of age-old *observation of the heavens* as the reflection of a universal human *compulsion*" (ibid. ["Prolegomena"]), 16.

[118] In a work published two years after the *Essays on a Science Mythology* (and ten years after the Berlin Seminar), Mircea Eliade reflected upon the symbolism of the center in line with Kerényi's reflections and recalling a series of phenomena—from the Theseus dance after slaying the Minotaur to medieval ceremonies in cathedral labyrinths and Navajo healing practices—remarkably similar to Jung's amplificatory itinerary in this seminar; he then commented on Jung's theories on "European mandalas" as further psychological evidence for the universal aliveness of mandala symbolism (Mircea Eliade, *Commentaires sur la Légende de maître Manole*, Fr. trans. from Romanian by Alain Paruit [Paris: L'Herne, 1994 [1943]], 144–45). See also Eliade's *Images and Symbols: Studies in Religious Symbolism*, trans. by Philip Mairet (Princeton, NJ: Princeton University Press), 1991 [1952], ch. 1; *Patterns in Comparative Religion*, trans. by Rosemary Sheed (New York: New American Library, 1974 [1958]), 367–87, and the chapter on mandalas in his *Yoga: Immortality and Freedom* [1933], trans. by Willard R. Trask (Bollingen Series LVI) (Princeton, NJ: Princeton University Press, 1958 [2nd edn 1969; repr., with a new Introduction by David Gordon White, 2007], esp. 226–27.

[119] *DAS*, 104–5. Jung would address the "subtle but enormous difference" between the Buddhist and the Christian mandalas in, e.g., "The Psychology of Eastern Meditation" [1948], *CW* 11, where he states, "The Christian attains his end *in Christ*, the Buddhist knows *he* is the Buddha. The Christian gets *out of* the transitory and ego-bound world of

From reflection on the similarities between the Eastern quest for spiritual enlightenment and the aims of Western psychotherapy, Jung was able to state that "[t]he Chinese form of Yoga is quite like the symbolism we get in dreams and from the unconscious in general. To speak of Yoga is to speak of a certain form of analytical method."[120] It is worth recalling here that Jung subsumed under the term "yoga" a variety of types of meditative orthopraxis from Taoism to Buddhism and Hinduism, which he regarded as highly developed systems of concentration and contemplation into the deep psyche originally stemming from "a natural process of introversion."[121]

The Berlin Seminar falls in the midst of Jung's involvement with Oriental spirituality and precisely when he was integrating his long-year investigation of Chinese Taoist philosophy with his interest in Hindu and Tantric symbolism, during a phase marked by a growing immersion in the study of alchemical imagery: a sulphurous subject indeed, until then largely considered by scholars abstruse, even suspect. Yet for him, as he would write in his memoir, being "grounded in the natural philosophy of the Middle Ages, alchemy formed the bridge on the one hand into the past, to Gnosticism, and on the other into the future, to the modern psychology of the unconscious."[122] Jung's in-depth exploration of alchemy got fully underway in 1934–35, the latter being the year when he undertook the painstaking compilation of an alchemical lexicon with the aim of orienting himself within the subject to which to he would contribute so much fundamental work in the following decades.[123] So, significantly, in

---

consciousness, but the Buddhist *still* reposes on the eternal ground of his inner nature" (§ 949).

[120] *DAS*, 117.

[121] "Yoga and Us" [1936], *CW* 11, § 873. In the words of Paul Bishop, "he saw meditation as primarily a one-sided attempt to withdraw from the world [. . .] as a surrender to the collective unconscious, an introverted journey into self-absorption" (Bishop, "Jung, Eastern Religion, and the Language of the Imagination," *The Eastern Buddhist* 18, no. 1 [1984]: 42–56 at 50; republished in Daniel J. Meckel and Robert L. Moore, eds, *Self and Liberation: The Jung/Buddhism Dialogue* [Mahwah, NJ: Paulist Press, 1992], 166–80]). For more on Jung's understanding of yoga, see J. J. Clarke's crucial book, *Jung and Eastern Thought*, cit., part 2 and passim.

[122] *MDR*, 201. See also "Paracelsus as a Spiritual Phenomenon" [1942], *CW* 13, § 237.

[123] See the series of reproductions from Jung's "alchemical excerpt books" in Sonu Shamdasani, *C. G. Jung: A Biography in Books* (New York: Norton, 2012), 172ff., and Thomas Fischer, "The Alchemical Rare Book Collection of C. G. Jung," *International Journal of Jungian Studies* 3, no. 2 (2011): 169–80. Jung's subsequent Eranos lectures in 1935 and 1936 were augmented in *Psychology and Alchemy* (1944), and the later *Psychology of the Transference* (1946), while to 1955 dates his magnum opus on the subject, *Mysterium Coniunctionis*.

the Berlin seminar he defines his method of working with dreams as "a kind of meditation,"[124] by way of affiliation with an ascetic praxis—rather than, as he would do some years later, pointing out its structural connection to alchemical *meditatio* ("an inner colloquy with one's good angel").[125] In fact this seminar includes only a few references to alchemy and its proto-chemical language which makes the alchemical *opus* "the immediate preliminary stage of this kind of psychology"; Jung furthermore addresses the awareness among "some later alchemists" that they were practicing "a peculiar kind of philosophy [. . .] in turning worthless matter [. . .] into gold." Hence his parallel with analytical work: in such psychological but also hermeneutical excavations along with his patients, he felt himself "in agreement with the secret workings of nature."[126]

8. West versus East, or the Berlin Seminar as a Phase in the Making of a Complex Psychology

Jung conceived his complex psychology as a discipline capable of engaging in interdisciplinary debate, on a comparative basis, to arrive at the widest possible gnoseological and practical perspective on the psyche.[127] At this juncture, Eastern thought provided him with crucial insights. In this Jung was in line with, and contributed outstandingly to, a developing contemporary transformation in the Western attitude toward the East, by considering the latter in a more differentiated manner, going beyond an idealized, primitivist, or post-romantic perspective. Around the turn of the twentieth century, along with an "orientalist wave" spreading by way of the syncretistic teachings of theosophy, and a developing interest in dialogue between different faiths (especially in the wake of the legendary 1893 World Parliament of Religions in Chicago, at which the Hindu monk Swami Vivekananda also initiated a successful popularization of Raja yoga

---

[124] Below [MT], 159.

[125] Quoted in "Archetypes of Collective Unconscious," [(1934) 1954], CW 9.1, § 85, following Martin Ruland's *Lexicon alchemiae* (1612). (This definition was added in the revised edition.) See also "Concerning Rebirth" [(1939) 1940/1950], CW 9.1, § 236. See also below [MT], n. 391.

[126] Below [MT], 291–96. It is perhaps unsurprising that Jung's exploration of the visionary, imaginal dimensions of the psyche should lead Antoine Faivre to consider him the last representative of Romantic *Naturphilosophie* (see his entry "Esotericism," in Mircea Eliade, ed., *The Encyclopedia of Religion*, 16 vols [New York: Macmillan, 1987], 5:156–63).

[127] On Jung's notion of "complex psychology," see in particular the sections on "Complex Psychology" and "The New Encyclopedia," in Shamdasani, *Jung and the Making of Modern Psychology*, cit., 13–22.

in the West), many hitherto unknown or previously untranslated classical
Oriental texts at last became available to scholars through the monumen-
tal collection of *Sacred Books of the East* edited by Max Müller.[128] On
the top of that, as observed by Susanne Marchand,

> The image of the Orient had changed. Nineteenth-century platitudes
> invoking oriental stagnation were repeatedly challenged by those
> who now admired the East's resilience as against the constant revo-
> lutions of fortune in the West. [. . .] The "primitivism" of the East
> had become a positive virtue, and the Orient no longer seemed weak
> or weird. It was now the West that was degenerate and idolatrous,
> abandoned by God and the *Weltgeist*. For some, the now established
> association of the Orient with antibourgeois knowledge made East-
> ern wisdom an essential element in a new sort of *Bildung*. For others,
> appreciation of the Orient figured in a campaign to save both the
> East and West from spiritual, or even biological, death.[129]

In his explorations, Jung avoided any temptation to demonstrate the
superiority of Christianity over Eastern doctrines, considered by many to
be an archaic stage of religious development, and equally any tendency
to contrast Western civilization belittlingly with the "true metaphysics of
Indian doctrines," regarded as a a sort of ancestral "cradle" of world culture. Instead, he pursued a "synthesis of the two cultural and religious
models."[130] While Jung's involvement with the East runs throughout his

---

[128] *The Sacred Books of East*, in fifty volumes, collected English translations of Asian
religious writings from Hinduism, Buddhism, Taoism, Confucianism, Zoroastrianism, Jain-
ism, and Islam. It was undertaken by Max Müller and others, appearing between 1879 and
1910, published by Oxford University Press. Jung owned the entire series.

[129] Susanne Marchand, "German Orientalism and the Decline of the West," *Proceedings of the American Philosophical Society* 145, no. 4: 465–73 at 472–73. For more on the
reception of Eastern thought in the West, including Jung's pivotal role, see Raymond
Schwab, *The Oriental Renaissance: Europe's Rediscovery of India and the East, 1680–
1880*, trans. by Gene Patterson-King and Victor Reinking (New York: Columbia University
Press), 1984; Karl Baier, *Yoga auf dem Weg nach Westen: Beiträge zur Rezeptionsgeschichte*
(Würzburg: Königshausen & Neumann, 1998); Clarke, *Oriental Enlightenment*, cit.;
Clarke, *Jung and Eastern Thought*, cit.; Shamdasani, "Introduction," cit.; Paul Bishop,
Dreams of Power: Tibetan Religion and The Western Imagination (Cranbury, NJ: Fairleigh
Dickinson University Press, 1994); and Harry Oldmeadow, *Journeys East: 20th Century
Western Encounters with Eastern Religious Traditions* (Bloomington, IN: World Wisdom,
2004).

[130] Christine Maillard, "L'apport de l'Inde à la pensée de Carl Gustav Jung," in Michel
Hulin and Christine Maillard, eds, *L'Inde inspiratrice: Réception de l'Inde en France et en
Allemagne (XIXᵉ–XXᵉ siècles)* (Strasbourg: Presses de l'Université de Strasbourg, 1996),
155–68 at 160.

life, the peak came in the years between the publication of his "Commentary on *The Secret of the Golden Flower*" (1929) and his trip to India (1937–38), which he realized constituted for him an "intermezzo" in his alchemical studies. The "Commentary" was reviewed as "instructive and impressive,"[131] spurring Western psychotherapy to achieve new epistemological depth, and welcomed as the inception of a transformation of psychology's canons, by moving away from the external observation of the object to "a looking inward and a recognition of timeless psychic symbols."[132] In 1930 Jung had managed to invite Richard Wilhelm to lecture at the psychotherapy conference, but the sinologist unfortunately died from a rare infection before he was able to attend. In his obituary tribute to Wilhelm, Jung stressed the value of their kind of collaboration, with a view to a more mature, cross-cultural, and thus complex, knowledge of the psyche, capable of building bridges between (and deriving benefit from) East and West, exhorting his readers to "[i]magine what it means when a practicing physician, who has to deal with people at their most sensitive and receptive, establishes contact with an Eastern system of healing!"[133]

Indeed, Jung's own engagement with Taoism was instrumental in his understanding of psychological wholeness, the integration of opposites, synchronicity, and the Self. Grounded thus in his collaboration with Wilhelm, "this connection between the 'European unconscious spirit' and Eastern eschatology became one of the major themes in Jung's work in the 1930s."[134] This led him progressively to involve himself with yoga, Tantra, and Buddhism through crucial collaborations not only with Jakob

---

[131] W. Stockmayer, "Weiteres Material zum Beleg der Archetypen im Sinne C. G. Jungs," *Zentralblatt für Psychotherapie* 3, no. 5 (1930): 314–17 at 314.

[132] Anonymous review of Richard Wilhelm and C. G. Jung, *Das Geheimnis der goldenen Blüte: Ein chinesisches Lebensbuch* (1929), in *Zentralblatt für Psychotherapie* 3, no. 2 (1930): 88–90 at 89.

[133] "Richard Wilhelm: In Memoriam" [1930], CW 15, § 90. Jung also highlighted the role on that occasion of Jakob Wilhelm Hauer, who lectured on "Der Yoga im Lichte der Psychotherapie" (in E. Kretschmer and W. Cimbal, eds, *Bericht über den V. Allgemeinen Ärztlichen Kongress für Psychotherapie in Baden-Baden, 26. bis 29. April 1930* [Leipzig: Hirzel, 1930], 1–21). In the same vein, Jung would invite Zimmer to open the psychotherapy congress of 1935. See below, 61–62.

[134] Shamdasani, "Introduction: *Liber Novus*," cit., 218. For more on this, see also Shamdasani, "Introduction: Jung's Journey," cit., esp. xxvii–xxxi. Jung had met Wilhelm at the "School of Wisdom" in Darmstadt, a pioneer institution for dialogue between East and East founded by Hermann Keyserling in 1920, which also proved to be a model for Eranos (see Hakl, *Eranos*, cit., ch. 3). The Taoist treatise was also instrumental in Jung's understanding of the circumambulation of the Self, which will be discussed in some depth in the present seminar. See Clarke, *Jung and Eastern Thought*, cit., 80–89.

Hauer and Heinrich Zimmer, but also with scholars such as Walter Evans-Wentz and Daisetz Taitaro Suzuki.[135] Zimmer in particular, as we shall see, considered Tantra the most archaic substratum of aboriginal India and the culmination of Indian thought, and contributed, as did Mircea Eliade, to its establishment in a central position in twentieth-century scholarly study of the history of religion, radically reversing a generally dismissive tendency on account especially of the Tantric emphasis on the body,[136] and eroticism.[137]

With regard to Jung's seminar on the psychology of Kundalini yoga in 1932, Shamdasani writes that

> at a time when psychology was characterized by the reign of behaviorism, the positivist experimental epistemology, and the growing dominance of psychoanalysis, and when developmental phases could hardly be associated with anything other than what was becoming the alpha and omega of the study of personality—the child—Kundalini yoga presented Jung with a model of something that was almost completely lacking in Western psychology—an account of the developmental phases of higher consciousness.[138]

---

[135] See Jung's "Psychological Commentary on *The Tibetan Book of the Great Liberation*" [1939/1954], CW 11. The "Bardo Thodol," commonly known as *The Tibetan Book of the Dead (Das tibetanische Totenbuch)* was published in English by the anthropologist, theosophist, and expert on Tibetan Buddhism W. Y. Evans-Wentz in 1927 (subsequently attaining vast success especially through the American counterculture of the 1960s). See also Jung's "Foreword to Suzuki's *Introduction to Zen Buddhism*" [1939/1949], CW 11, which equally deals, from a Western, psychological perspective, with the book by the great popularizer of Mahayana and Zen Buddhism in the West, Daisetz Teitaro Suzuki: *Die große Befreiung: Einführung in den Zen-Buddhismus* (Leipzig: Weller, 1939), which was translated into German by Zimmer.

[136] For Feuerstein "the body, according to Tantra, holds all the secrets of the universe. It is a miniature replica of the macrocosm. The Tantric adepts also understood the body as a temple of the Divine—an idea present already in ancient Upanishadic times but not followed through" (*Deeper Dimension*, cit., 330).

[137] Hugh B. Urban, *Tantra, Sex, Secrecy, Politics, and Power in the Study of Religion* (Berkeley, CA: University of California Press), 169; see also ch. 5. "For European colonizers, Orientalist scholars, and Christian missionaries of the Victorian era, Tantra was generally seen as the worst, most degenerate and depraved example of all the worst tendencies in the 'Indian mind', a pathological mixture of religion and sensuality that had led to the decline of modern Hinduism" (Hugh B. Urban, *The Power of Tantra: Religion, Sexuality, and the Politics of South Asian Studies* [London: I. B. Tauris 2010], 1). "Yet for most contemporary New Age and popular writers, conversely, Tantra is now celebrated as much-needed affirmation of physical pleasure and sexuality, as a 'yoga of sex' or 'cult or ecstasy' that might counteract the hypocritical prudery of the Christian West" (ibid.).

[138] Shamdasani, "Introduction: Jung's Journey," cit., xxiii–xxiv.

Indeed, Jung pointed out, for instance, that

[t]he analytic process thus occasions a broadening of consciousness, but the relation of the ego to its objects still remains [. . .]. Only in the continuation of the analysis does the analogy with yoga set in, in that consciousness is severed from its objects (*Secret of the Golden Flower*). This process is linked up with the process of individuation, which begins with the self severing itself as unique from the objects and the ego. It is as if consciousness separated from the objects and from the ego and emigrated to the non-ego—to the other center, to the foreign yet originally own [. . .]. It is a psychical experience, which in practice is expressed as a feeling of deliverance [. . .]. One does not become apathetic but is freed from entanglement. Consciousness is removed to a sphere of objectlessness.[139]

In his pondering of the connections between Eastern soteriological techniques and the paths of Western depth psychology, which will be further discussed in section XI of this Introduction, Jung's high esteem for the teachings of Taoism, Hinduism, Tantrism, or Buddhism did not extend to a promotion of their methods. Assuming that "any religious or philosophical practice amounts to a psychological discipline; in other words, it is a method of psychic hygiene,"[140] he interpreted yoga from a psychological and functional, rather than a philosophical or metaphysical, perspective. His approach to the East was thus marked by a fundamental dichotomy. On the one—more epistemic—hand, he sought to investigate morphological affinities, with the aim of identifying recurring transpersonal archetypal structures and thereby demonstrating the validity of his thesis of the collective unconscious. On the other—more practical—hand, he discouraged Westerners from engaging in adopting Eastern yoga practices, invoking their different historical, cultural, and psychological background. Already in 1923 he had stated, in the previously cited letter to Oscar A. H. Schmitz,

The products of the Oriental mind are based on its own peculiar history, which is radically different from ours. Those peoples have gone through an uninterrupted development from the primitive state of natural polydemonism to polytheism at its most splendid, and beyond that to a religion of ideas within which the originally magical

---

[139] Jung, *Psychology of Kundalini Yoga*, cit., 82–83.
[140] "Psychotherapists or the Clergy" [1932], *CW* 11, § 531; "Yoga and the West" [1936], *CW* 11, § 866.

practices could evolve into a method of self-improvement. These antecedents do not apply to us.[141]

He even called "poisonous" the use by Westerners of Eastern "preformed" meditation instruments—including traditional sacred devices such as yantras and mandalas—in the persuasion that the Westerner would do this "only in order to increase his specific lie."[142] Jung argued, conversely, that the Western psyche's *production* (of its own symbols), rather than *imitatation* (of Eastern symbols), should remain paramount.

In his essay "Yoga and the West" (1936), he insisted that "[t]he spiritual development of the West has been along entirely different lines from that of the East and has therefore produced conditions which are the most unfavorable soil one can think of for the application of yoga."[143] The Western attitude toward yoga, Jung assumed, had developed along two separate lines: on the one hand it was regarded as a strictly academic science, and on the other it became "something very like a religion," especially on account of the influence of theosophy.[144] Eastern teaching, in his judgment, was understood through the filter of "a specifically Western phenomenon—really set in with the Renaissance": namely, the split between "science and philosophy"; to which he added that the "syncretistic outgrowths" and "the importation on a mass scale of exotic religious systems" in the nineteenth century could be seen as roughly corresponding to "the Hellenistic syncretism of the third and fourth centuries A.D."[145] Convinced of the virtually irreconcilabile distance between the West and "the unity and wholeness of nature which is characteristic of yoga," Jung argued that

[t]hrough his historical development, the European has become so far removed from his roots that his mind was finally split into faith

---

[141] Jung, C. G. *Jung: Letters*, cit., 1:39 (May 26, 1923).

[142] *DAS*, 493. In this Jung's position differed from that of Avalon, who maintained that "[w]e, who are foreigners, must place ourselves in the skin of the Hindu, and must look at their doctrine and ritual through their eyes and not our own" (quoted in Shamdasani, "Introduction: Jung's Journey," cit., xxvi). For a critical analysis of Jung's differentiated stance toward the East, and more specifically toward Buddhism, see respectively Clarke, *Jung and Eastern Thought*, cit., chs 8 and 9, and Polly Young-Eisendrath and Shoji Muramoto, eds, *Awakening and Insight: Zen Buddhism and Psychotherapy* (Hove: Brunner-Routledge), 2002. For further valuable insights, see Leon Schlamm, "C. G. Jung's Ambivalent Relationship to the Hindu Religious Tradition: A Depth-Psychologist's Encounter with 'The Dreamlike World of India'," *Journal of Indian Philosophy and Religion* 4 (1999), 49–74.

[143] "Yoga and the West" [1936], CW 11, § 876.

[144] Ibid., § 859.

[145] Ibid., § 860 and 861.

and knowledge, in the same way that every psychological exaggeration breaks up into its inherent opposites. He needs to return, not to Nature in the manner of Rousseau, but to his own nature. His task is to find the natural man again.[146]

So despite the "important parallels" he found with the symbolism of Tantric yoga, Lamaism, and Taoistic Yoga, he maintained that "in the West, nothing ought to be forced on the unconscious" for "usually, consciousness is characterized by an intensity and narrowness that have a cramping effect, and this ought not to be emphasized still further." To mitigate such conditions, Jung listed a variety of Western methods, such as Schultz's autogenic training and his own method, which "like Freud's, is built up on the practice of confession." Contextually, he defined the active imagination as "special training for switching off consciousness, at least to a relative extent, thus giving the unconscious contents a chance to develop." And after recalling Ignatius of Loyola's spiritual exercises, he concluded his article on the subject by asserting, "In the course of the centuries the West will produce its own yoga, and it will be on the basis laid down by Christianity."[147]

This hermeneutical stance, aimed at establishing a psychology on a comparative, multidisciplinary basis, was profoundly endorsed by the founder of Eranos, Olga Fröbe-Kapteyn, who situated Jung's analytical psychology itself in line with previous "Yoga traditions" of the West such as those of the Gnostics, Hermetics, and alchemists. Consequently, the

---

[146] Ibid., § 867 and 868. And shortly after, "Western man has no need of more superiority over nature, whether outside or inside. He has both in almost devilish perfection. What he lacks is conscious recognition of his inferiority to the nature around and within him. He must learn that he may not do exactly as he wills. If he does not learn this, his own nature will destroy him" (§ 870).

[147] Ibid., § 874–876. In the ETH Lectures Jung would further discuss this nexus (C. G. Jung, *Psychology of Yoga and Meditation: Lectures delivered at ETH Zurich, Volume 6: 1938 to 1940*, ed. and introduced by Martin Liebscher, trans. by Heather McCartney and John Peck [Philemon Series] [Princeton, NJ: Princeton University Press, 2020], 164 and passim); while in his "Psychological Commentary on *The Tibetan Book of the Great Liberation*" (1939), he would advocate for "get[ting] at the Eastern values from within and not from without, seeking them in ourselves, in the unconscious. We shall then discover how great is our fear of the unconscious and how formidable are our resistances. Because of these resistances we doubt the very thing that seems so obvious to the East, namely, the *self-liberating power of the introverted mind*" ("Psychological Commentary on *The Tibetan Book of the Great Liberation*" [1939/1954], CW 11, § 773). Such a self-liberating power, he argued, is far from being encouraged in "our religion [. . .]. Yet a very modern form of psychology—'analytical' or 'complex' psychology—envisages the possibility of there being certain processes in the unconscious which, by virtue of their symbolism, compensate the defects and anfractuosities of the conscious attitude" (§ 779).

search for correspondences and differences between Eastern and Western traditions, which she considered the core aim of Eranos's congressional project, could pave the way for a synthesis capable of redeeming Europe in its crisis. In this vein she created in the 1930s, in parallel with the summer meetings and at Jung's suggestion, an impressive archive of archetypal images which was to become the basis of the New York Archive for Research in Archetypal Symbolism (ARAS).[148] Years later, Fröbe-Kapteyn would recall that "C. G. Jung's participation [at Eranos] revealed the gateway to the archetypes, and that established the foundation and background for all our other efforts."[149]

Eranos's cross-cultural project for a new *Bildung* aimed at an amplification of Western consciousness was consonant with Jung's commitment "to know everything that is of vital importance to the life of the psyche," for "psychotherapy is an intermediate field of research which requires the collaboration of many different branches of learning."[150] In Eranos Jung met and conversed with an impressive range of scholars, such as the sinologist Erwin Rousselle, the mythologist Károly Kerényi, the Islamologist Henry Corbin, the zoologist Adolf Portmann, and the philosopher and professor of Jewish mysticism Gershom Scholem. The opening lecture at the first Eranos meeting in August, 1933, dedicated to "Yoga and Meditation in East and West," was given by Heinrich Zimmer; which brings us to an appropriate point to introduce this eminent Indologist.

9. Heinrich R. Zimmer: A Biographical Profile

Heinrich Robert Zimmer was born in Greifswald in 1890, and went to Berlin and Munich to study Sanskrit philology, art and literary history, and comparative linguistics. After graduating in 1913, he served in the military during the First World War as an interpreter of French. This experience was a real rite of passage: "Four year of human, sub- and superhuman experience, initiations in the trenches, staffs, meeting more people

---

[148] See https:/aras.org/about (accessed March 24, 2024). For the archive's history, see Bernardini, *Jung a Eranos*, cit., part 3.

[149] Olga Fröbe-Kapteyn, "Vorwort," *Eranos-Jahrbuch XXV/1956* (Zurich: Rhein, 1957), 7. See also Giovanni Sorge, "Love as Devotion: Olga Fröbe-Kapteyn's Relationship with Eranos and Jungian Psychology," in Fabio Merlini, Lawrence E. Sullivan, Riccardo Bernardini, and Kate Olson, eds, *Love on a Fragile Thread/L'amore sul filo della fragilità: Eranos Yearbook/Annale di Eranos 70, 2009-2010-2011* (Ascona and Einsiedeln: Eranos Foundation/Daimon, 2012), 388–434. For a thorough examination of the larger context, see Hakl, *Eranos*, cit., and Bernardini, *Jung a Eranos*, cit.

[150] "Editorial" [1935], CW 10, § 1046.

than I ever did before and watching them in all kinds of revealing attitudes. The initiation of Life (including Death) playing its own symphony with the fullest possible orchestration."[151] Afterwards he took up the study of Sinology, Indology, Buddhist and Tantric texts, mythology, and symbolism, and in 1920 was appointed associate professor in Indian philology at the University of Greifswald. In 1926 he was named professor of Indology and mythological research in Heidelberg and published *Kunstform und Yoga im indischen Kultbild*:[152] a work which served, as the leading perennialist and historian of Indian art Ananda Coomaraswamy stated in a 1932 review, to change the direction of the study of Indian art in the West, for "no more valuable book for the understanding of Indian art [. . .] has yet been published."[153] This book brought Zimmer a recognition that continued to grow with his other major studies from the interwar period.[154]

In 1928, Zimmer married Christiane von Hofmannsthal (only daughter of the celebrated Austrian writer and poet), who gave him three children; after his father-in-law's death in 1929, he served as co-publisher of the *Nachlass* and anonymously edited some of Hofmannsthal's posthumously published works.[155] He also had another lifelong relationship, known to his wife, with Mila Esslinger Rauch, with whom he had three more children.

---

[151] Heinrich Zimmer, "Some Biographical Remarks about Henry R. Zimmer" (January 1943), in Heinrich Zimmer, *Artistic Form and Yoga in the Sacred Images of India*, trans. and ed. by Gerald Chapple and James B. Lawson with J. M. McNight (Princeton, NJ: Princeton University Press, 1984), 243–60 at 250.

[152] Heinrich Zimmer, *Kunstform und Yoga im indischen Kultbild* (Berlin: Frankfurter Verlags-Anstalt A.-G., 1926); translated as *Artistic Form and Yoga in the Sacred Images of India*, cit. (henceforth *Artistic Form and Yoga*).

[153] Quoted in Mary F. Linda, "Zimmer and Coomaraswamy: Visions and Visualizations," in Case, *Heinrich Zimmer*, cit., 119–41 at 122. Hauer too praised Zimmer's "geistreiche[s] Buch" (ingenious book) in his [Foreword to] *Der Yoga als Heilweg: Nach den indischen Quellen dargestellt* (Stuttgart: W. Kohlhammer, 1932), v–xvii at v–vi.

[154] These were: *Ewiges Indien: Leitmotive indischen Daseins* (Eternal India: leitmotifs of Indian existence) (1930), probably "Zimmer's least translatable" book because of the difficulty of its "philosophical language and verbal pyrotechnics" (Gerald Chapple, "Heinrich and Henry R. Zimmer: The Translator Translated," in Case, *Heinrich Zimmer*, cit., 61–86 at 69 n. 17); *Maya: Der indische Mythos* (Maya: the myths of India) (1936), Zimmer's magnum opus along with *Artistic Form and Yoga*; and *Weisheit Indiens: Märchen und Sinnbilder* (Wisdom of India: fairy tales and parables) (1938), his last book to appear in Germany, dedicated to Jung ("al maestro di color che sanno in Verehrung und Dankbarkeit" ["to the *maestro* of those who know"—recalling Dante's epithet for Aristotle: *Inferno*, canto IV, 132—"with admiration and gratitude"]).

[155] Lukas Rauch, personal communication (Winterthur, 2010). For a sensitive sketch of Zimmer by Hofmannstahl, see Chapple, "Heinrich and Henry R. Zimmer," cit., 68–69.

Zimmer was not only a brilliant scholar and popularizer of Indian thought and myths, but also a prolific translator from Sanskrit.[156] He maintained, moreover, a vast network of relationships with intellectuals, philosophers, and literati, including Karl Jaspers, Emil Nolde, Hermann Hesse, Alfred Weber, Alfred Kubin, and Raymond Klibansky.[157] He was also a friend of Thomas Mann, and provided inspiration for Mann's novella of 1940 *Die vertauschten Köpfe—Eine indische Legende* (*The Transposed Heads: A Legend of India*). Zimmer's personal connection with Jung from 1932 was a turning point in his life, which also opened up to him several new possibilities and audiences[158] (though not everyone welcomed Zimmer's growing closeness to analytical psychology: several "head-shaking colleagues," such as Ludwig Edelstein, "regretted these diversions of his activity, valuable as they were in themselves. We wished that he would devote himself entirely to his own field, in which he could make a unique contribution.")[159] The Eranos conferences proved moreover to be a further occasion for the meeting of kindred spirits, and Zimmer's opening lecture on Tantric yoga at the first conference was followed by three more appearances, in 1934, 1938, and 1939.[160]

---

[156] See Heinrich Zimmer, *Karman: Ein buddhistischer Legendenkranz* (Munich: Bruckmann, 1925); then came *Anbetung mir: Indische Offenbarungswörte* (Munich: Oldenburg, 1929); *Spiel um den Elefanten: Ein Buch von indischer Natur* (Munich: Oldenburg, 1929); Sir George Dunbar, *Geschichte Indiens von den ältesten Zeiten bis zur Gegenwart* (Munich: Oldenburg, 1937); Daisetz Teitaro Suzuki's *Die große Befreiung*, cit. (1939); and *Zeichen der Liebe: Wie man lieben und die Liebe dichten soll; Ein indisches Lehrbuch für Liebende und Dichter* (Stuttgart: Hatje, 1948).

[157] Klibansky, a German-Canadian historian of philosophy, remembered Zimmer as being "a genial fellow [...]. He set forth a profound interpretation of Hindu mythology and Hindu philosophy. His work was not merely philological, and was not confined to India alone, especially because he was equally interested in psychology, which sharply distinguished his India, whose burden he carried lightly, invoked from the India of other Indologists" (Raymond Klibansky, *Erinnerung an ein Jahrhundert: Gespräche mit Georges Leroux* [Frankfurt am Main: Insel, 2001], 60).

[158] Margaret Case also notes that after Zimmer met Jung "his writing became more cautious, as if some doubts had finally found a voice. The early years of their friendship coincided with the increasingly difficult times in Nazi Germany, and it was as if Jung had given him the language to express at least indirectly the discouragement that he must have felt even as he continued to exude good-humored vitality" (Margaret H. Case, "Introduction" to Case, *Heinrich Zimmer*, cit., 3–13 at 9).

[159] From Edelstein's "Preface" to Henry R. Zimmer, *Hindu Medicine* (Baltimore: Johns Hopkins Press, 1948), xxii–xxiii. On the question of Zimmer's "Jungianism," see also Case, "Introduction," cit., 3 and passim.

[160] See Heinrich Zimmer, "On the Significance of the Indian Tantric Yoga" (1933), in Joseph Campbell, ed., *Papers from the Eranos Yearbooks, Eranos 4: Spiritual Disciplines* (Bollingen Series XXX) (Princeton, NJ: Princeton University Press, 1960), 3–58. His subsequent Eranos lectures were entitled, respectively, "Indische Mythen als Symbole" (Indian

Yet for political reasons, and because of the non-Aryian ancestry of his wife, from 1933 Zimmer came under surveillance by the Heidelberg police. Despite "severely reduced publishing possibilities" he continued to publish also in non-academic journals on various topics ("ranging from Indian temples to Richard Wagner to a playful treatment of Sisyphus as a neurotic"), but, as Gerald Chapple notes, "the tragedy was that Zimmer, still in his forties and at the height of his powers, was being cut off from communicating his discoveries at a time when he was profitably expanding his interests in so many directions."[161] After escaping arrest in November 1938, in March 1939 he emigrated to Oxford where he received an (unpaid) position as visiting professor at Balliol College. Subsequently, on June 1, 1940, he entered the United States, where despite difficulties he adapted well, moving his family to New Rochelle in New York state. Already in November he gave three lectures on Hindu medicine at the Johns Hopkins Institute of the History of Medicine in Baltimore (to be published in 1948).[162] In the United States as in Germany, however, a degree of ostracism, on the part of two powerful established academic figures in particular, contributed to the obstruction of his career,[163] so that only in early 1943, a few months before his sudden death, was he eventually appointed visiting lecturer in Hindu Philosophy and Religion at Columbia University.

Jung's acquaintances in the United States proved to be important for Zimmer. Upon his arrival in New York he had entered into contact with the American Jungians Kristine Mann, Eleonore Bertine, and Esther Harding, and he gave several lectures and courses at the Analytical Psychology Club. Furthermore, "from the outset [Zimmer became] the principal advisor, mentor, and guide" of Mary Conover Mellon,[164] who with her husband, the multimillionaire philanthropist Paul Mellon, established the Bollingen Press, and subsequently, from 1942, the Bollingen Foundation, an educational

---

myth as symbols), "Die indische Weltmutter" (The Indian world-mother), and "Tod und Wiedergeburt im indischen Licht" (Death and rebirth in Indian light).

[161] Chapple, "Heinrich and Henry R. Zimmer," cit., 73.

[162] Zimmer, *Hindu Medicine*, cit.; the above-mentioned "Editor's Preface" by Zimmer's student and friend Ludwig Edelstein provides a vivid portrait of Zimmer. For more on Zimmer's study of Indian medical lore, see Kenneth G. Zysk, "Magic, Myth, Mysticism, Medicine," in Case, *Heinrich Zimmer*, cit., 87–107, who credits him with being "perhaps the first scholar to notice in Vedic hymns the importance of the myths associated with healing plants" (92).

[163] See in this regard Zimmer, "Some Biographical Remarks," cit., 247; Chapple, "Heinrich and Henry R. Zimmer," cit., 63, 77–79, and passim.

[164] William McGuire, "Zimmer and the Mellons," in Case, *Heinrich Zimmer*, cit., 31–42 at 34.

institution which awarded over three thousand fellowships, to among others Eranos lecturers such as Kerényi, Eliade, Neumann, and Paul Radin. Appointed to the editorial board of the Bollingen Series, Zimmer exercised a "manifold" influence in planning the Bollingen editorial series, which came to include, besides Jung's works and a selection of Eranos lectures (in six volumes originally shepherded by Zimmer and edited by Joseph Campbell), important studies, often beautifully illustrated, in cultural and art history, archaeology, religion, anthropology, philosophy, mythology and symbolism, psychology, aesthetics, literary criticism, and poetry. It was Zimmer who proposed that the series be published by Pantheon Books of New York, founded by his friend Kurt Wolff, who had emigrated from Germany to New York as an exile in 1941.[165] On the editorial board was Ximena de Angulo Roelli, who together with her mother, Jung's translator and confidante Cary Fink Baynes, already knew Zimmer from the Eranos meetings during the 1930s. She remembers him thus:

> Zimmer? Oh my, who could ever forget him! He was 'come una fontana che sprizza parole' [like a fountain spouting words]. His manuscripts were built just like a musician's score, marked up in different colors, red and green and "crescendo" everywhere, etcetera. And he also spoke in the same way. He was fabulous. He was a thrilling lecturer. The tragic thing was that he was an Indologist who never had a penny, he never got to India. It was terrible.[166]

Indeed Zimmer's three attempts to visit India failed.[167] And on March 20, 1943—he was during those weeks lecturing on the Tarot in

---

[165] Ibid., 39. The Bollingen Series (see its impressive list in McGuire, *Bollingen*, cit., 295–309) was transferred in the late 1960s to Princeton University Press.

[166] Ximena de Angulo Roelli, personal communication (Cavigliano, 2008). Cf. Joseph Campbell's recollection of how Zimmer worked on his lectures in Chapple, "Heinrich and Henry R. Zimmer," cit., 75, and the enthused recollection of a listener in Ascona in 1938: "When he is excited or has had a glass of Italian wine, he spouts vocabulary like a geyser or like James Joyce. To hear him is like watching Shankar dance. It is mythology orchestrated" (quoted by McGuire, *Bollingen*, cit., 30–31).

[167] After an unsuccessful plan for a trip in the Twenties, Zimmer was asked by his friend Helmuth von Glasenapp, professor of Indology at Königsberg and of Comparative Religion at Tübingen, to guide an academic tour to India in January 1937, planned through a Viennese travel agency (Zimmer to Esslinger, October 7, 1936, BHZME J14 1936b/423, DLA); the plan was repeatedly postponed and finally canceled. "Maybe the travel agency in Vienna is not purely Aryan," he commented bitterly (Zimmer to Esslinger, undated [November/December 1936], BHZME J14 1936b/222, DLA). Chapple instead reports a refusal from the ministry due to Zimmer's wife's Jewish ancestry ("Heinrich and Henry R. Zimmer, cit., 73). A later planned Indian journey in the autumn of 1939 "probably with

the apartment of artist and anthropologist Maud Oakes—he was fatally struck down by a brief bout of pneumonia, at the premature age of fifty-two. At the end of that year the first volume of the Bollingen Series, a "Navaho war ceremonial" entitled *Where the Two Came to Their Father*, was published, dedicated to the scholar "who was so greatly instrumental in the founding of the Bollingen Series, and whose generous advice and help will be missed by those who carry it forward."[168] The following year, as a kind of memorial to him, Jung edited his translation of writings by Sri Ramana Maharshi,[169] the Indian guru whom Jung—to Zimmer's regret—had chosen not to visit on his trip to India (preferring to see a humble disciple of his).[170] Zimmer's premature decease not only left a great void in Oriental studies, but also cut short a remarkable dialogue with Jung's depth psychology. Zimmer's life-long friend, the writer Herbert Nette, pondered "how thoroughly he had absorbed and taken to heart the teachings of the East" (and, we might also add, those of Jung), quoting one of Zimmer's last letters to him, in which he had written,

> [O]nly when you sense that everything that happens to you, and is somehow connected with you, is something that really belongs to you as a unified whole—only then can you understand your own fate and sense its meaning; that is *amor fati*. A fate we cannot transform into meaning will crush us.[171]

Left to deal with the great scholar's legacy was his pupil Joseph Campbell, the influential author and mythologist (known especially for *The Hero with a Thousand Faces* [1949], which he reportedly composed "in a baroque style 'à la Zimmer'")[172] who, with the assistance of Ananda Coomaraswamy, went on to edit Zimmer's unpublished works in four volumes.[173]

Peter Baynes [...] was—alas, as always—prevented by the outbreak of war" (Hannah, *Jung: His Life and Work*, cit., 217).

[168] McGuire, "Zimmer and the Mellons," cit., 40. See also McGuire, "Foreword to the Third Edition," in Joseph Campbell, ed., *Where the Two Came to Their Father: A Navaho War Ceremonial Given by Jeff King* (text and paintings recorded by Maud Oakes) (Bollingen Series I) (Princeton, NJ: Princeton University Press, 1991 [1943]), vii.

[169] Heinrich Zimmer, trans., *Der Weg zum Selbst: Lehre und Leben des Heiligen Shri Ramana Maharshi aus Tiruvannamalei*, ed. by C. G. Jung (Zurich: Rascher, 1944).

[170] See Jung's Introduction to Zimmer, *Der Weg zum Selbst*, cit.: "The Holy Men of India" [1944], CW 11, § 952–53.

[171] Herbert Nette, "An Epitaph for Heinrich Zimmer," [1948], in Case, *Heinrich Zimmer*, cit., 21–30 at 30.

[172] Chapple, "Heinrich and Henry R. Zimmer," cit., 81 n. 55.

[173] See ibid., 80–84. The books are: *Myths and Symbols in Indian Art and Civilization* (Bollingen Series VI) (New York: Pantheon, 1946); *Hindu Medicine*, cit.; *The King and the*

10. Amazed and Amused: Impressions from the Jung–Zimmer Encounter

Having read Zimmer's "fascinating book" *Kunstform und Yoga* (*Artistic Form and Yoga*) only after publishing his "Commentary on *The Secret of the Golden Flower*," Jung had not been able "to make use of Zimmer's material, which was of the greatest value for me."[174] He also reports that his commentary initially provoked anger in Zimmer, who "dashed the book against the wall [. . .]. Like so many others, he reacted to the word 'psychological' as a bull does to the red cape."[175] Yet a letter of Zimmer's to Herbert Nette of the end of 1931 sheds a different light on Jung's recollection:

> As you know, I [. . .] don't quite share Jung's views, although I consider his notion of aggregation suitable for merely private-personal orientation and communication. The fact that he only saw a copy of my old book [*Artistic Form and Yoga*] so late explains why he didn't mention it in *The Golden Flower*; at the time, I did feel somewhat offended, just as that circumstance troubled him. Naturally, R. Wilhelm was familiar with it, and I never quite understood why he never mentioned the sheer fact that the technique, structure, and material of the Chinese treatise can be understood only as a special Buddhist form of general-Indian Tantrism in a blended late-Buddhist, and Chinese Taoist atmosphere. Besides, Jung, notwithstanding his many great aspects, is probably a power-happy fox, inflated by a warranted self-assurance about his intellect, his vigor, and the adoration he has received from solvent and devoted females. Which is not meant as an objection. Someone who no longer fears his godlikeness can be quite a respectable fellow, as matters stand nowadays.[176]

Corpse: Tales of the Soul's Conquest of Evil (Bollingen Series XI) (New York: Pantheon, 1948); *Philosophies of India* (Bollingen Series XXVI) (New York: Pantheon, 1951); and *The Art of Indian Asia: Its Mythology and Transformations*, 2 vols (Bollingen Series XXXIX) (New York: Pantheon, 1955). See also the "Selected Bibliography of Zimmer's Works" in Zimmer, *Artistic Form and Yoga*, cit., 261–67. For a discussion of Campbell's own editorial role with regard to Zimmer's English corpus, see Wendy Doniger, "The King and the Corpse and the Rabbi and the Talk-Show Star: Zimmer's Legacy to Mythologists and Indologists," in Case, *Heinrich Zimmer*, cit., 51–59.

[174] C. G. Jung and Aniela Jaffé, *Erinnerungen, Träume, Gedanken*, recorded and ed. by Aniela Jaffé (Zurich and Stuttgart: Rascher, 1962), 385.

[175] Ibid.

[176] Zimmer to Nette, undated (ca. November/December 1931), BHZDR 17-5, DLA. Quoting the same passage, Chapple notes that, "[o]ne irony here is that, after he landed in

It may be noticed that the reference to "godlikeness" [*Gottähnlichkeit*], surely familiar from Zimmer's longstanding involvement with the super-humanity of countless deities of the Hindu pantheon, perhaps hints at his reading of Jung's *The Relations* (1928), in which this term, borrowed from Goethe's *Faust* and soon to become a leitmotif in Jung's work, was largely used with reference to the issue of psychic inflation or "self-deification"; more specifically, it was linked with the figure of the "mana personality" which Zimmer, as we shall see, was to discuss again.[177]

For sure, Zimmer's initial perplexities regarding Jung's method would rapidly recede in the wake of their encounter, which "marked a watershed in Zimmer's life."[178] In that period he came into contact with Gustav Richard Heyer, a close friend of Jung and among the first popularizers of his ideas in Germany.[179] A successful neuropathologist and psychotherapist, who stands among the fathers of psychosomatics, Heyer led, in the lively Munich cultural scene buzzing with a bewildering mixture of artistic vanguards, literary ferments, and occultist interests, a psychoanalytic working group notable for its female contingent—so much so that it was called "der Amazonenclub."[180] With Heyer's mediation, Zimmer first met Jung in June 1932 and spent some days, with his wife, as a guest in Jung's home in Küsnacht. "When I first met him," he reminisced later in New York, "and observed him during a weekend I spent in his house, at

New York, Zimmer was to become very dependent on the financial support of the ladies from the Jungian circle, nicknamed with a pun *die Jungfrauen*, 'the virgins'" (Chapple, "Heinrich and Henry R. Zimmer," cit., 71; see also n. 22).

[177] See below [ZL], 96ff.; see also Jung, *The Relations between the I and the Unconscious* [1928], CW 7, chs 1.2 and 2.4. Jung treated this theme thoroughly in his Zarathustra Seminar: *Nietzsche's Zarathustra*, cit.

[178] Case, "Introduction," cit., 7. See also Henry R. Zimmer, "The Impress of Dr. Jung on My Profession," in Case, *Heinrich Zimmer*, cit., 43–48.

[179] Zimmer wrote to Esslinger, "I had lunch with a well-known psychoanalyst. We had a good talk about my yoga plans, and I very much hope he secures my participation at a conference on yoga next summer, at C. G. Jung's in Zurich, where I could give a talk on Hatha Yoga to a gathering of initiates. This could help different kinds of people and things take a step forward" (Zimmer to Esslinger, undated [November 1931], BHZME AN 669, DLA).

[180] Lockot, *Erinnern und Durcharbeiten*, cit., 162. For more information, including an autobiographical portrait of Heyer, see ibid., 161–72. Heyer can be accounted, together with Kranefeldt, among the more fervent defenders of Jung's psychology and fierce opponents of psychoanalysis. He lectured at Eranos in 1933, 1934, 1935, and 1938. In 1937 he joined the Nazi Party, and in 1939 moved to Berlin to work at the Göring Institute. Heyer is credited with having protected some Jews, among them Käthe Bügler and Max Zeller (see Thomas B. Kirsch, *The Jungians: A Comparative and Historical Perspective* [London: Routledge, 2000], 96; also 125–26). Before the war's end Jung had broken off all relations with him.

his table and in his garden, in his boat on the lake and while meeting people, he struck me as the most accomplished embodiment of the great medicine man, the perfect wizard, the master of Zen initiations."[181] Zimmer's successful debut lecture on June 18, 1932 at the Psychological Club on "Einige Aspekte des Yoga" (Certain aspects of yoga) initiated a collaboration that lasted until his emigration.[182] In October, Zimmer attended the previously mentioned seminar on Kundalini yoga, while on January 14, 1933, he was invited to Berlin by the Society for Analytical Psychology. He gave a lecture with the title "Yoga und Wir" (Yoga and us), in which he discussed the relationships between yoga and Western psychotherapy. "It is most amusing," he wrote to Mila Esslinger, "that the psychoanalysts were rather baffled, since they had expected Jung to receive either outright affirmation or homage [. . .], but what became evident were the distances!," while "others found his way of handling matters, without any gossip and secretiveness whatsoever, very appealing."[183] Indeed, Zimmer found sympathetic ears, and would be invited to lecture again in the years to follow.[184] Käthe Bügler reminisces that

---

[181] Zimmer, "Impress of Dr. Jung," cit., 45.

[182] As proved by his subsequent lectures listed in the register of the Zurich Psychologischer Club: "Hathayoga. Mit Lichtbildern" (On Hatha yoga, with transparencies), June 10, 1933; "Die Geschichte vom Indischen König und dem Leichnam" (The Indian tale of the king and the corpse), June 23, 1934 (which was particularly appreciated by Jung, and published, with a few changes, in the volume in honor of his sixtieth birthday edited by the Club the following year: *Die Kulturelle Bedeutung der Komplexen Psychologie* [Berlin: Springer, 1935], 171–94); "Indische Schlangensymbolik" (Indian serpent symbolism), June 1, 1935; "Der autonome Komplex in der Dingwelt oder Abu Kasems Pantoffeln (The autonomous complex in the thing-world, or Abu Kasem's slippers), March 28, 1936; "Der seelische Hintergrund indischer Volkserzählungen" (The spiritual background of Indian folktales), October 30, 1937; and "Abenteuer zweier Artusritter (Gawain und Owain) als Individuationssymbole des Atruskreises" (The adventures of two Arthurian knights, Gawain and Owain, as individuation symbols from King Arthur's Round Table), June 25, 1938.

[183] Zimmer to Esslinger, January 17, 1933, BHZME AN 683, DLA.

[184] On November 11, 1933, Zimmer lectured on "Die praktische Psychologie des Tantra Yoga" (The practical psychology of Tantra yoga) and the following spring he held a six-day seminar (see Zimmer, *Bericht über das Berliner Seminar "Indische Mythen im Lichte neuerer Psychologie,"* vom 19. bis 24. März 1934, Berlin, 1934, cyclostyled typescript, 85 pp., JA) at which he presented a wide array of tales and comments on various Hindu gods, extending also to Greek mythology and concluding, notably, with the Chassidic story of Rabbi Löb of Prague: an impoverished Rabbi who, following a dream, travels to Prague to look for a treasure, only to discover that the treasure lies beneath his own stove at home (cf. "The Treasure" in Martin Buber, *Tales of the Hasidism: The Later Masters* [New York: Schocken Books, 1975 (1948)], 245–46). On January 22, 23, and 24, 1936, he would give the following lectures: "Die Geschichte von Abu Kasems Pantoffeln" (The story of Abu Kasem's slippers); "Der autonome Komplex in der Dingewelt" (The autonomous complex

Zimmer's lectures were more than a translation from Sanskrit. These lively stories suddenly became our own life, our essence and passion. We could not discern how brilliantly the language became living speech, and mythical story turned into visible action. These fabulous experiences flowed over us in vivid colors, becoming intelligible to our European outlook [. . .]. Afterwards as usual one of the members had us join Zimmer for discussion over wine and a buffet until the wee hours. In Zimmer's presence everything turns light and lively.[185]

Likewise Zimmer's lecture on Hatha yoga on June 10, 1933 at the Psychological Club was much appreciated by Jung.[186] It was likely on this occasion that Jung invited Zimmer to introduce his Berlin Seminar with a lecture. Bügler described the latter as a "brilliant performance,"[187] while Barbara Hannah found it

> a masterpiece. Zimmer held the audience spellbound, a feat, since he was little known at that time and the large audience, which had collected from all over Germany and abroad, had come primarily to hear Jung and were disappointed that he was not lecturing that first Sunday evening. But Zimmer had not only an excellent knowledge and grasp of his subject but also a very creative mind and an extremely lively delivery: in short, he was one of the best lecturers I have ever heard.[188]

Zimmer's voiced his satisfaction to Mila Esslinger: "I can take only a few quick minutes to write to you, so great is the whirl I'm swimming in. The talk went over in a big way, more so than the earlier one [i.e., his lecture of January 14], many here are telling me so," adding that "Jung's Seminar is excellent and very stimulating, one can't help meeting a lot of notables."[189] He then reported that his presentation had been "a big hit and very well-attended" and the audience "quite warm and forthcoming." Furthermore,

---

in the world of things); "Symbolische Züge in Vishnu Mythos" (Symbolic characteristics in the Vishnu myth); and "Schiva als Animus Figur" (Shiva as an animus figure).

[185] Bügler, "Die Entwicklung," cit., 30.

[186] "Zimmer was here in Zurich recently and gave us an extremely beautiful lecture on Hatha yoga. Zimmer has achieved a deepening and a ripening. On this occasion he and his wife stayed with us" (Jung to Jacobi, June 19, 1933, JA [Hs 1056:2142]).

[187] Bügler, "Die Entwicklung," cit., 29.

[188] Hannah, *Jung: His Life and Work*, cit., 210.

[189] Zimmer to Esslinger, June 29, 1933, BHZME AN 973, DLA.

in every respect the stay here was gratifying, my standing with Jung and his circle obviously seems to be rather high. What he himself presented on that occasion was extremely fine and altogether eminently intelligent, as only he could have brought it off. I'm still feeling strongly, after the meeting with Jung, that we have both, I believe, gotten a solid feeling for each other, and in meeting me he really heard the microcosmic human being seated in the last row. He's a man from whom I can always take repeated nourishment, as in the Hindu myths.[190]

From these spur-of-the-moment impressions, it is clear that Jung's personality and mind exerted considerable charm over Zimmer, who according to his son Lukas Rauch "found in Jung an ideal conversation partner."[191] Years later Zimmer even wrote, "The mere fact that nature allowed this unique, mountainous example of the human species to come into existence, was, and is, one of the major blessings of my spiritual and my very earthly life."[192] Jung in turn reminisced: "I found him to be a congenial man of the most animated temperament. He was voluble and spoke rapidly, but could also listen intently and with keen attention."[193] He also wrote that in their work together "he gave me invaluable insights into the Oriental psyche, not only through his immense technical knowledge, but above all through his brilliant grasp of the meaning and content of Indian mythology."[194]

11. Yoga, Individuation, and Western Psychotherapy According to Zimmer

Zimmer recalled that when he began his studies, the idea of Indian philosophy itself was widely seen as a contradiction in terms. However, thanks to the cultural ferment in religious studies engendered by key figures such as Wilhelm Dilthey, with his reappraisal of the place of the human sciences, and Rudolf Otto, with his influential religio-phenomenological perspective (highly important both for Zimmer and Jung), the existential meaning of the life experience (*Erlebnis*) began to take pride of place over a merely positivistic, philological perspective. This had a remarkable

---

[190] Zimmer to Esslinger, June 30, 1933, BHZME AN 259, DLA.
[191] Lukas Rauch, personal communication (Winterthur, 2010).
[192] Zimmer, "Some Biographical Remarks," cit., 260.
[193] Jung and Jaffé, *Erinnerungen*, cit., 385.
[194] "The Holy Men of India" [1944], CW 11, § 950.

impact, far before his collaboration with Jung, on Zimmer's unique, masterly approach to his sources. As Wendy Doniger notes,

> Zimmer chose, loved, illuminated, and compared in detail a number of great myths that continue to inform our appreciation of Indian mythology. He resurrected undeservedly neglected texts and provided us with a basic corpus of myths chosen with an unfailing eye for a truly wonderful story, impeccably translated, and wisely glossed.[195]

Facing the huge Hindu mythological heritage,

> he preferred to ignore the usual star players: the *Rigveda*, the *Ramayana* and the *Mahabharata*, and the much overcited and overrated Bhagavad Gita, or even the slightly less clichéd Brahmanas and Upanishads. Instead, indulging in a kind of literary affirmative action, he rescued from academic obscurity the Puranas, more particularly the late (and therefore devalued) Puranas.[196]

Zimmer's encounter with Jung's depth psychology led him progressively to integrate certain aspects of Jung's hermeneutics into his own religio-phenomenological perspective, which was already informed by a comparativist approach. Moreover he found himself compliant with Jung's—and for that matter Eranos's—purpose of seeking a confrontation between Eastern soteriological systems and the deepest questions of Western psychotherapy. (In so doing, the meanings of some of the terms he used, such as "individuation," and "Self," sometimes overlap without fully

---

[195] Doniger, "King and the Corpse," cit., 54. On Zimmer's recollection of the intellectual climate at the time of his studies, see Henrich Zimmer, *Philosophies of India*, ed. by Joseph Campbell (Bollingen Series XXVI) (New York: Pantheon, 1951; Princeton, NJ: Princeton University Press, 1969), 27–34.

[196] Doniger, "King and the Corpse," cit., 55. The Puranas primarily deal with the liturgy, rituality, and iconography of religious ceremonies. Zimmer translated some of them in his voluminous *Maya, der indische Mythos*, cit. Moreover, "[h]e gave relatively short shrift to the already trite avatars of Vishnu (which Christian Indologists tended to favor because of their apparent resemblance to another Incarnation). He gave pride of place to a mythology that now seems mainstream—but only because he made it so. He went for riddles, philosophical puzzles, animals, goddesses, everyday people. He liked the things that were not like Greek mythology" (ibid., 56). Yet at the same time he kept always an eye on the latter, and on *Naturphilosophie*. Doniger laments, however, that Zimmer's body of work (by which she intends primarily his posthumously edited corpus) might be said to have engendered an "often infuriating psychological reductionism," although this was counterbalanced by "his way of deliberately shunning the heroic dragon-slaying myths with their happy endings" which in her judgment "goes against the usual healthy-minded trend of Jungian interpretations" (57).

corresponding to those of analytical psychology). For instance, in 1932 he pointed out in relation to the Jaina-Yoga that

> unlike our psychotherapy, which seeks to establish a healing balance between the realms of the ego and the unconscious, this yoga embraces both spheres and sets them apart. It separates a person's individual husks from our supra-individual core, that is, life itself within us. It teaches us to realize that their exposure to involuntary participation is an outward process, one that does not touch the core of our essence. A person's full sacrifice gives us an anonymous, sovereign Self.[197]

In a review of the same year, Zimmer declared that analytical psychology aims at "an autodiagnosis of the adept for the purpose of uncovering the personality complex with its deeper layers to make it run more smoothly in greater harmony," while

> Yoga is actually about the reduction of individuation and its unconscious "willingness," whose tangibility is steadily made to dwindle; it is a matter of laying bare and empowering an anonymous person-in-depth, which in many ways resembles Jung's mana personality. The goal of classical yoga *kaivalya* in the individual, that is, totality and integration as opposed to particularity and differentiation, is a magical life from the collective unconscious, in which space and time, individual memory and character, have been overcome or are felt to be a *quantité négligeable*, as a mask for the world which no longer sticks to the wearer, having lost its suggestive power to convince him that he is really the figure of that mask.[198]

These passages serve to demonstrate the specific way in which Zimmer sought, already in the year he met Jung, to find a point of convergence between, and in some manner integrate, the final goals of yogic meditation and those of analytical psychology. While yoga pursues a state of complete separation of *purusha* from *prakrti*, by overcoming both the conscious and

---

[197] Zimmer, "Yoga und Wir. Zimmer, Sommer 1932," HZMK 85, DLA, cit., 20–21. Cf. Zimmer, "Yoga und Wir [January 1933]," typescript, APC, 23, and *Indische Sphären*, cit., 2nd edn (1963), 83.

[198] Heinrich Zimmer, review of J. H. Schultz, *Das autogene Training (Konzentrative Selbstentspannung): Versuch einer klinisch-praktischen Darstellung* (1932), and of J. W. Hauer, *Der Yoga als Heilweg* (1932), in *Der Nervenarzt*, year 6, no. 4, April 15, 1933, 203–4. It is worth recalling here Patanjali's opening sentence: "Yoga consists in the (intentional) stopping of the spontaneous activities of the mind-stuff" (*Yogasutra* 1. 2; see Zimmer, *Philosophies of India*, cit., 283–94 at 284).

the unconscious, Jung's psychology tends to their dynamic, progressive integration towards the attainment of the Self. For the yoga practitioner, unconscious life is an obstacle to the achievement of absolute freedom and aloneness (*kaivalya*); consequently, the Self is a state completely devoid of images. Analytical psychology supplies no concept corresponding to the yogic supreme liberation: at that point, the crossover breaks down. In analysis, healing is nowhere conceived as a dimension utterly free and detached from images, but rather as a responsible integration, and higher awareness, of conscious and unconscious elements, for the sake not of perfection (*Vollkommenheit*), but of wholeness (*Vollständigkeit*)—in tune with the Latin concept of *integritas*.[199] Yet Jung's conception of the (collective) unconscious was far from that of psychoanalysis: he saw it not only as a sort of repository of ancestral, biologically as well as culturally determined images which host and reproduce recurrent experiences of mankind, but also as a prospective, synthetic, and teleological dimension that surrounds (or is surrounded by) the archetypal principle and the *primum movens* of individuation, the *Selbst*. Thus it is in the collective unconscious, not the Self, that Zimmer looks for—and sees—a connection between the goals of the two systems; and, in this connection, he pays special attention in Jung's notion of the mana personality: that is, the psychic phase characterized by a sense of superiority yet replete too with megalomanic, inflationary, daimonic risks, that anticipates the Self. Indeed, in Jung's theorizations of the collective unconscious Zimmer found a remarkable heuristic link to the Hindu concept of inheritance, as evidenced by his above-mentioned Berlin lecture of January 1933:

> What depth psychology establishes as a generic heritage, engrams of mythical-magical states of thought and being, is at the same time reproduced individually by every human being. This view coincides with the Indian notion that while self-fashioned heredity is individual for and in each of us, possessing it constitutes the most

---

[199] *Integritas* was addressed by, for example, Cicero in *De officiis*; and by Thomas Aquinas, as brilliantly explored in Umberto Eco, *The Aesthetic of Thomas Aquinas*, trans. by Hugh Bredin (Cambridge, MA: Harvard University Press, 1988), 98–102. See the inspiring book by John Beebe, *Integrity in Depth* (College Station, TX: Texas A&M University Press, 1992). As has been observed, "Jung does not see the self beyond opposites, but sees the self as conjunction or reconciliation of opposites and points out that this is expressed in mandala symbols in both the East and the West" (Shoji Muramoto, "Jung and Buddhism," in Young-Eisendrath and Muramoto, *Awakening and Insight*, cit., 119–32 at 126).

universal phenomenon of all: namely, the very circumstance that creates individuation.[200]

Likewise, Jung's notion of archetypes soon took on weight in Zimmer's reflections. For instance, in his first Eranos lecture, he claimed that modern Western (and especially analytical) psychology "is a science [*sic*] born of our time and our predicament," capable of putting us "into relation with the very same reality which speaks through the fading hieroglyphic system of other ages." Referring to such "suprapersonal sphere of our inner life," he argued that "our deeper dreams raise this deposit up to us, and it is nothing other than what the myths of all times have preserved in their figures, situations, and symbols."[201] Such images from the unconscious, similarly to the sacred images used by the yogin and tantra practitioner, serve as "a ruler [or regent] within us," that is, "a guiding principle from the unconscious."[202] He also affirmed (echoing in a way Freud's famous epigraph to his *Interpretation of Dreams*, taken from Vergil's *Aeneid* 7.312: "Flectere si nequeo Superos, Acheronta movebo" [If I cannot bend the higher powers, I'll stir up Hell]),

> If we cannot raise our spiritual eye and find our regent in the heavens, we can discern it in inward voices and signs, as [did] Socrates; and, after all, is its situation in space anything more than imagery, a metaphoric visualization? What is the deeper relation of outward to inward, of macrocosm to microcosm? The merit of the new depth psychology is that it unearths that which is timeless in us, in a form appropriate to our time, so that we can comprehend it and live by it.[203]

Yet the path toward such attainments might also entail magic, mana, and demonism, as Zimmer recalls in the Berlin Seminar. In a similar vein, in his opening lecture at the psychotherapy conference at Bad Nauheim,[204] he affirmed that the relation of the Indian "guru" with his pupil "is a

---

[200] Zimmer, "Yoga und Wir [January 1933]," cit., 16–17, and "Yoga und Wir. Zimmer, Sommer 1932," cit., 15.

[201] Zimmer, "On the Significance of the Indian Tantric Yoga," cit., 8 (translation modified; cf. the original: "Zur Bedeutung des Indischen Tantra-Yoga," *Eranos-Jahrbuch* I/1933 [Zurich: Rhein, 1933], 9–94 at 17).

[202] Ibid.

[203] Ibid., 28–29.

[204] Heinrich Zimmer, "Indische Anschauungen über Psychotherapie," *Zentralblatt für Psychotherapie* 8, no. 3 (1935): 147–61; its English translation "Indian Views on Psychotherapy," quoted here, appeared in *Prabuddha Bharata or Awakened India* 16, no. 2 (February 1936): 177–89; this special issue of the Indian philosophy journal dedicated to the

priestly one, full of magic and sacraments, witchcraft and transformation," while the Western psychotherapist, "looked at from the Indian viewpoint," appears "handicapped by the fact that he is not allowed to use more witchcraft." So the latter can only "let the patient's unconscious itself effect the transformation," or else perform a conjuring trick on the patient by bringing *mana* to bear. He then addressed the difference between classical psychoanalysis and analytical psychology: while the unconscious processes in the psychoanalytic system seem to "appear as diabolical forces," in analytical psychology the same have largely assumed the direction "already anticipated by the German Romantics: they have become what might be called the great demonic powers which both threaten and bless, being dark and light at the same time," in such a way as to reconcile them to the dominant "cult of rational thinking and conscious discipline of will-power."[205] Consistently with Jung, Zimmer then stated that "from the analysis and observation and collection of psychological facts we may evolve a method somewhat similar to the Indian one, which would grow up from our own Western soil and material into a kind of synthetic dietetics of the psyche." Yet in concluding, he spelled out his view with a vivid metaphor: "Compared with this extensively consolidated Indian system of guiding of the soul, analytical psychology in all its ramifications still looks like an infant beside a man who has already achieved the greater part of his life."[206]

12. MANDALA, YANTRA, AND THE "MACHINE": ZIMMER'S INFLUENCE UPON JUNG

Jung came across Zimmer's *Artistic Form and Yoga* (1926; English translation 1984) in 1930, as can be deduced both from the (personally) handwritten inscription in his copy of the book and, as we shall see, from the appearance of the yantra theme in *Dream Analysis*.[207] His acquaintance

---

hundredth anniversary of Sri Ramakrishna's birth also included also Jung's article "Yoga and the West" (CW 11).

[205] Zimmer, "Indian Views," cit., 181–83.

[206] Ibid., 189. Cf. Jung's affirmation of same year: "We Europeans [. . .] are just a peninsula of Asia, and on that continent there are old civilizations where people have trained their minds in introspective psychology for thousands of years, whereas we began with our psychology not even yesterday but only this morning" (*The Tavistock Lectures* [1935], CW 18, § 139).

[207] Jung's copy of Zimmer's *Kunstform und Yoga* exhibits a certain number of underlinings and lines at the margin. In the first January session of the Dream Analysis Seminar (January 22, 1930), Jung showed the participants "photographs and plaster impressions of

with the work allowed him to enrich his understanding of the symbolism, and the proper cultic functioning, of Hindu sacred images. *Artistic Form and Yoga* in fact dealt with Hindu meditative and the ritual practices in a remarkably innovative way, distinct from the dominant contemporary studies marked either by the prevailing academic philological-positivist, or by an exclusively spiritualistic-occultist, approach one.[208] With an uncommon command of Hindu literature, Zimmer brilliantly penetrates, in a richly inflected literary style, a conceptual, philosophical, and cosmological universe. The book comes with an iconographic apparatus of thirty-six plates (supplemented with as many as fifty-four illustrations in the English edition), including a series of yantras on copper which until Avalon's publications had remained "terra incognita for the art historian."[209]

Zimmer conceived the Tantra "as both the most profound aspect of Indian thought and the most sophisticated aesthetic tradition"[210] and put the *Kultbilder*, or sacred cult images—in particular the yantra and the mandala—at the heart of Indian art, which he regarded as the quintessential expression of Indian metaphysics. Hindu sacred images, he explained, are not "self-contained constructs, nor pure expressions of a religious *Weltanschauung* that our knowledge can appropriate; they are purposeful links in a chain of spiritual, ritualistic procedure."[211] Contextually, the yantra, a simple structure whose geometric, aniconic equation undergirds the mandala, is "an instrument [*Gerät*] or tool [*Werkzeug*], a device [*Apparat*] or mechanism [*Mechanismus*] employed for carrying out

---

designs, seals, engraved on the surfaces of Babylonian cylinders made of amethyst and jade," which he found "in looking through the cellars of the East Asiatic department of the Museum in Berlin" (Staatliches Museum) during an otherwise undocumented stay (*DAS*, 439 and n. 1). Perhaps Jung came upon Zimmer's book too in the German capital. In what follows I partly take up my article, Giovanni Sorge, "Jung's Hermeneutic of the Yantra and the Mandala: The Influence of Heinrich Zimmer's Book *Kunstform und Yoga im indischen Kultbild*," *Spring Journal* 90 (2013): 77–104.

[208] As Zimmer himself put it, "I did not write it for professionals, nor as a contribution to specialized studies. I had to write it, to realize my self and to come into my own," and to do away with "the then prevailing purely aesthetic approach to Indian art on the one had, and with the barren classistic criticism on the other" ("Some Biographical Remarks," cit., 255).

[209] Zimmer, *Artistic Form and Yoga*, cit., 236.

[210] Urban, *Power of Tantra*, cit., 169. In Zimmer's words, Tantra is "a period as grand and rich as the Vedas, the Epic, Purânas, etc.: the latest crystallization of Indian wisdom, the indispensable closing link of a chain, affording keys to countless problems in the history of Buddhism and Hinduism, in mythology and symbolism" ("Some Biographical Remarks," cit., 254).

[211] Zimmer, *Artistic Form and Yoga*, cit., 5 (cf. Zimmer, *Kunstform und Yoga*, cit., 8).

a specific task. The sacred image is a device of very efficient construction used for both magical and ritualistic spiritual functions."[212] Likewise, "the figurative image (*pratima*) of a divine being" has the same technical function, one which must also inform our understanding of the "*cakras* ('circle-shaped' designs), *mandalas* ('ring-shaped' designs), and, related to them, exclusively geometric linear figures that have no other name but *yantra*."[213] The latter is also defined as a "'tool' for many purposes, [whose] lowest function is a magical one."[214]

That said, as Martin Liebscher has observed, Jung was already familiar with Tantra texts thanks to the works of Avalon, and Zimmer's *Artistic Form and Yoga* served to provide further insights into the influence of the Tantric body of thought on "those heterodox Buddhist and Jaina sects that flourished and declined in the midst of orthodox Hinduism," and the influence of orthodox Brahman Tantric texts in "the general formal aspect of Buddhist sacred images."[215] Jung specifically benefited from Zimmer's commentary to the rather obscure Tibetan (Lamaistic) Tantra text *Shri-chakra-sambhara*, which Zimmer regarded as "the most complete directions for developing before the inner eye linear yantra with figurative decoration that is alive with the dynamics of enfolding and unfolding."[216] Jung would deal repeatedly with this text, which was instrumental to his psychological understanding of the mandala;[217] and similarly with the *Amitayurdhyana-sutra* (Treatise on Amitabha meditation), one of three texts basic to Pure Land Buddhism, which deals with meditations based on complex visualization procedures. In Jung's eyes, the *Amitayurdhyana-sutra* rises to become a paradigmatic example of the Eastern sacred image and its practice. Contextually, "in his differentiation between the religious and spiritual practices of the West and the East Jung followed Zimmer's dichotomy of outward sight and inward vision."[218] Zimmer's interpretations of yoga, along with those of Avalon and Hauer, Liebscher concludes, were instrumental in Jung's first clarifi-

---

[212] Ibid., 28 (cf. *Kunstform und Yoga*, cit., 25–26; in Jung's copy, this passage is marked by a line in the margin).

[213] Ibid., 28–29 (cf. *Kunstform und Yoga*, cit., 26).

[214] Ibid., 68 (cf. *Kunstform und Yoga*, cit., 57).

[215] Ibid., 23, quoted by Martin Liebscher, "Introduction" to Jung, *Psychology of Yoga and Meditation*, cit., xlv–lxix at lxi.

[216] Zimmer, *Artistic Form and Yoga*, cit., 81, quoted by Liebscher, "Introduction," cit., lxv.

[217] Liebscher, "Introduction," cit., lx; see also lx, lxiii, lxviii (and Jung, *Psychology of Kundalini Yoga*, cit., 12).

[218] Liebscher, "Introduction," cit., lxii.

cation of "his concept of active imagination against the background of Eastern as well as Western meditative religious and spiritual practices and suggested the psychology of the unconscious as their twentieth-century equivalent."[219] Moreover Jung's understanding of the *vajra* was clearly indebted to Zimmer's detailed account of Tantric meditation.[220] The Lamaist *vajramandala*, used by Jung as a frontispiece for *The Secret of the Golden Flower*, and discussed in *Dream Analysis*, was also commented on in the Berlin Seminar (although, as Liebscher too concedes, no sources so far can prove definitively that Jung also displayed it to his Berlin audience).[221]

Jung's attention seems also to have been drawn by Zimmer's cross-cultural reflections on the Indian lotus as "seat of the Absolute, the Divine (of Brahma or Buddha)" regarded as the source of the many symbolic variants which are to be found in Eastern and Western religious traditions.[222] In particular, he drew a parallel between the symbolism of Dante's ascent of Mount Purgatory on his way to Paradise in the *Divine Comedy* and the sacred world-mountain Sumeru. In Zimmer's words,

> Dante climbs up to the heavens that, in the cosmic architectonics of his vision, correspond to the ever more spiritual heavenly worlds experienced in Indian yoga, although the names and the numbers in his vision are indebted to classical astronomy. Passing beyond the spheres of the sun, moon, and the five planets, beyond the heaven above them that rotates most quickly of all around the earth, he enters the empyrean. There, transformed by the fountain of divine light that floods his sight, he beholds the rose of Heaven, a perfect replica of those many-petaled lotuses—here reshaped by the Christian tradition—that form the center of so many mandalas and purely linear yantras, are the symbolic dwelling place of the Absolute, and serve as a frame for Its unfoldings into the world of illusion. The Holy Trinity floats above the center of this infinitely great flower,

---

[219] Ibid., lxiii.

[220] Liebscher, "Introduction," cit., lxv. *Vajra* means "thunderbolt," "diamond," or "adamantine," and is also a secret name for the penis (Georg Feuerstein, *The Encyclopedia of Yoga and Tantra* [Boulder, CO: Shambhala, 2022 (1997)], 389).

[221] Ibid., lxvii–lxviii, and main text (Jung, *Psychology of Yoga and Meditation*, cit.), 57. See *DAS*, February 19, 1930, 479, and below, n. 228 and [MT], 231–32. Jung's reliance on Zimmer's research is also evidenced by "his explanation of the mystic syllables *yam* (air), *ram* (fire), *vam* (water), and *lam* (earth) in the lecture of 10 February 1939" (Liebscher, "Introduction," cit., lxv).

[222] Zimmer, *Artistic Form and Yoga*, cit., 239ff. See also below [MT], 221 and n. 482.

but each of its petals, in ring upon ring, serves as the seat for a saint or a blessèd soul.[223]

Zimmer connected the rose as "grand amphitheater of deified essence" and the "rings of petals in mandalas and yantras" where Buddhas and bodhisattvas are placed for the initiate's contemplation, affirming that "[a]s a symbol, Dante's heavenly rose must be considered as the equal of the gleaming rose windows of the cathedrals which (at Chartres, for example) portray Christ surrounded by rings of saints and martyrs, and encircled by angels, thrones, and powers."[224] These considerations on "mandalic" affinities may well have resonated for Jung, and seem to anticipate reflections revealed in his "Commentary on the *Secret of the Golden Flower*" and later developed in, for instance, the ETH Lectures.[225]

To return now to the predecessor of the Berlin Seminar, the 1928–30 seminar recorded in *Dream Analysis*: Zimmer's book finally provided Jung with new hermeneutic inputs for the understanding of a recurrent oneiric motif, the machine, as evidenced by a closer look at this in the seminar sessions from January 1930 onward. This motif first appears in the third dream (considered also in the Berlin Seminar), and recurs in subsequent dreams in *Dream Analysis* (the analysand was a passionate driver with a weakness for cars). While the first mention of a yantra appears in an earlier session, Jung's first direct reference to Zimmer occurs on May 7, 1930, in relation to a dream (no. 25) in which a machine appears; in this regard, he remarks that he had "often emphasized the fact that a symbol functions like a machine in our psychology."[226] Three months earlier, he had defined the yantra as "any kind of device or symbol that serves the purpose of transforming or centering the libido of the one who is concentrating upon it. It has been translated by a German scholar [i.e., Zimmer, unnamed] by the term 'machine.'" He went on to consider ritual practices meant to transform psychic energy, citing the Indigenous Australian's *churinga* as "the most primitive form of worship and the most primitive

---

[223] Zimmer, *Artistic Form and Yoga*, cit., 239. On ascension myths and symbolism, see Mircea Eliade, *Myths, Dreams and Mysteries: The Encounter Between Contemporary Faiths and Archaic Realities* (New York: Harper & Row, 1975 [1957]), and Eliade, *Patterns in Comparative Religion*, cit., ch. 2, esp. 102–8.

[224] Zimmer, *Artistic Form and Yoga*, cit., 239, 240. For more on this, see the magisterial Joseph Campbell, *The Mythic Image* (Princeton, NJ: Princeton University Press, 1974), ch. 3 ("The Lotus and the Rose").

[225] See Jung, *Psychology of Yoga and Meditation*, cit., e.g., 171–73, 279–80.

[226] *DAS*, May 7, 1930, 580.

form of yantra."[227] Ten days earlier Jung had showed "the reproduction of a Tibetan mandala" calling it "a *yantra*, used for the purpose of concentration upon the most philosophical thought of the Tibetan Lamas,"[228] then adding that he had "another mandala where instead of a thunderbolt in the center, there is the god Mahasuka, one form of the Indian god Shiva, in the embrace of his wife Shakti."[229] But it was in the session of January 29, 1930 that Jung interpreted the functioning of the symbol itself as a machine.[230] On this occasion he commented on a dream, the twenty-second, in which a sort of grinding machine appears. This device links back both to a stripe motif in a preceding dream (no. 18), and to the aforementioned third dream (analyzed the previous year, on January 30, 1929), in which a steamroller laid down a labyrinthine figure. Here Jung pauses to integrate (and change) the principal meaning he had previously ascribed to that mechanism: namely, the sexual component embedded in "the realm of automatic processes, the instincts."[231] Turning his attention to that labyrinthine-mandalic figure, Jung now associates it with the typical structure of a Buddhist mandala, bringing it into wider connection with "a sort of map or ground plan of the structure of the psyche or self, the expression of what man is as a psychical entity."[232] This resonantly echoes Zimmer's definition of the mandala as "the road map of Truth."[233] Jung then goes on to move the focus, so to speak, from space to time, interpreting "the main body or the virtual center" as the expression of the Self, surrounded by its components "as the months or

[227] Ibid., February 29, 1930, 492. On the *churinga* (or *tjurunga*) see, e.g., *Psychological Types* [1921], CW 6, § 325 and § 496; and "On Psychic Energy" [1928], CW 8, § 119, where Jung refers to it as an "energic concept," and § 92, where rhythmic rubbing of the *churinga* is put in connection with the line of thought inherent in the totemism idea.

[228] *DAS*, February 19, 1930, 479; also n. 1. The *vajramandala* is reproduced in ibid., 477, in the frontispiece to *The Secret of the Golden Flower*, in *Psychology and Alchemy* [1944], CW 12, Fig. 43, and in "Concerning Mandala Symbolism" [1950], CW 9.1, Fig. 1, with an analysis in § 630–39. See also Jung, *Psychology of Yoga and Meditation*, cit., 56–57, 63ff.

[229] *DAS*, February 19, 1930, 479. Cf. Zimmer, *Artistic Form and Yoga*, cit., 90ff., and Figs 22 and 23. Zimmer speaks of Mahasuka embracing the female figure and represented in the flower's center by four heads which signify the four elements "in their immaterial, supra-sensory state" and, simultaneously, the four eternal feelings which constitute the adept's "ever-increasing maturation towards Nirvana" (96).

[230] *DAS*, January 29, 1930, 444ff.

[231] Ibid., 449.

[232] Ibid., 452.

[233] Zimmer, *Artistic Form and Yoga*, cit., 108. Equally, mandalas are "like road maps in their origin and purpose. In them there can be no free play of an active imaginative fantasy; they are designed like maps" (107).

days are the constituents of the year." He also singles out "an analogy in the early Christian idea that Christ's body was the Church year," maintaining that "the *quadratura circuli* was the problem of the Middle Ages, the problem of psychological completeness," and claims that "this idea of the mandala expressing the totality of the human being and the [human being's] right position in the universe, is the fundamental idea underlying the motif of the machine."[234]

The connection between the former and the new connotation is not unproblematic, as voiced on February 5, 1930, when after reiterating the link of the machine-motif to the sexual function as the "*partie inférieure*," Jung adds that the dream image must be reconnected to the mandala—"a circular symbol which no one would associate with sex."[235] The mandala in fact "is used for the transformation of energy, as in certain rituals the *yantra* is used"; contextually, he affirms that "certain temples have the form of typical mandalas, like the famous temple of Borobudur in Java, which is a circle in a square."[236]

Now, it is noteworthy that Jung's very first mention of the yantra, in the context of his inquiries about mandalic meanings, is tied to a reference to the famous Indonesian temple, for Zimmer was among the first to argue comprehensively for the correspondence between its structure and the mandala.[237] In a chapter entitled "Borobudur: An Architectural Mandala of the Pilgrim's Path" he marvels:

> Borobudur must be regarded as the most impressive mandala the art of Buddhism has ever created in the visible world as a symbol of its truth [. . .]. The purpose of Borobudur is to release a spiritual process in the pilgrim during his ascent through sculpture-embellished terraces to the unadorned summit, to bring about a complete transformation of his sense of being, a transformation by nature related to the activity of meditation as practiced by the adept of the Tibetan mandala.[238]

---

[234] *DAS*, January 29, 1930, 453.

[235] *DAS*, February 5, 1930, 458 and 461.

[236] Ibid, 462.

[237] As acknowledged by the French orientalist Paul Mus in *Barabudur: Esquisse d'une histoire du bouddhisme fondée sur la critique archéologique des textes*, 2 vols (Hanoi: Imprimerie d'Extreme-orient, 1935), 1:52ff. See also Mus, "The Problematic of the Self, West and East, and the Mandala Pattern," in Charles A. Moore, ed., *Philosophy and Culture, East and West: East–West Philosophy in Practical Perspective* (Honolulu: University of Hawaii Press, 1962), 594–610.

[238] Zimmer, *Artistic Form and Yoga*, cit., 114. See also his *Art of Indian Asia*, cit., 1: ch. 7.

The planimetry of Hindu temples, traditionally construed as the abode of the god's body and representing a *mahupurusha* (great man)—a definition of the cosmos, that is, already attested in the *Rigveda* (10.90)— corresponds in fact to yantric diagrams, and further studies have successively reinforced this point.[239] In the session of February 12, 1930, Jung reaffirmed "the peculiar fact that the machines were connected with the mandala" which, in turn, "has to do with [the analysand's] sexuality" and "is linked up with [his] inferior function."[240] He then observed, "We see that the mandala symbolism comes in to show him that it really is his own individual *yantra*, that mechanism which ought to function and to transform him—as if he were walking round and round at Borobudur."[241] Then he admits,

> The mandala really is an effect of some fundamental idea in man for which I find no explanation. [. . .] So this machine represents an underlying fact of an ideal nature and is the means by which he can transform himself.[242]

And again with reference to the analysand,

> By being identical with the machine he would arrive at his goal. The Eastern idea is demonstrated by its chiefly circular character, in which the cross is not so obviously represented; the idea is that man should enter the centre, and there he should become identical with the god that occupies it. Our Western mandalas on the other side show a tendency to represent the cross in the center in the following fashion: ⊕ This would mean a differentiation of the most central thing and that does not exist in the East. It is probably what

---

[239] Mircea Eliade insists on this theme, for instance in "Centre du monde, temple, maison," in *Le Symbolisme cosmique des monuments religieux: Actes de la conference [. . .] à Rome, avril–mai 1955 [. . .]* (Serie Orientale Roma 14) (Rome: Is.M.E.O., 1957), 57–82 at 67ff., and *The Myth of the Eternal Return: Cosmos and History*, trans. by Willard R. Trask (Princeton, NJ: Princeton University Press), 12ff., and compares the structure of Borobudur with a mandala, as well as with the heavenly Jerusalem (in *Commentaires sur la Légende de maître Manole*, cit., 143). See also Stella Kramrisch, in *The Hindu Temple: An Introduction to Its Meaning and Forms* (New York: Harper and Row, 1977) and Alex Wayman, "Reflections on the Theory of Barabudur as a Mandala," in Luis O. Gómez and Hiram W. Woodward Jr., eds, *Barabudur: History and Significance of a Buddhist Monument* (Berkeley, CA: Asian Humanities Press, 1981), 139–72.

[240] *DAS*, February 12, 1930, 472 and 474.

[241] Ibid., 473.

[242] Ibid.

they criticize in us because it is missing in themselves, and that is why the East is coming to the West.[243]

Jung went on to comment upon the recurrence of the cross at the center of "European mandalas," addressing also the topic of the circumambulation—which would be explored in depth in the cross-cultural amplifications of the Berlin Seminar—yet assuming in this instance that "working round in a circle is characteristic of the East."[244] He also noted that the Western mandala works "entirely differently from the Eastern way. It is not a finished temple in which there is a definite ritual, it is only an attempt"[245]—an "attempt" nonetheless replete with a paramount epistemic and heuristic value, and instrumental in showing the centripetal tendency of the collective (that is, transcultural and transgeographical) psyche.

In sum, these passages from *Dream Analysis* show that Jung borrowed from Zimmer's work elements which concurred in leading to a reinterpretation of the dream motif of the machine. Subsequently, in the Berlin Seminar, this reinterpretation is treated as established, although Jung mentions the yantra only once, and that, somewhat surprisingly, without any reference to Zimmer's *Artistic Form and Yoga*. Zimmer's explanations on the psycho-spiritual function of the Hindu symbol (*Sinnbild*) as simultaneously a representation (*Abbild*) of God, a ritual image (*Kultbild*), and a tool (*Werkzeug*) for transforming psycho-spiritual energy—specifically in the yantra—might also have been influential, as recounted above, in Jung's understanding of the symbol *per se*, the theory of which he had formulated comprehensively a decade before in *Psychological Types*, where the symbol, in antithesis to the univocal intelligibility of the sign, expresses an unrepresentable (nor conceivable in its essence) *quid* that transcends human understanding.[246] Although in 1928 it was had already been described as "the psychological mechanism that transforms energy,"[247] in 1930 it acquires a further connotation of a certain deliberate instrumentality; that is, a tool "to control the unconscious factors," for "it always

---

[243] Ibid., 473–74.

[244] Ibid., 474.

[245] Ibid., 469.

[246] *Psychological Types* [1921], CW 6. See "Symbol" in the section on Definitions.

[247] "On Psychic Energy" [1928], CW 8, § 88 and passim. With reference to the spring ceremony of the Australian Wachandi, for magical fertilization of the earth, Jung wrote, "Just as a power-station imitates a waterfall and thereby gains possession of its energy, so the psychic mechanism imitates the instinct and is thereby enabled to apply its energy for special purposes" (§ 83).

has been of the utmost importance for man to have a certain magic control of the gods."[248] Turning to Christianity, and notwithstanding the doctrine of divine grace grounded in the assumption that salvation does not lie in man's own hands, but comes through Christ (e.g., Ephesians 2:8–10), Jung defined the Catholic sacraments as "magic devices, magic methods, of forcing the grace of God."[249] And two sessions later, apropos the Eastern influence on the early Church, he went so far as to say that "the rosary is an Eastern yantra."[250]

Throughout his work, Jung's exegesis of the yantra-mandala theme is characterized, and aims at integrating, two poles, as it were: one being the spontaneous psychic emergence of geometrical configurations, and the other the instrumental function inherent in the classical use of these sacred diagrams, whereby their *trait d'union* inheres in the symbolic ability of the psyche to channel the individuation process. Immediately after Jung's reading of Zimmer, however, the instrumental function of the symbol seems to assume a certain preeminence. One could even say that the ultimate meaning of the yantra-mandala as explained by Zimmer has been worked out by complex psychology along the lines of the individuation process, in a way as a *function of analysis itself*. Indeed, analysis can be conceived as a kind of yantra: a modern ritual device capable of enabling the *circumambulatio* around—and toward—the deeper center of the personality. The (so to speak) machine of analysis,—as a potent reshaper of psycho-spiritual energies, as it were lays down an Ariadne's thread guiding one through the labyrinth, gradually relating a person to the axial point, the transpersonal focus of an individual, unique pattern.

---

[248] *DAS*, May 7, 1930, 581.

[249] Ibid. Besides, one can clearly perceive here Jung's interest in the Catholic liturgy, from his own post-Reformation stance (the belief in salvation through *sola gratia*—grace alone, from God, and *sola fide*—faith alone, in the individual, being two of the five basic theological tenets of Protestantism). Later in the Berlin Seminar he defines the faith the symbolon par excellence (see below [MT], 201–2).

[250] *DAS*, May 21, 1930, 607. See also a rather unusual passage in the *Deutsches Seminar* (1931) which states that the mandala must not be seen as "purely a product of nature [*blosses Naturprodukt*]," but as "a special ritual instrument fabricated by man: a yantra" (Jung, *On Active Imagination*, cit. [forthcoming]). Elsewhere, and recurrently, Jung maintains that the "spontaneous occurrence in modern individuals [. . .] in conditions of psychic dissociation or disorientation" of geometric, circular images such as mandalas represents "*an attempt at self-healing* on the part of Nature, which does not spring from conscious reflection, but from an instinctive impulse" ("Mandalas" [1955], *CW* 9.1, § 714). On this nexus, cf. Giuseppe Tucci, *The Theory and the Practice of the Mandala, with Special Reference to the Modern Psychology of the Subconscious*, trans. by Alan Houghton Brodrick (Mineola, NY: Dover, 2001 [orig. Ital. edn 1949]), 36–37.

A pivotal "device" within the psychotherapeutic *opus* is the active imagination, "an active engagement," Jung will poignantly say, "with otherwise passive fantasy."[251] Yet here in the Berlin Seminar "contemplation" lies at the core of this operation: a dream "should be contemplated from all sides, and filled with libido so that it begins to live and finally gives birth to its inner meaning and tendency. In that way, one performs what the mandala and circumambulation mean."[252] Jung points out that *betrachten* (to observe) derives etymologically from *trächtig machen* (to make pregnant); thus to contemplate a dream or a fantasy means to "concentrate on it until it becomes pregnant."[253] By means of such dynamic processes of the reception and engaged expression of imaginal content on a concrete, three-dimensional level ("Since I did not understand this dream-image," as Jung wrote in his memoir about his dream of Philemon, "I painted it in order to better visualize [*veranschaulichen*] it to myself"),[254] unconscious contents are not mastered, but rather become conducive to an expansion of consciousness.

Toward the end of the seminar, by way of making a daring connection, Jung refers to the legendary biblical episode of Joshua's taking of Jericho on the seventh day of the march around its walls, and states, "The machine is [...] a mechanized capacity for effort that we must exert in the *circumambulatio*, so to speak, to perform the conversion or transformation."[255]

13. CONCLUDING REFLECTIONS

The Berlin Seminar may be seen as a contribution to the ongoing epistemological debate over hermeneutic values in any discourse having to do with the psyche. That is, it specifically supports a psychology (ψυχή λόγος, a discussion about the psyche or soul) embedded in a larger framework alongside Oriental meditative systems. Yet this puzzling journey in a soul's story somehow says more about analytical psychology than about the dreamer himself. So, we are left wondering how the analysand brought his remarkable dreams to bear upon his situation. We may also

---

[251] Jung, *Psychology of Yoga and Meditation*, cit., 178.

[252] See below [MT], 256.

[253] Below [MT], 251. In *The Psychology of Kundalini Yoga*, Jung connected this operation of *betrachten* with the *dharana* (Appendix 2, 81), thus paralleling the basis of active imagination with the vigilant attention and concentration that characterize Buddhism in particular, but in fact meditation in general. See also Jung, *Visions*, cit., 2:661.

[254] *MDR*, 183 (translation modified). *Veranschaulichen* also means "illustrate," and "realize" (cf. Jung and Jaffé, *Erinnerungen*, cit., 186).

[255] See below [MT], 256.

wonder to what extent his participation to the previous seminar, on dream analysis, influenced his oneiric life.

This question goes hand in hand with the perennial one of the suggestive influence exerted, or even spell cast, upon the analysand by psychotherapeutic systems, along with their hermeneutics.[256] Jung was aware of the impact his publications would have on the genuineness of the symbolic unconscious manifestations of his patients, and frequently recalled the long period he had waited before first revealing his discoveries in his "Commentary on the *Secret of the Golden Flower*" (1929);[257] while no more than two years after the present seminar, at the psychotherapy congress in Bad Nauheim, psychiatrist Fritz Kunkel would caution against the propensity of both psychotherapists and patients to get caught or "encapsulated" in semantic, conceptual, and imaginative automatisms. He half-amusedly remarked,

> For example, if [the psychotherapist] is a Freudian, he and his followers will very often dream church towers and walking sticks as well as jewelry boxes and handbags. But if he is a Jungian, you will find in his environment primarily dreams of snakes uncoiling from the interior of a lotus flower, or embryos that are still locked inside the earth. There seem to be sanatoriums where dreams exclusively serve one or the other language. This doesn't depend in the first instance on race, blood, or the collective unconscious, but above all on positive academic transference.[258]

Indeed, we cannot address these questions without first taking into account a certain optical illusion created by the seminar situation, insofar as the extensive amplifications laid out for a public audience are set-pieces of demonstration, ill suited to the intimate relationship built up in a therapeutic dialogue. With that distinction in mind, one can say that in both scenarios, Jung's focus on the peculiar richness of dream language carries another kind of meaning still: one that discloses new perspectives and opens onto otherwise inaccessible horizons. It is up to the individual to

---

[256] On this nexus, see Shamdasani, *Jung and the Making of Modern Psychology*, cit., 152–57.

[257] See, e.g., below [MT], 241, and n. 526.

[258] Fritz Künkel, "Lehrbarkeit der tiefenpsychologischen Denkweisen," *Zentralblatt für Psychotherapie* 8, no. 4 (1935): 235–48 at 241. On Fritz Künkel (1889–1956), a leading representative of Adlerian psychology in Germany who created his own system called "We-Psychology" which included sociological and markedly religious elements, see John A. Sanford, ed., *Fritz Kunkel: Selected Writings* (Mahwah, NJ: Paulist Press, 1984).

welcome and integrate such acquisitions with a responsible attitude toward the unconscious. The same responsible attitude furthermore enjoins us, nowadays, to (re)consider cautiously the risks inherent in the extension of psychic dynamics to national(istic) and collective scenarios.

The present edition wishes to privilege the connection, as experienced by the original participants, between Jung's seminar and Zimmer's lecture, in the hope that this can contribute to promoting knowledge of the still largely "*un*translated" or "not yet translated" German orientalist, of whom there has long been, and remains, a "great need for a personal and intellectual biography."[259] The intensity of dialogue between the two is captured in an impalpable arabesque of subtle—whether or not deliberate—links between the lecture on yoga and the seminar sessions. That is to say, approached from one side by Jung along heterogeneous lines of resonance, and from the other by Zimmer in terms of the cosmic dance of Shiva, creator and destroyer of worlds in his union with Shakti, the Berlin Seminar affords a stereoscopic view not explicitly framed by its protocols. Near the end of his lecture, Zimmer evokes the state in which the yogi, at the apex of his ascetic path, recognizes himself as a yantra or mandala, and his own body as a cultic instrument (*Kultgerät*), in a culminating awareness of the *coniunctio oppositorum*. One reaches behind the secret of the world process, to how the transcendental One (the supraworldly, auspicious Shiva, the Supreme God) becomes the manifoldly ephemeral, the world, due to the infinite agility of his creative force, Shakti, who is the Kundalini within us. This evocation of the unity of opposites appears here as Shiva and Shakti: transcendence, resting in itself, and the immanence of the world that comes into play from and within itself, beckon toward two aspects of existence, resonating with the constitutive presence of Death in (every single) Life—and with the deep meaning of (not only Western) philosophy itself, as recalled at the incipit of the lecture. Later in the seminar, Jung underlines the importance of *Geschehenlassen*, or "letting-happen," in observing fantasy processes. This is not the meditative composure alluded to by Zimmer, but a non-interfering, nonprejudicial posture on the part of the rational mind toward the products of fantasy, which constitutes a crucial (therapeutic) attitude for Jung: a "phenomenological stance" vis-à-vis the imaginal realm. In the meantime, however, he accommodates that propensity within the historical-existential horizon of individual typology and the specific personal equation. That complicates any hard and fast strictures about the appropriateness of

---

[259] As Chapple puts it ("Heinrich and Henry R. Zimmer, cit., 85).

"letting-happen," for the latter must be coupled with (and integrated into) the compensating direction: one "must learn will-power." Such flexible learning must go on constantly if consciousness is to remain open to our higher, largely unknown aims, for

> the entire psychic disposition puts all of these things into play simultaneously, and [. . .] we stand in their midst. This is something in which we are completely contained. We represent consciousness, and are surrounded by the unconscious. We are the ball that is thrown back and forth. How, for instance, is the tennis ball supposed to decide the outcome of the game? That makes no sense whatsoever.[260]

The key compass to navigate in life's seas remains, Jung maintains, a responsible awareness of and toward the language of the unconscious; an openness to the intermediate realm of images. Commenting upon the images—to conclude with a last "spoiler"—of the conclusive dream, Jung finds in "the breath of the sea" the expression of a new, wider understanding of the rhythm of life that the psyche announces to the analysand. This hint of such a natural, transforming pulse may find a resonance with the Hindu myth of the Hamsa, or great swan, evoked in Zimmer's lecture and elsewhere linked to "the condition of cosmic night, in which the world is dissolved once again in the primordial waters of the unconscious, in the formless body of cosmic being."[261] By referring to such ample "up-and-down movement," Jung aligns this kindred harmonic in the dream with the corresponding rhythm in the Hindu mandala, which takes its rise from both the smallest within and the greatest around it.[262] This rhythm articulates the paradoxical dimensions of the soul and at the same time unlocks an intimation of the primordial *purusha*. And here Jung mentions the "personal and impersonal life of the soul" of the dreamer, evoking the ineffable linkage within everyone between temporality and timelessness, *prakrti* and *purusha*, and their interregnum: that "strange world of images" which, as pointless or even ridiculous as it might sometimes appear, "fills us with the sacred, brings us healing." Yet behind the threatening power of the waves which alludes to "a future that outgrows him," also lies a danger, a potential "explosive effect." Thus Jung does not refrain from warning against the fundamental

---

[260] Below [MT], 250. This well resonates with the famous Heideggerian concept of "*Geworfenheit*" or "Thrownness" into the world. The initial dazzling scene of Woody Allen's movie *Match Point* (2005) also comes to mind.

[261] Zimmer, *Indische Sphären*, cit., 119 (2nd edn, 115).

[262] Below [MT], 298ff.

bipolarity at the bottom of each archetypal force—between inflation and salvation, development and regression, new life and death. Before and beyond a conclusive reference to the Promised Land (that in the eyes of posterity might be reworked with a sinister, involuntary predictive connotation), Jung points out that dreams in themselves can bring one nowhere. They are only a glimpse, a working direction, one might say. Therefore, after all, now "comes the detail work": a grounded concluding note of humility that affirms the care necessary at every step of the analytical process.

Heinrich Zimmer

On the Psychology of Yoga

Editor's Note to Zimmer's Lecture

The present text appears to issue from a work in progress beginning at least with Zimmer's lecture given in Berlin on January 14, 1933 to the Society for Analytical Psychology, "Yoga und Wir" (Yoga and us), which in turn reproduces, with a few changes, one given the previous year at the Psychology Club ("Einige Aspekte des Yoga" [Certain aspects of yoga], June 18, 1932), as attested by two typescripts: "Yoga und Wir. Zimmer, Sommer 1932" (39 pp.), HZMK 85, DLA, and "Yoga und Wir [January 1933]" (45 pp.), APC.[263] Both redactions would be reworked and published in 1934 under the title "Yoga und Maya" (i.e., the article published in "Corona," year 4, no. 4 [1934], 378–400, not the homonymous book [Munich: Oldenburg, 1940]), and further revised and enlarged in a homonymous chapter in Zimmer's *Indische Sphären* (Munich: Oldenburg, 1935 [2nd edn: Zurich: Rascher, 1963]). I have indicated pertinent modifications and revisions in the notes.

Zimmer's distinctive written (and teaching) style, with its literary allusions and overtones (particularly challenging for translators), will be apparent to readers of this valuable translation, even though certain nuances turn out to be untranslatable. Some literary references could not be traced, moreover, in part because Zimmer left them unidentified.

---

[263] Moreover, an untitled and undated typescript preserved in DLA (beginning, "Philosophieren ist nach Sokrates," 6 pp., HZMK 130) reproduces—with a few modifications and a different layout—the first part of this lecture.

On the Psychology of Yoga

HEINRICH ZIMMER

FOR SOCRATES, philosophy concerns itself with death[264]—just as yoga grows from the shadow cast over life by impermanence.

This is the meaning of the legend of that Indian prince, at whose birth the interpreters of signs read on his body that he was destined to become either a king who would rule the world or a Buddha who would overcome the world. His father sought to sequester him from all messengers of impermanence, so that he would remain in the world and be king, rather than venture into solitude to become a yogi. The father confined his son to a park,[265] where he surrounded him with all his friends so that his gaze could not fall upon the painful features on the face of life. A cool palace for the summer heat, a warm one for winter, and a third for the rains turn the admonishing passing of the seasons for the prince into a play of alternating pleasures; well-guarded gates tightly seal the prince's idyll, and women, flowers, the sound of lutes, and many other pleasures stave off the dreadful reality of life.

---

[264] Plato, *Phaedo* 64a–68b. This dialogue retells the conversations of Socrates with his friends, Crito and Simmias, before his death. Socrates's teaching—underlying Plato's entire philosophy—revolved around the soul's immortality, (physical) death as (soul) liberation, and philosophy (literally, "love of wisdom") as a proper preparation for death. The doctrine of anamnesis or recollection plays a decisive role: see *Phaedo* 72–75, and *Meno* 81d. One may recognize a profound resonance with the Christian dictum "initium sapientiae timor Domini" (The fear of the Lord is the beginning of wisdom) (Psalm 110:10 [Vulgate Psalm 111:10]).

[265] Shuddhodana (Pali: Suddhodana), the father of the Prince Siddhartha (Siddhartha Gautama, in Pali Siddhattha Gotama, the birth name of the Buddha), was a *raja* or king of the warlike dynasty of the Shakya. Anxious about the sage Asita's prophecy, Shuddhodana tried to isolate his son among worldly pleasures. Zimmer's summary follows the account in the second-century CE epic poem the *Buddhacarita*: see Edward B. Cowell, ed. and trans., *The Buddha-Carita, or the Life of the Buddha* [1894]; repr. (with Cowell's Sanskrit text [1893]) (New Delhi: Cosmo, 1977).

But the future Buddha reaches across the walls of his paradise, as if he desires the nourishing bitterness of the real. On three excursions into the realms of human life, he encounters a very old man, a man who is ill, and a dead man lying on a bier. Thus, impermanence speaks to him three times. His fourth encounter is with a yogi, who symbolizes the surpassing of the previous three figures. On his return, moved by this figure and his words, impermanence strikes him with its subtlest, sharpest arrow, for he hears the following message: "A son has been born unto you!" On hearing this, a shadowy word crosses his lips: "Rahulam jatam, bandhanam jatam"—"a small Rahu is born, a fetter is born";[266] this bitter joke, this play on words, transforms the sound and meaning of the message.

Rahu is a daimon, who chases the moon across the heavens and threatens to devour it during the lunar eclipse.[267] But the moon is both the source and symbol of all vegetative forces of life. While the moon is the receptacle containing the nectar of the gods, above all it is a term of endearment for

---

[266] Siddhartha's exclamation when his wife Yashodhara announces Rahula's birth: from the *Jataka* (ca. 380 BCE), a collection of 547 poems narrating previous human and animal incarnations of the Buddha: see T. W. Rhys Davies, trans., *The Jātaka Together with Its Commentary, Being Tales of the Anterior Births of Gotama Buddha*, ed. by Viggo Fausbøll, vol. 1 (London Pali Text Society/Luzac, 1962 [1877]), 60. "Fetter" strongly qualifies the joy deriving from a child's birth. Rahula later became a disciple of the Buddha. According to the Mulasarvastivada Vinaya (an early school of Buddhism in India, 2nd–7th century CE), Rahula was born on the day of Siddhartha's enlightenment, thereby linking his name to the lunar eclipse caused by Rahu (see the following note).

[267] See Heinrich Zimmer, *Myths and Symbols in Indian Art and Civilization*, ed. by Joseph Campbell (Bollingen Series VI) (New York: Pantheon, 1946), 175ff. Hindus give great importance to lunar and solar eclipses. Rahu is the monster demon who at the time of the Churning of the Milky Ocean—a major Hindu myth concerning the ever-continuing struggle between the *deva*s (gods) and the *asura*s (demons, or titans) elaborated in the *Vishnu Purana*—disguised himself as a *deva* and drunk the nectar of immortality. "When the Sun and Moon called this to the attention of Vishnu, the god immediately cut off Rahu's head, but since he had drunk the nectar, his head became immortal. The head was placed in the heavens and Rahu was allowed as a means of revenge on the Sun and the Moon to approach them at certain times and thus render them unclean so that their bodies at those times would become thin and black. The duration of an eclipse is therefore regarded as a time of uncleanness, and orthodox Hindus purify themselves at that time by bathing in the holy rivers of India" (Kenneth W. Morgan, *The Religion of the Hindus* [Delhi: Motilal Banarsidass, 1987 (1953)], 94–95). In art, Rahu is shown as a dragon; in Vedic astrology, it is one of the nine planets, and inauspicious. "Physically speaking, it is one of two points in the heavens where the moon crosses the ecliptic or path of the sun. The point where the moon crosses the ecliptic moving from south to north is Rāhu, the north node. The south node is Ketu. Rāhu and Ketu are depicted as a serpent demon who encircles the Earth; Ketu is the dragon's tail and Rāhu is the head" ("Rāhu," in Satguru Sivaya Subramuniyaswami, *Dancing with Śiva: Hinduism's Contemporary Catechism*, 6th edn [Kapaa, HI: Himalayan Academy Publications, 1979], 815. See also Rajesh Kochhar, "Rāhu and Ketu in Mythological and Astronomological Contexts," *Indian Journal of History of Science* 45, no. 2 (2010), 287–97.

"son," epitomizing both pleasure and refreshment.[268] Just as the moon gifts the gods with their potion of immortality, the sight of his son affords the father the comforting certainty that he will live on after his death: in offering sacrifices to the ancestors, the son feeds the deceased members of the family in the otherworld, to prevent them from falling into a bottomless abyss. Therefore, another word for son is "joy"—*nandana*—for he is the utmost joy of both his father and the entire family. With trepidation, he awaits and delivers both his living and his departed mountain ancestors, without further offerings to the dead having to be made in vain.

"A small moon is born, joy is born": this must have been the call that reached the prince's ear, but he playfully reversed sound and sense into their opposites. Thus, the words did not announce one who would give him life beyond death, but instead one who took life away from him, and thereby reminded him of his own impermanence. Neither did they announce one who would be a "joy" (*nandana*) onto him, but instead a "fetter" (*bandhana*), which would tie him to his house and to the world through love and need.

For the son is the sweetest and most relentless messenger of impermanence. He reveals this mortality to the father by appearing as his reborn I, as his second self, appointed to survive him and to inherit the sphere of life from him. Even without being destined to slay his father, Oedipus is symbolically both the death of Laios and the bringer of that death.[269] Equally, the grain that falls from the ear of corn in order to sprout deceives the ear which claims its continued existence by standing upright, precisely by falling to the ground. This is the meaning of the myth of Kronos, the father who devoured his children as soon as they glimpsed the light of day;[270] he understood that their mere existence meant his death, and that next to their lives his continued existence amounted to the merest reprieve. Their entering the world pronounced upon him the death sentence of impermanence and already extinguished him before nature's

---

[268] Indirect reference to Candra ("Shining"), a lunar deity in Puranic mythology identified with the Soma, the nectar of immortality, a plant called "God for the Gods" (*Rigveda* 9.42).

[269] Reference to Sophocles's *Oedipus Rex* (ca. 430 BCE). Oedipus (literally "swollen foot," from the piercing and binding of his feet by his parents when they abandoned him) becomes king of Thebes but is destined to murder his father Laius and marry his mother Jocasta.

[270] Kronos: a pre-Olympian god of Greek religion and mythology, son of Ouranos (Heaven) and Gaia (Earth). After eviscerating his father Ouranos, Kronos devoured all his sons, to forestall the prophecy that one of them would overthrow him. See Hesiod's *Theogony* (7th–6th century BCE).

Medusa countenance,[271] which must take from him what it will give to his children.

Thus, the future Buddha, the yogi of India who is revered worldwide, is assailed by all the messengers of impermanence; admonished by them, he undertakes the path to "illumination" in yoga, which is meant to transport him beyond transience. Therefore, when he proffers the fruit of his illumination to all beings of all worlds, he utters the words, "Open is the gate to immortality for those with ears to hear—may they have faith!" Among his earliest disciples a pair of friends stands out, one endowed with miraculous powers, the other with knowledge; when they went forth in quest of a true teacher, they promised each other that "whoever first attains immortality shall announce it to the other."[272]

This immortality cannot be the fruit of the timebound. Everything that has become must become undone, and thus also all the fruit that this can bring forth. "Everything that has become is fragile," says the Buddha. Thus, everything that has shape slips away. No sacrifice and no deed can burst asunder the miracle of change, the spell of finitude. What defines the limits of all ritual, all magic, and all moral greatness is that they are finite, and thus capable of attaining only limited ends.

Against the face of impermanence, the Greeks set the golden shield of fame, of "inextinguishable fame," which Achilles chose over longevity to secure the immortality of a great name.[273] Christians overcame the fear

---

[271] In Greek mythology, Medusa (or Gorgo) is an iconic figure who, after incurring Athena's wrath because of her ill-fated love affair with Poseidon, was transformed into a vicious monster with writhing snakes entwining her head. Medusa's gruesome gaze, capable of petrifying any living being, commonly represents an apotropaic symbol and was interpreted by Freud as an emblem of the castration complex in his posthumously published essay "The Head of Medusa" (1922). Cf. the intriguing essay by Alexander M. Schlutz, "Recovering the Beauty of Medusa," in *Studies in Romanticism* 54, no. 3 (2015): 329–53.

[272] Reference to Sariputta (Śāriputra; Pali: Sāriputta) and Moggallana (Maudgalyayana; Pali: Moggallāna), the Buddha's foremost followers. At first wandering mendicants seeking truth and the deathless, then followers of an ascetic, they became Buddhist. Their conversion is narrated in the *Shariputraprakarana*, a drama ascribed to Ashvaghosha (2nd century CE).

[273] Fame in battle—a topos in the *Iliad*—originates from the intimate wish and iron will of the hero deliberately to choose to go toward his destiny and premature death. The term *kleos*, originally "what has been heard," therefore comes to signify the hero's glory conferred by oral epic (see Rosalind Thomas, "The Place of the Poet in Archaic Society," in Anton Powell, ed., *The Greek World* [London: Routledge, 1995], 104–29, esp. 113–17); Bruce Lincoln, *Death, War, and Sacrifice: Studies in Ideology and Practice* [Chicago: Chicago University Press, 1991], 15 and passim). As for the shield of Achilles, used in his fight with Hector and famously described in Book 18 of the *Iliad* (478–608), see the truly evocative poem by W. H. Auden "The Shield of Achilles" (1952) and the recent study of its

of death through their faith in the bodily resurrected one, who bestows eternal life upon those who believe in him. The Indian knows that he is entirely mortal, and thus exposed to the universal fate. Spellbound by involuntary participation in the flow of life, which condenses into figures within and around him and endlessly allows these to fade away, he lives his impermanence even while it chases him—and so how does he escape it? Where would he find immortality, which has neither come into existence nor will pass away?[274]

This everlastingness appears and manifests itself in impermanence: its playful involvement in the alteration of forms is what endures forever. Within and behind every shape that blossoms and withers, as pure process, like a soap bubble—be it a tree or a human body, which assumes no permanent shape whether short-lived or long-lasting—an intangibility that inheres in all formations, physical and psychic, reveals itself by bringing forth every tangible thing. By this measure, everything tangible is a semblance, because it evaporates like a cloud. Neither was it present, nor will it be present. As ephemeral as a gesture, as irretrievable as an impression, it is already crumbling. But to us it seems to be made for eternity. To Indian thinking, the terms "real" and "unreal," "actual" and "unactual," gain color and substance from the conceptions of immortality and impermanence. Because everything tangible in the world is transient, like any impression, sensation, or feeling, it is no more than "appearance"—that is, Maya.[275] Where the world at every instant is a mutable gesture of the

---

spectacular detailed imagery by Andreas Jung, *Der Schild des Achilleus: Spiegel von Schöpfung und Schicksal* (Einsiedeln: Daimon, 2021).

[274] Zimmer considers the figure of Achilles as a paradigmatic Western model also in his *Indische Sphären* (Munich: Oldenburg, 1935 [2nd edn: Zurich: Rascher, 1963]), 67ff. (2nd edn, 68ff.), where he stresses its distance from an Oriental conception of heroism. He points out that "in our country, this type of yogi, and his antisocial self-engrossment, corresponds to a madman and his delusions of grandeur." He adds, "Gandhi, a highly social and political figure whose program for national and economic independence is oriented toward the West, still derives his effect on the broad masses of the Indian population to a large extent from a renewal of this type of ascetic in himself: he drives himself relentlessly toward suffering, not only as a political martyr—but just as much by constantly exhibiting and exemplifying the self-torment of self-imposed, voluntary abstinence, and of the purposeless mortification of his body" (*Indische Sphären*, cit., 66 [2nd edn, 67]).

[275] "*Māyā*, from the root *mā*, 'to measure, to form, to build,' denotes, in the first place, the power of a god or demon to produce illusory effects, to change form, and to appear under deceiving masks" (Heinrich Zimmer, *Philosophies of India*, ed. by Joseph Campbell [Bollingen Series XXVI] [New York: Pantheon, 1951; Princeton, NJ: Princeton University Press, 1969], 19 n. 11). For the Vedanta, the whole Cosmos as an "effect of nescience" is *maya*, an illusion that mocks "the perceiving, cogitating, and intuitive faculties at every turn" (414). "God" perhaps "exists only in association with the power (*Shakti, maya*), of

divine, its alteration of fortune over aeons is the playful, raving dance of its force, self-intoxicated. Given this character, however, the deathless can never become manifest. Instead, whatever appears and becomes tangible is a semblance of the deathless. Thus, immortality resides in all tangible transience as an intangible interior, as a hereafter—but who indeed grasps this? Yoga is the path across this threshold of the afterlife that exists within us. Its teachings impart the techniques for ascending this path through the tangible layers of the person, through the realm of the senses and thought, which contributes its substance to the unity of personal consciousness, inwardly through the sphere that is more encompassing than the conscious I, through layers of pure inner experience and realization, into a depth that consumes everything just as it carries everything.

Yoga struggles with the questions that exercise all philosophy: Who am I? Where do I begin, where do I end, and where is my foundation?[276] Is the person the overarching thing, as consciousness likes to believe? Or is

---

self-misrepresentation" (427). Though accepting this principle, Tantrism interprets *maya* as a showing of the dynamic aspect of the divine. *Maya* then represents "the unsubstantial, phenomenal character of the observed and manipulated world, as well as of the mind itself—the conscious and even subconscious stratifications and powers of the personality. It is a concept that holds a key position in Vedantic thought and teaching, and, if misunderstood, may lead the pupil to the conclusion that the external world and his ego are devoid of all reality" (19). Moreover, Zimmer points out that "the Hindu view and symbolism of the macrocosmic universal maya is based on millennia of introspection, as a result of which experience the creative processes of the human psyche have been accepted as man's best clues to the powers, activities, and attitudes of the world-creative supramundane Being. In the process of evolving a dream world of dream scenery and dream people—supplying also a heroic dream double of our own ego, to endure and enjoy all sorts of strange adventures, we do not suffer the least diminution, but on the contrary realize an expansion of our personal substance" (390–91). *Maya* is also a "deceit, fraud, any act of trickery or magic, a diplomatic feat" (122), which Zimmer connected with the "atmosphere of danger, suspicion, and threat" that pervaded ancient Indian affairs, just as during the decline of the Roman Empire famously described by Edward Gibbon, "as well in the Greek accounts of Achaemenids of ancient Persia, the Moslem records of the caliphates at Baghdad, Cairo, and elsewhere, and the histories of Ottoman power in Constantinople. It is the atmosphere that is general today, particularly in the sphere controlled by the totalitarian states" (112–13). The *maya* notion was popularized in the West by Arthur Schopenhauer, a philosopher dear to Zimmer, who created the influential, highly evocative, yet to a large extent fallacious, notion of the "Veil of Maya" (see Douglas L. Berger, *"The Veil of Maya": Schopenhauer's System and Early Indian Thought* [Binghamton, NY: Global Academic Publishing, 2004]). "Maya," as Wendy Doniger O'Flaherty puts it, "not only deceives people about things they think they know; more basically, it limits their knowledge" (*Dreams, Illusion, and Other Realities* [Chicago: University of Chicago Press, 1986], 119).

[276] Later he adds, "Wo sind die Grenzen des Ichs?" (Where are the boundaries of the I?) (Zimmer, "Yoga und Maya," *Corona*, year 4, no. 4, 1934, 378–400 at 382; and *Indische Sphären*, cit., 105 [2nd edn, 102]).

it merely the bright blossom on a dark tree,[277] a shapely wave on calm, deeper waters—a surface structure, hollowed from another that withdraws from it, leaving it to do as it pleases in evanescence? The necessary thing is to grasp this other, this afterlife within us, and to understand and master the order of dependency which it sustains, because it is a matter of attaining sovereignty within the house of our body, our world.[278]

The aphorisms in the *Yoga Sutras* describe yoga's classic path toward attaining this objective.[279] One central concept of this system concerns

---

[277] I have substituted the term *Laub* (foliage) in the original transcription with *Baum* (tree), which appears in the other redactions of the text ("Yoga und Maya," cit., 382; *Indische Sphären*, cit., 105 [second edn, 102]; also in Zimmer's typescript beginning "Philosophieren ist nach Sokrates" HZMK 130, DLA, p. 5), which suggests a transcription error. [Ed.]

[278] This last sentence alludes simultaneously to the notion of the human body as a temple (in both the East and the West: cf. Saint Paul's "your body is a temple of the Holy Spirit" [1 Cor. 6:19]), and Freud's famous maxim, that "the ego is not master in its own house," being opposed by both superego and id (Sigmund Freud, "A Difficulty in the Path of Psycho-Analysis" [1917], collected in *The Standard Edition of the Complete Psychological Works of Sigmund Freud*, ed. and trans. by James Strachey with Anna Freud [London: Hogarth Press, 1953–74], vol. 17: *1917–1919* [1971 (1955)], 143). From this point on, in "Yoga und Maya" as well as in *Indische Sphären* (and in the typescript HZMK 130, DLA), the theme shifts, turning on the impossibility of human self-understanding ("We do not wholly possess ourselves; we have nothing other than what we are aware of—a flowing, a slipping from our grasp—which is swallowed by the night, and dispersed by the day" ["Yoga und Maya," cit., 382; *Indische Sphären*, cit., 105 (2nd edn, 102); HZMK 130, DLA, 5]). Subsequently "Yoga und Maya" addresses breathing techniques (382ff.), and *Indische Sphären* aspects of Hatha yoga physiology (105ff. [2nd edn, 102ff.]).

[279] The *Yoga Sutras*, in four books attributed to Patanjali, form—together with their ancient commentary *Yoga-bhashya*—the canonical basis of classical yoga, supplying the first coherent system for the Yoga-darshana, one of the six orthodox philosophical systems of Hinduism based on the Vedas (the Nyaya and the Vaiseshika, the Sankhya and the Yoga, and the Mimamsa and the Vedanta). *Yoga Sutras*, also known as the Eightfold Path or Limbs (different from the Buddha's noble Eightfold Path) that correspond to the rules leading to the liberation, is divided into four *pada* or books named Samadhi, Sadhana, Vibhuti, and Kaivalya. Zimmer indicates elsewhere that the first three can be tied to the second century BCE and the last—the most lyrical—to the fifth century BCE (*Philosophies of India*, cit., 283). Recent studies tend to date this last to between 325 and 425 CE (see P. A. Maas, *Samādhipāda: Das erste Kapitel des Pātañjalayogaśāstra zum ersten Mal kritisch ediert* [Aachen: Shaker, 2006]). The *Yoga Sutras* system, retroactively known as Raja yoga (Royal yoga) had for long fallen into relative obscurity until its rediscovery in the early nineteenth century by British orientalists and successive popularization by the Theosophical Society and Swami Vivekananda (who regarded it as the "supreme contemplative path to self-realization" [David Gordon White, *The Yoga Sutra of Patanjali: A Biography* (Princeton, NJ: Princeton University Press, 2014), xvi–xvii]). Jung commented on the *Yogasutras* in his pre-war ETH Zurich lectures (C. G. Jung, *Psychology of Yoga and Meditation: Lectures delivered at ETH Zurich, Volume 6: 1938 to 1940*, ed. and introduced by Martin Liebscher, trans. by Heather McCartney and John Peck [Philemon Series] [Princeton, NJ, Princeton University Press, 2020], 213–14 and 219ff.).

the "impairments"—that is, the so-called "hindrances" or "afflictions" (Sanskrit refers to these as *kleshas*).[280] The term denotes everything in general language that impairs the ideal condition of an object or person: a withered floral decoration or a threadbare or soiled garment is afflicted with *klesha*, and so, too, is the moon when it is wrapped in cloud; trying business affairs and commitments are *kleshas*, since they obstruct perfect happiness.[281]

In the psychology of yoga, there are five "impairments" of our original splendor: partiality or "ignorance"; the feeling of "I am I"; desire and aversion; and lastly, the drive toward creaturely existence.[282] These five "impairments" are "perversions" of our true essence. By "swaying" or "oscillating," they consolidate the workings of the psychic and physical "world substance [*Weltstoff*] from which we are formed"—and which psychically manifests itself in three aggregate forms: as utter clarity or the

---

[280] "*Kleśa*, a common word in everyday Indian speech, is derived from the root *kliś*, to be tormented or afflicted, to suffer, to feel pain or distress" (Zimmer, *Philosophies of India*, cit., 294; Cf. his treatment of the "hindrancens" in the homonymous chapter in Part III, ch. 2.3).

[281] The *klesha*s are generally held responsible for perpetuating *samsara* and therefore stand as the chief obstacle to recognizing one's essential identity with eternal and unchangeable essence. See also *Indische Sphären*, cit., 146 (2nd edn, 140). Surendranath Dasgupta translates *klesha* as "affliction" (*Yoga as Philosophy and Religion* [London: Kegan Paul, 1924], 104; cf. C. G. Jung, *The Psychology of Kundalini Yoga: Notes of the Seminar Given in 1932*, ed. by Sonu Shamdasani [Bollingen Series XCIX] [Princeton, NJ: Princeton University Press, 1996], 4–5 and n. 4); and Georg Feuerstein as the "causes of suffering" (*The Deeper Dimension of Yoga: Theory and Practice* [Boston, MA: Shambhala, 2003], 280, 284).

[282] On the five "impairments" (treated in the second book of the *Yoga Sutras*, *Sadhana Pada*, *Yogasutra* 2. 3–17), see also Zimmer's *Philosophies of India*: "1. *Avidyā* [. . .] is the root of all our so-called conscious thought. 2. *Asmitā* (*asmi*='I am'): the sensation, and crude notion, 'I am I;' cogito ergo sum; this obvious ego, supporting my experience, is the real essence and foundation of my being. 3. *Rāja* [*sic*]: attachment, sympathy, interest; affection of every kind. 4. *Dveṣa*: the feeling contrary to rāja [*sic*]; disinclination, distaste, dislike, repugnance, and hatred. *Rāja* [*sic*] and *dvesa*, sympathy and antipathy, are the root of all the pairs of opposites (*dvandva*) in the sphere of human emotions, reactions, and opinion [. . .]. 5. *Abhiniveśa*: clinging to life as to a process that should go on without end: e.g., the will to live. These five hindrances, or impairments [. . .] are generated involuntarily and continuously, welling in an uninterrupted effluence from the hidden source of our phenomenal existence. They give strength to the substance of ego, and ceaselessly build up its illusory frame" (*Philosophies of India*, cit., 295). Cf. Larson's translation of the "five afflictions" as "ignorance, egoity, attachment, aversion and the clinging to conventional life" (G. J. Larson, "Introduction," in Gerald James Larson and Ram Shankar Bhattacharya, eds, *Encyclopedia of Indian Philosophies, Volume 12: Yoga: India's Philosophy of Meditation* [Delhi: Motilal Banarsidass, 2008], 1–153 at 116). Jung also dealt with the *klesha*s. For instance, in the seminar *The Psychology of Kundalini Yoga*, cit., he compared the *klesha* of discrimination (*asmita*) to "what one would describe in Western philosophical terms as an

calm ocean of mind; as the whirling dust of the affects, which clouds clarity; and as creaturely stupor, a dull and bestial complacency.[283]

Partiality, the first "impairment" of our splendor, is the "motherly soil" of the other four conditions, which are the forms into which it changes. Partiality, which pervades all of these forms, seems to create the impression that the ephemeral, impure, suffering, and insubstantial character of both our natural existence and the world are its opposites: namely, everlasting, pure, joyous, and substantial. This compelling impression leads to mistaken "naming," such as "the earth is everlasting, the heavens occupied by the moon and stars are everlasting, and the gods inhabiting the heavens are immortal"—whereas the opposite is true. The Indian notion of impermanence does not start from the self-evident transience of human existence, but instead throws it into total relief against the insubstantiality and transience of the cosmic, astral, geological, and theological.[284]

If someone says, "This young girl, who is as sweet as the thin waxing crescent of the new moon, her limbs formed as if from honey and nectar, seems to have broken out of and descended from the life-giving lunar globe; her eyes, curved long like the petals of a dark-blue lotus, her eyes, full of alluring playfulness, allow the world to heave a deep sigh of joy"—if someone says this, caught by sensual allure, then the yogi considers this a reversal of the pair of terms "pure–impure," because in the

---

urge or instinct of individuation" (4); notably, he suggested calling it "the *kleśa* of individuation" (5) arguing its close connection with the *entelechia* (ἐντελέχεια), the recurring term in Aristotle's work (and commonly associated with energy) indicating the capacity of every form of life to reach its goal, or make actual what is potential—consequently, the *quid* able to lead the human being to her/himself. (On this topic, there comes to mind James Hillman's *The Soul's Code: In Search of Character and Calling* [New York: Random House, 1996]). The *klesha*s, Jung said elsewhere, "produce ultimately, in the long duration of development, what we call consciousness" and are like "the great forces" constituting "the formidable structure of the collective unconscious," called *archaí* by the Gnostics, and by Saint Paul also *thronoi*, or "powers and principalities" (C. G. Jung, *Visions: Notes of the Seminar Given in 1930–1934*, ed. by Claire Douglas, 2 vols [Bollingen Series XCIX] [Princeton, NJ: Princeton University Press, 1997], 2:1024). Later in "The Psychology of Eastern Meditation" [1948], Jung affirmed that the *klesha*s "would correspond, in the language of Saint Augustine, to *superbia* and *concupiscentia*" (CW 11, § 912).

[283] Zimmer refers to the constitutive qualities of the three *guna*s that according to the Samkhya system—which constitute the theoretical foundation of the *Yoga Sutras*—drench the whole of physical and psychic nature (*prakrti*): 1) purity, equilibrium (*sattva*); 2) activity, passion (*rajas*); and 3) inertness, obscurity (*tamas*). Jung discusses the three *guna*s in *Psychology of Yoga and Meditation*, cit., 218ff.

[284] Cf. this last statement modified in *Indische Sphären*, cit.: "Do not begin with the infirmity of the creature, which throws into relief both the profound insubstantiality of the creature and its cosmic-astral insubstantiality, as well as the impermanence of the gods" (147 [2nd edn, 140–41]).

eyes of the wise the body is "impure." It is so because its origin, seed, main-
tenance, excretions, and its end all require that it be cleansed.

Our "impairments" cling to everything that takes on objective form
for us. They are at work in all instances of our "mistaken notions." But
for the yogi, who perfects himself, "they wane with diminishing partiality." Owing to his efforts, they grow ever smaller in him; in his state of
rapture, they have fallen asleep for a time; in the child of the world, however, they are immense, readily shattered when they break each other, for
instance when conflicting in the interplay between desire and aversion. But
if they are taken together, then our impairments are what we must call our
personal life: the entanglement in the world as it presents itself to us; the
clinging to our I as it is naively given to us; affection and aversion, which
guide us on all paths throughout our lives; and the drive toward creaturely
existence, which unites us with the worm; the primal scream uttered when
danger confronts us and when we come to die by drowning, "I do not
not want to be—I want to be!"[285] It guides us through our whole life and
thrusts us into future existence; at first it flows out of itself, "for how else
would a newborn creature, which has never tasted death, have an aversion toward the strangeness of death?"

This primal sound also means, "I do not *not* want to *become*! I want
to *become*." It implies the urge of life flowing out of its given shape to
form new shapes, in order not to be devoured by the death of the person.[286] This urge represents the first mythical stirring of world genesis;
in the ancient myths, the "Lord of Spawn,"[287] the world creator in the

---

[285] *Yogasutra* 2. 9. In his *Philosophies of India*, Zimmer cites the same plea thus: "'May
I not cease to exist! May I go on increasing!'," as well as the commentary of James Haugh-
ton Woods, who translated this sutra as follows: "The will-to-live (*abhiniveśa*) sleeping on
[by the force of] its own nature exists in this form even in the wise" (Haughton Woods, *The
Yoga-System of Patañjali, or The Ancient Hindu Doctrine of Concentration of Mind* [Har-
vard Oriental Series 17] [Cambridge, MA: Harvard University Press, 1927 (repr. Mineola,
NY: Dover, 2003)], 117, cited in Zimmer, *Philosophies of India*, cit., 299); cf. also G. J.
Larson's translation: "Clinging to ordinary life in the sense of going with the flow of one's
ordinary impulses is characteristic even of a learned person" (Gerald James Larson and
Ram Shankar Bhattacharya, eds, *Encyclopedia of Indian Philosophies, Volume 12: Yoga:
India's Philosophy of Meditation* [Delhi: Motilal Banarsidass, 2008], 168).

[286] See Heinrich Zimmer, *Ewiges Indien: Leitmotive indischen Daseins* (Potsdam: Müller & Kiepenheuer, 1930): "The creature's cry of fear, which pleads for never-ending, timeless existence, is no different from the cry of pleasure of the swelling urge of life—'I want
to be more! I want to give birth!'—which demands existence without bounds. It only
sounds different if it is fraught with the shudders of death or begetting" (123).

[287] Reference to Prajapati, a creator god like Vishnu but presiding over procreation as
well, is prominent in the Upanishads, the Brahmanas, and the sections on sacrifice in the
core of the *Rigveda* ("Hymn in praise of sacred knowledge"), the oldest of Indian sacred

Vedic scriptures, is induced by a twofold motive to give birth to creatures from himself: he feels "lonely, destitute, and fearful," and so he generates a world of life—or rather, he wants "to be many [. . .] and to give birth in abundance."[288] He is not spirit [*Geist*] but creature, the *creatura totius* from which all creatures come, and his two impulses are a single urge, the urge of every creature: to exist, in whichever way that may be, and not fall away; to exist beyond itself and vanish within itself.[289]

All these "impairments" or "hindrances" are a timeless legacy of our creaturely nature. They lie at the bottom of the impregnations of our core, which have no beginning, and which afflict us from earlier lives as an unconscious treasury of readiness to react and behave. They need to be overcome and diminished. Not only the person but also the creature within us should be cast off, and the beyond of both, which lies within us impalpably, should be given to us.

The "impairments" of our essence prevent the human being from attaining a single-minded disposition, an imperturbable concentration of stillness within, where the illuminating knowledge of our essence, which brings the inner hereafter within our reach, becomes possible. In this respect, the three means of "active" yoga come to our assistance: asceticism, learning, and a devotion to the divine.[290] Asceticism burns away

---

text (ca. 1200 BCE), dealing with the *asura*s and the supreme gods. In the post-Vedic era, Prajapati is identified with Brahma.

[288] Cf. *Philosophies of India*, cit., 300, where Zimmer translates, "May I give increase; may I bring forth creatures!" referring to "*Śatapatha Brāhmaṇa* 2. 2. 4; 6. 1. 1–9; 11. 5. 8. 1. Compare *Bṛhadāraṇyaka Upaniṣad* 1. 2, 1–7 and 1. 4, 1–5" (300, n. 23). See also his *Ewiges Indien*, cit., 122–23, and cf. *Indische Sphären*, cit., 148 (2nd edn, 141). Zimmer comments, "The elemental cry and craving to expand, even to multiply in new forms in order to circumvent the inevitable doom of individual death," as taken from a Brahmana-period myth (ca. 900–600 BCE) describes "the first, world-creative impulse of Prajapati, the 'Lord of Creatures'" (*Philosophies of India*, cit., 300).

[289] "This ancient god-creator," Zimmer writes further, "was not an abstract divine spirit, like the one in the first chapters of the Old Testament, who, floating in the pure void, beyond and aloof from the confused welter of the dark world of matter, created the universe by the sheer magic of the commands of his holy voice, summoning all things into being by the mere utterance of their names. Prajapati, rather, was a personification of the all-containing life-matter and life-force itself, yearning to develop into teeming worlds" (*Philosophies of India*, cit., 300). For an overview of this god in Hindu literature, see Jan Gonda, "The Popular Prajāpati," *History of Religion* 22 (1982): 129–49.

[290] According to *Yogasutra* 2. 1, "the structured practice of disciplined meditation involves ascetic exercises (for the body), sacred study/recitation (for speech and voice) and focusing on God (as a content of meditation) (for the mind)" (Larson and Bhattacharya, *Encyclopedia of Indian Philosophies*, cit., 167). The "yoga of action," or Patanjali's Kriya yoga, refers to a combination of the three main yogic paths, i.e., *karma*, *jnana* and

the impurity within us. Stemming from self-determination without a beginning, this impurity darkens us with the many hues of our infinite conduct in
earlier lives. This natural impurity of our inner condition rests upon both
the "impairments" and the impregnations of our unconscious by earlier
conduct. Meanwhile, "the mesh of things and situations outside draws
near" to incite its unconscious readiness to permit ever-new events to occur
spontaneously in consciousness. With regard to its material nature, this
plentiful impurity consists of the two inferior aggregate states of the
physical-psychic world substance within us: the clouding dust of the affects, and animal stupor. Both should be dispersed and dissolved by ascetic
fervor, just as the blazing heat of the sun devours the clouds and dissolves
them into the ethereally clear calm of the firmament. "Learning," the second means of "active" yoga, involves the recitation of purifying, healing
syllables and words, whose effect is magical and suggestive. It also implies
the repeated recitation of the teachings. Freed from the spell of self-
consciousness, this promotes the passage of self-transformation. Thirdly,
"dedicating oneself to the divine" means that whatever one does, together
with all the fruits it bears, does not serve oneself, nor should one expect
anything from it for oneself. Instead, one sacrifices this to God, according
to the motto, "whatever I do willingly or unwillingly, whether pure or impure, I commit unto you, since my actions follow your guidance." This
amounts to perfect self-abdication,[291] as taught by the *Bhagavadgita*:[292]

---

bhakti, indicated in sequence of sutras in the second book of the *Yoga Sutras* concerning
exercises that facilitate attaining the level of consciousness described in the first book
(*Yogasutra* 2. 1–27; Larson, "Introduction," cit., 11). The threefold path, or the three core
practices of yoga known as Kriya yoga, comprises: *tapas* (heat, "ascetic fervor"), *svad-
hyaya* (*jnana*, study, knowledge), and *ishvarapranidhana* (*bhakti*, devotion to God); see
Yogasutra 2. 1 and 2. 43–45. In this way Patanjali magisterially distilled and systematized
in the of the Raja ("royal") yoga the earlier message included in the *Bhagavadgita* (see
below, n. 292).

[291] Zimmer adds in *Indische Sphären*, cit.: "in faithful devotion (*bhakti*)" (149 [2nd edn,
142]). Cf. his *Philosophies of India*, cit.: "Complete surrender to the will and grace of God"
(302).

[292] The *Bhagavadgita* (Song of the divine, or Song of the venerable) is the famous devotional epic about Krishna and Arjuna, concerning a crucial battle supposed to have taken
place ca. 1000 BCE, and part of the encompassing epic cycle, the *Mahabharata*—ascribed,
like the Homeric epics, to a legendary bard (Vyasa), but in fact composed by various hands
and in several stages, coming to a written form between 50 BCE and 50 CE and further
attaining its current form (as reconstructed by Vishnu S. Sukthankar in the Pune critical
edition) during the Gupta Empire (ca. 330–ca. 500 CE). On the composition and date of
the *Mahabharata*, see J. L. Fitzgerald, "The *Mahābhārata*," in Knut A. Jacobsen et al., eds,
Brill's Encyclopedia of Hinduism, 6 vols (Leiden: Brill, 2009–14), vol. 2 (2010): *Sacred*

that is, "it is not I who acts."[293] In a life dedicated wholly to such conduct, the aim is to reduce the "impairments" to a small size; that is, the epitome of person and creature within us, and to sterilize[294] and "burn" its seed, which sprouts ever-new flowers of individuation, ever-new fruits of destiny, in the fire of asceticism.[295] Then, untouched by all impairments and afflictions, a "most refined knowledge" will be able to become functional. It will manifest itself as a purely discerning knowledge: that is, that an innermost, purely observant core stands within us beyond the tangible person and the deepest darkness of its unconscious readiness, beyond the psychic-physical matter of the world, which has purified itself from its inferior aggregate forms of the clouding affects and animal stupor to become purest clarity and calm. Gathered into a clear mirror by way of concentrating and unifying the incessantly whirling stream of outer impressions and inner reactions, the disposition reflects this otherworldly greatness; it

---

Texts, Ritual Traditions, Arts, Concepts, 72–94. In the *Bhagavadgita* the tripartition (*Trimarga*) of yoga's soteriological paths occurs in teachings imparted by Krishna to Arjuna. Thus Karma yoga, Jnana yoga and Bhakti yoga are introduced as pathways respectively privileging askesis, the study of the sacred texts, and devotion to God, consistent with different temperaments yet aiming at one supreme goal, the soul's reunion with its universal essence. Yet devotion in the *Bhagavadgita* becomes the instrument of salvation par excellence. It has been noted that in the *Mahabharata* "the Yogin is better than the jnanin, the karmin and the tapasvin—the wise, the active and the ascetic. The *Yoga* here spoken of is not the *Yoga* of Patanjali or *Yoga* in any technical sense, but it indicates the union with the Divine, or what the Gita mentions as *Brahmic* consciousness, that resting and living in the Divine, in the Absolute, which is the sum and substance of spiritual realization" (Nalini Kanta Brahma, *Philosophy of Hindu Sādhanā* [New Delhi: Asoke K. Ghosh, Prentice-Hall of India, 2007 (orig. London: Kegan Paul, Trench, Trubner, 1932)], 251).

[293] The renunciation of attachment to the fruits of action is the leitmotif of the whole text. It epitomizes Krishna's teaching in *Bhagavadgita* 14.25: "The same in honor and dishonor, / not taking sides from friend or foe, / who gives up all undertakings, / is said to transcend the qualities [*gunas*]"; and 13.29: "And he who sees that actions are / entirely performed by matter, / the Self having no agency, / that one truly sees, Arjuna" (Gavin Flood and Charles Martin, eds, *The Bhagavad Gita: A New Translation* [New York: Norton, 2012], 117 and 110). See also 18.26: "One who is free from attachment, / firmly resolved, self-effacing, / unmoved by success or failure, / is said to be a pure agent" (ibid., 139).

[294] The original term (Ger. *sterilisieren*) will be substituted with "to kill its seeds" ("Yoga und Maya," cit., 397; *Indische Sphären*, cit., 149 [2nd edn, 142]. See also *Ewiges Indien*, cit., 123).

[295] According the *Bhagavadgita*, "Who in action sees non-action, / and sees in non-action, action, / is a wise man, is disciplined, / whole in all actions he performs. // In all his undertakings, he / ignores desires and volition. / The wise see him as wise, whose acts / have been burned in wisdom's fire" (Flood and Martin, *Bhagavad Gita*, cit., 4.19 [38]. See also Zimmer, *Philosophies of India*, cit., 302–3).

sets itself apart and sees itself as an unwoven, unentangled otherworld, which is mirrored only by the completely purified person.

This "discerning knowledge"[296] of the person and the otherworldly core signifies the introduction of a "backward process" within us:[297] the constant soaring of person and world from our depth, allowing our unconscious readiness to take shape in inner and outer actions, turns into a diminishing of the spheres which, fulfilled by the dynamics of these actions, are constantly regenerated. Being separate or distinct, known as the condition of *kaivalya*, takes the place of an entanglement with the world and the I. But the condition and notion of *kaivalya* covers only one side of the meaning of "singleness," namely, only that of the process of impairment or diminishment, which dissolves the person and creature as the focal point of our existence.[298] By moving beyond both and returning inward, we become not poorer, but richer. This is what the notion of impairment or affliction captures, which defines the nature of both person and creature within us. Their impairment or diminution means the falling away of something that oppresses us and that impedes us from enjoying our true and full splendor.

Freed from the impairments, the way to our ideal, unburdened being proceeds by way of practicing a series of exercises of wonderful miraculous experiences, through which we gradually approach the fullness of our

---

[296] Reference to a nexus including the *jnana* or *gnana* (Pali: *ñāṇa*) or *jnana-dyaka* (transcendental knowledge) and *viveka* ("discriminative insight"), representing "the enemy of our usual ignorance, i.e., of *avidyā* and therefore the chief instrument to disentangle us from the force of the gunas" (Zimmer, *Philosophies of India*, cit., 302; see also Heinrich Zimmer, trans., *Der Weg zum Selbst: Lehre und Leben des Heiligen Shri Ramana Maharshi aus Tiruvannamalei*, ed. by C. G. Jung [Zurich: Rascher, 1944], 247). Cf. for instance *Yogasutra* 2. 28: "When impurities have been destroyed as a result of the systematic practice of the limbs (or levels) of disciplined meditation, knowledge comes to include the full illumination of discriminative awareness" (quoted in Larson and Bhattacharya, *Encyclopedia of Indian Philosophies*, cit., 170). Zimmer also evokes the *buddhi*, human intelligence as the highest faculty of the substance (*prakrti*) to comprehend, by reflection, the spirit (*purusha*) according to the Samkyha.

[297] Cf. *Yogasutra* 4. 4–34, "reversal or turning back of the guṇas" (Larson and Bhattacharya, *Encyclopedia of Indian Philosophies*, cit., 183). See also Zimmer, *Indische Sphären*, cit., 90 (2nd edn, 143).

[298] "*Kaivalya* is the state of one who is *kevala*, an adjective meaning 'peculiar, exclusive, isolated, alone; pure, simple, unmingled, unattended by anything else; bare, uncovered (as ground)'; and at the same time, 'whole, entire, absolute, and perfect' [. . .]. *Kaivalya*, consequently, is 'perfect isolation, final emancipation, exclusiveness, and detachment', and at the same time, 'perfection, omniscience, and beatitude'" (Zimmer, *Philosophies of India*, cit., 305). Larson renders the term as "spiritual liberation": "When the *sattva* and *puruṣa* have been clearly distinguished, [there arises] spiritual liberation" (*Yogasutra* 3. 55, in Larson and Bhattacharya, *Encyclopedia of Indian Philosophies*, cit., 179).

splendor, so-called *kaivalya*. An entire book of the *Yoga Sutras* is devoted to the unfolding of miraculous powers.[299] They have been understood as "supernormal powers"—which they are, so long as one considers the diminished, burdened condition of the child of earth as "normal." But if one follows the *Yoga Sutras* and their terminology, for which our personal and creaturely aspects are the abnormal condition because they constitute a burden, then one is speaking as if from the standpoint of the *kaivalya*—and from where else should yoga name things?[300] For these "miraculous powers" are the happiest extensions of our secret and full splendor. Therein, the powers of our deeper essence, which are imprisoned and shackled in the child of earth, are liberated.[301]

The inner concentration on an object, which forgets itself and becomes wholly absorbed by that object and is no more than its own emanation, transforms the focused adept into the object and bestows upon his sphere a suprasensory, supralogical knowledge.[302] This concentration proceeds

---

[299] *Yogasutra*'s third book, called *Vibhuti Pada* (The path of realization) focuses on the advanced stages of the eight-limbed yoga practice, and therefore deals with the *siddhis* (variously translated as "[spiritual] super-human powers, perfection, ability, success"), the term appearing for the first time in the *Mahabharata*. The *Yoga Sutras* classify sixty-eight *siddhis*. *Siddhi* can also signify "paranormal attainment" and "magic ability" (Georg Feuerstein, *The Encyclopedia of Yoga and Tantra* [Boulder, CO: Shambhala, 2022 (1997)], 348).

[300] Cf. Jung: "The Hindu thinks in terms of the great light. His thinking starts not from a personal but from a cosmic *ajna*. His thinking begins with the brahman, and our with the ego," for "to the Indian, the brahman or the *purusha* is the one unquestioned reality; to us [it] is the final result of extremely bold speculation" (*Psychology of Kundalini Yoga*, cit., 67, 69).

[301] Such powers—Zimmer observes in *Philosophies of India*, cit.—are not extraneous to humans, but "man's original property"; they appear "supernormal from the point of view of our naive and worldly, 'normal' life," but in fact they are "attributes of the pristine reality of our nature that in the course of yoga become restored to us" (307).

[302] Zimmer likely alludes here to *dharana*, one of the eight paths of Patanjali's yoga system, meaning "focused concentration," or "fixation of the mind (on one object or thought)"; he might also refer to *samyama* (restraint), meaning the inner discipline engendered by the combination of simultaneous practice of *dharana*, *dhyana* (meditation) and *samadhi* (union), namely the three final stages of Patanjali's Raja yoga). Moreover, it brings to mind the *ekagrata* (Skr. and Pali; "one-pointedness," "unification of the mind"). In this regard cf. Eliade, who states that "[a]ny ascetic practice begins by *establishing*, by holding a particular body posture (asana) so as to enlarge the states of consciousness and to focus these states on a single image. Such practice also delineates the passage from a profane condition (which is psychologically and metaphysically 'nomadic,' dissipated, frivolous, irresponsible) to one prerequisite to salvation" (Mircea Eliade, *Commentaires sur la Légende de maître Manole*, Fr. trans. from Romanian by Alain Paruit [Paris: L'Herne, 1994], 138). Moreover, "religious meditation and concentration always mean freeing oneself from chaos, so as to root oneself in a stable place; they impede dissipation, chaotic fluctuation, and becoming embroiled in the amorphous. [. . .] In mystical and ascetic practices, the new

by way of as many steps as the objects of fixation to which it can adhere. It concerns itself both with the organs of the body and with the elements of the world. Such concentration on the aspects of time as past, present, and future, freed from the binding of the creature to the immediate and the present, provides a knowledge of the past and the future. Put differently, the unconscious opens itself up with its treasure of impregnations and readiness, whereas the yogi intuitively reads therein his past in previous lives, which have precipitated into this wealth of signs and seeds within him. The concentration on the immediate impression of a person leads to a knowledge of their thoughts.

Knowledge and powers grow that lie slumbering in the child of earth. Concentration on the nature of the sun affords an intuition of the structure of the cosmos, whose all-seeing, wandering eye is the sun. Concentration on the moon opens up the world of the stars, whereas concentration on the polestar affords an intuition of that star's orbit of the pole. Concentration on the navel, the pole of the body, gifts us with intuitive knowledge of the workings of the organs.[303] Wonderful powers follow upon wonderful knowledge; among those mentioned are the loss of bodily heaviness, power over the elements, and lightning-fast movement of thought through space.

Without doubt, here one finds oneself in the sphere that C. G. Jung's analytical psychology has revealed as the "mana personality" within us. (On the characteristics of this sphere, it suffices to refer to Jung's well-known study *The Relations between the I and the Unconscious*.)[304] On

---

life inaugurated by spiritual exercises is, as in ordinary religious practices, a path taken toward the center, that is, a return to the ultimate reality" (139).

[303] "Knowledge of (various) universes"; "of the orderly arrangement of the stars"; "of the motion of the stars"; "of the orderly arrangement of the body" (*Yogasutra* 3, respectively 26, 27, 28, 29; quoted in Larson and Bhattacharya, *Encyclopedia of Indian Philosophies*, cit., 176).

[304] Jung's book published in 1928 distills the core of his psychodynamic system, presenting the main archetypal phases of the individuation process: i.e., the persona, the dyad animus–anima, the shadow, the mana personality, and the Self. Borrowing the Melanesian notion of "mana," generally considered to indicate an impersonal, mysterious, supernatural power associated with special persons, places, and spirits (see Mircea Eliade, *Patterns in Comparative Religion*, trans. by Rosemary Sheed [New York: New American Library, 1974 (1958)], 19–23), he conceives the "mana personality" as "a dominant of the collective unconscious, the recognized archetype of the powerful man in the form of hero, chief, magician, medicine man, and saint, the lord of man and spirits, the friend of Gods," while for women, it usually manifests as "a sublime, matriarchal figure, the Great Mother" (*CW* 7, § 377 and § 379). With its magical power, the mana personality predominantly entails a specific risk of the possession and enchantment of the ego, through the sense of a superiority derived from new acquisitions from the (collective) unconscious, in what Jung names a

the path of yoga, this sphere stands precisely where analytical psychology has found it within us: namely, behind the dissociation of the personality, the dismantling aimed at deliberately by yoga. In Western patients, this sphere breaks out of the unconscious as complexes which have become autonomous, in spontaneous inflation.[305]

Yoga also fully grasps the danger of inflation.[306] While all these extensions of the person are not goals in themselves, they are signposts along

---

condition of "godlikeness" (Gottähnlichkeit). Such a condition essentially represents an unwarranted, megalomanic amplification of the ego: that is, a psychic inflation. (For a survey of this notion, see Giovanni Sorge, "The Construct of the 'Mana Personality' in Jung's Works: A Historic-Hermeneutic Perspective," Parts 1 and 2, Journal of Analytical Psychology 65, no. 2 [2020]: 136–52, and 65 no. 3 [2020]: 519–37, respectively.) Jung proceeds in the same text to evoke, in tones of fascination, the next, supreme stage of the individuation process: "The dissolution of the mana-personality through conscious assimilation of its contents leads us, by a natural route, back to ourselves as an actual, living something, poised between two world-pictures and their darkly discerned potencies. This 'something' is strange to us and yet so near, wholly ourselves and yet unknowable, a virtual centre of so mysterious a constitution that it can claim anything—kinship with beasts and gods, with crystals and with stars—without moving us to wonder, without even exciting our disapprobation" (CW 7, § 398). Jung is speaking here, namely, of the Self, which he regards as "no more than a psychological concept, a construct that serves to express an unknowable essence which we cannot grasp as such, since by definition it transcends our powers of comprehension. It might equally well be called the 'God within us.' The beginnings of our whole psychic life seem to be inextricably rooted in this point, and all our highest and ultimate purposes seem to be striving towards it. This paradox is unavoidable, as always, when we try to define something that lies beyond the bourn of our understanding" (§ 399).

[305] Notably, in the corresponding paragraph beginning "Without doubt," as reworked in Indische Sphären, Zimmer will erase the mention to Jung's book, choose the more generic "Sphäre der Magie des Unbewussten" (magical sphere of the unconscious), and omit the reference to the "autonom gewordene[r] Komplexe" (complexes that have become autonomous) and to the mana personality itself, even replacing this last with the psychoanalytic "Über-Ich" (superego or over-I), and retaining the reference to "[psychic] inflation." See the passage in full in Indische Sphären, cit., 151–52 [2nd edn, 144–45]. (The same will be the case for the corresponding reworked passage in "Yoga und Maya," cit., 399). Zimmer's subsequent change in terms of a major—at least lexical—approximation to the Freudian theory invites us to consider the parallel drawn in the present lecture between siddhi powers and the mana personality as a kind of working hypothesis, which we may speculate he discussed with Jung on some occasion.

[306] At the 1935 psychotherapy congress, Zimmer would insist on the differing approaches of yoga and analytical psychology toward the Self: "When the Westerner thinks of unconscious, he is inclined to feel as though he were contemplating an abyss. To the Indian, on the other hand, his own psychology appears rather as having the structure of an onion, skin after skin: there is the rind belonging to the senses, the next one belonging to the intellect, the one of the consciousness of his own being, then his own unconscious—all these are layers, but when he is going to reach the kernel of life? In all Indian psychology and metaphysics the word 'Self' takes on an incredible emphasis, the 'Atman' does not correspond to the Latin 'ipse,' which flavours of a pompous inflated personality, rather does it correspond to the word 'Self,' to the reflective pronoun in its connotation of 'reaching one's self,' 'to

the onward path toward *kaivalya*. *Kaivalya* stands at the end of this passage of experience, but only for that particular adept who does not let a vestige of his ego become entangled in these allurements while practicing this path. These temptations of the suprapersonal sphere draw near the yogi. The "mana sphere," which he enters in order to cross rather than to populate, consolidates itself within him as figures resembling his divine splendor and attraction. They afford him heavenly pleasures, the drink of immortality, the fulfillment of all wishes, the abolition of spatial barriers, and divine insight. But without heeding them he says to himself, "Roasted on the frightening embers of the timeless cycle of birth and death and deeply entangled in its darkness, I have struggled to attain the illuminating light of yoga, which destroys the gloom of the 'afflictions.' These things of the senses, whose womb is desire, are hostile, just as blowing winds are to the light of the lamp. I have beheld the light of this lamp. How could I be seduced by this fata morgana, succumb to it, and become firewood for the newly kindled embers of the cycle of life? Farewell, ye cousins to dream images, the desired goal of the deplorable!" Nonetheless, whoever decides thus can still fall victim to concentration in yoga. The wise yogi will not be proud even of having encountered and declined these great allurements. Such pride would be one last sublime inhibition, a final triumph of an inflated ego destined to become a mana person.

Tantric literature vividly depicts the daimonic side of this passage into the otherworld of the person.[307] The character of these experiences as

---

return to one's self.' The very essence of the ascetic therapy of Yoga, therefore, could be expressed in the idea of returning into the very heart of the onion, and leaving behind all the outer skins thus to achieve salvation and unity" (Heinrich Zimmer, "Indian Views on Psychotherapy," *Prabuddha Bharata or Awakening India* 16, no. 2 [February 1936]: 177–89 at 187). Cf. Coomaraswamy's argument against the inappropriateness of translating *atman* as "Self," because of a certain "selfishness" and the reflexive connotation inherent in the term, which "hardly occurs in the Rig Veda" (Ananda Coomaraswamy, "Vedic Exemplarism," *Harvard Journal of Asiatic Studies* 1, no. 1 [1936]: 44–64 at 55 n. 23).

[307] The Tantras are a textual corpus from a variegated movement of spiritual teachings and esoteric traditions that influenced the whole of Hinduism beginning around the sixth century CE. Apart from its various currents within Hinduism, there are also a Buddhist, a Jaina, and even a Sufi or Hindu-Islamic Tantrism. The homologizing of bodily organs and functions with cosmic regions and rhythms, already present in the Vedas, was transformed by Tantrism into a coherent system. The Tantric tradition gave a new centrality to the female energy, and stressed the relevance of Shiva and Shakti, whose union represents the universal energy inherent in every individual. This tradition also elaborated a mystical physiology focused on the relationships between microcosm and macrocosm and based on the *chakra*s, literally "wheels," or energy centers. The doctrines and practices of Tantrism are traditionally considered transgressive, because of their insistence on "the holiness and purity of all things; the 'five forbidden things' [. . . that] constitute the substance of the

temptations is brought to life. Everything as yet purely daimonic in his nature obstructs the yogi's path, time and again. When he has overcome the outer sphere of the senses, time and again it approaches him within, just as from above, as a supernatural temptation, as cosmic powers threatening to tear him asunder, because he strives to rise above them, or because they seek to lure him with blazing lust, tearing him backward through involuntary participation in desire. The vast realm of black magic, replete with devilry and she-devils, glittering with the magic of omnipotence, erupts like the seemingly already extinguished volcano of suppressed daimonism, which has not yet been leached. Its rain of ashes clouds the firmament of pure brightness, into which the yogi thought he had already emerged, with embers and darkness. Here lies the zone of true crisis, here one decides either to fall headlong like Icarus into the daimonic inflation of one's ego, feeling like a great magician, and fall like a tyrant into the depths of one's own flames—or indeed to see through one's own fraud to recognize that everything, that all these worlds and powers, are nothing other than explosions of our own daimonism.[308]

This temptation increases until the final stage of the way. If the yogi turns away from it, to concentrate on diminishing the passions and stupor in the daimonic, then he enters "singleness"—he experiences himself as detached from everything that defines his consciousness, and that could affect his unconscious.[309] By describing this state as "singleness," however, one has obviously named only the negative aspect of *kaivalya*, the detachment

---

sacramental fare in certain Tantric rites: wine (*madya*), meat (*mamsa*), fish (*matsya*), parched grain (*mudra*), and sexual intercourse (*maithuna*)" (Zimmer, *Philosophies of India*, cit., 572). Certain esoteric Tantric schools also envisage the consumption of forbidden substances such as semen, menstrual blood, urine, and even human flesh. For an excellent introduction to Tantra, see David Gordon White, *Tantra in Practice* (Delhi: Motilal Banarsidass, 2001); for a study of its Western reception, Hugh B. Urban, *The Power of Tantra: Religion, Sexuality and Politics of South Asian Studies* (London: I. B. Tauris, 2010).

[308] In *Indische Sphären*, cit., Zimmer broadens his phrasing after "daimonism" as follows: "this concerns all this: worlds and forces are nothing other than creatures born from our own daimonism, the unleashed games played by the unconscious, who devours those who abuse it" (92 [2nd edn, 91–92]).

[309] After "singleness" (*Ledigsein*) Zimmer continues in *Indische Sphären*, cit., "and experiences within himself a detachment, which leaves behind, husk-like, everything that besets others." And again, "The Buddha does not call this state of detachment from involuntary participation in the ego, the world, and the unconscious 'Being as such' but 'Expiration'; from the point of view of depletion, he understands this as the final process. For us, who are still 'inflamed,' the Buddha has 'expired.' With this term, he avoids the danger of ontological statement, of objectifying this 'Being,' which is distinct from everything that is, and thereby he avoids the danger of all theories of yoga degenerating into ontology or metaphysics" (92 [2nd edn, 91]).

from the personal and creaturely; only, that is, the diminishment of these determinations. The ever-widening stock of capabilities, which are wonderful when seen from the standpoint of the child of the world, at the same time describes various facets growing within us, including the light-heartedness and undiminished scope of our nature, which develops into full splendor in this final state. This final state integrates all such individual gain into itself; it gathers within itself what in differentiated individual growth represents itself as the powers of cognition and effect in the particular spheres of both the microcosm of our body and the macrocosm of the world. It merges everything attained in the final state of splendor into an undifferentiated calm and an indefinable unity.

This final state is considered in positive terms: to be and to be effective as the integration of all possibilities, a totality of all individual determinations beyond all opposition, to relate to something and to conduct oneself well. Thus, it [the final state] is removed from any destiny, which always proves to be the consequence of individuation and differentiation.[310] Here one experiences an existence beyond all determination and dependency. It is a floating abundance of all possibilities, relieved from the creaturely compulsion to carry oneself into effect in any tangible reality. This state of being knows that it is infinite and devoid of all qualification, just as the preexistence of the divine, fateless and immortal, has still not overflowed its calm to differentiate itself and become the world's plaything. For our nature, *kaivalya* means the *restitutio in integrum*.[311]

---

[310] A difference between Jung's notion of the mana personality and the *siddhi*s in the *Yoga Sutra*s may be perceived in the fact that Jung focuses on the presumption of the subject's ego to possess—risking identification with—a superior, divine-like knowledge and power; for him it is a hint of the Self, while the *siddhi*s represent a hint of immortality, if the demonic temptation is overcome. Daemonism and black magic winds for Jung in the subject's incorrect relationship to the mana-figure, for the *Yoga Sutra*s in the *siddhi*s themselves. Employing psychologically inflected language, Zimmer sets up a dialogue between yogic and analytical-psychological goals: for instance, by defining the state of an ultimate detachment, singleness and separation (*kaivalya*) as consequence of both "individuation and differentiation," thereby combining one term (the first) germane to the fundamental psychological process according to Jung's depth psychology with another (the second) liable to echo a crucial act in the individuation process itself—that is, the differentiation from the collective, as explained, for example, in *Psychological Types* (CW 6).

[311] Lat.: "reinstatement of integrity, restoration of the original state." The phrase was used in medicine to signify the rehabilitation of diseased organs. In *Indische Sphären*, cit., Zimmer adds, "the ascending Kundalini performs the gradual integration of the self-differentiating elements into the all-integrating ether and on making its journey home beyond that ether, toward the transcendent recumbent God, achieves the total integration of the world body" (153 [2nd edn, 146]). In *Philosophies of India*, cit., he renders *kaivalya* both as "perfect isolation, final emancipation, exclusiveness, and detachment" and as

The exercises of Hatha yoga,[312] which exerts physiological force (*hatha*)[313] by means of breathing, point in the same direction. Hatha Yoga teaches many ways of seizing immortality, which exists both within and behind everything that comes to pass. Its technique gives rise to other yogic methods, which in turn build on it. Therefore, it refers to itself as the "staircase"[314] that ascends to these other methods, just as stairs lead

---

"perfection, omniscience, and beatitude" (305), simultaneously observing that the Samkhya-yoga system shares many features with the ancient pre-Aryan philosophy preserved in Jain beliefs, and also that the *Yoga Sutras* retain the double meaning of isolation and perfection for *kaivalya* (306–7). This differs from the monistic Vedanta, where "there is, finally and fundamentally, but one essence, Brahman, and that this unfolds into the world-mirage of the visible multitude of beings" (306). Defining *kaivalya* as both the discovery of true nature and a "discarding [of] the mask" of personality, Zimmer facilitates the impression of a connection with the aims of psychoanalysis (310) and specifically analytical psychology, which insists on a differentiation from identification with the "persona" or mask; and despite the absolute detachment of *kaivalya* he tries to link the final state in Indian philosophy, as well as for Western thinkers such as Plato, with a state that transcends the "power and task of reason" (312). Thus the final condition corresponds to transcendence as a union of opposites that one can pursue through "an everlasting dialectical process" and "a constant transformation of things into their antitheses" (313).

[312] Hatha yoga, going back to the legendary figures of Matsyendranatha and Gorakshanatha (or Gorakhnath) around 1000 CE, and essentially based on the teaching of the *Yoga Sutras*, makes physical discipline central, with the body as an instrument of awareness and self-overcoming. Its meditative method or orthopraxis, which is in fact a form of Tantra yoga, consists in a complex discipline of *asana* (body posture), *pranayama* (vital breath control), and practices of inner cleansing or *shatkarman* (the six "[purifying] acts"). The nearly one thousand *asanas*—psycho-physical exercises originating in Hindu initiatic traditions—take their names from mythology and nature, particularly animals. They are often performed together with *mudras* (symbolic hand gestures), *pranayama* (breathing techniques), and *mantras* (sounds), so as to enhance their effects. Hatha yoga "approaches Self-realization through the vehicle of the physical body and its energetic (pranic/etheric) template" and "it means to transubstantiate the body into a 'divine body' (*divya-deha*) or 'adamantine body' (*vajra-deha*), which is endowed with all kinds of paranormal capacities" (Feuerstein, *Deeper Dimension*, cit., 41). In other words, Hatha yoga is "the forceful channeling and control of the vital breaths (*pranas*) and of the thermal energy (*tapas, yogagni*) of the subtle body" (David Gordon White, *The Alchemical Body: Siddha Translations in Medieval India* [Chicago: Chicago University Press, 1996], 39). See also the entry "Hatha Yoga" by James Mallinson in Jacobsen et al., *Brill's Encyclopedia of Hinduism*, cit., vol. 3: *Society, Religious Specialists, Religious Traditions, Philosophy* (2011), 770–81.

[313] "The term '*hatha*' is from the root *hath* meaning 'to treat with violence' or 'to oppress,' and hence, the expression 'Hatha Yoga' means something like 'the discipline of (bodily) exertion'" (Larson, "Introduction," cit., 141). Several texts assign a different etymological value to the term *hatha*, by splicing "*ha*" (sun) with "*tha*" (moon), or by correlating masculine and feminine, sound and silence, micro- and macrocosms, and Shakti and Shiva (ibid., 142).

[314] As the very first sentence of the classical Hatha yoga text *Hatha Yoga Pradipika*, compiled in the fifteenth century CE from earlier sources, reads (I.1): "Hatha yoga, [. . .] like a staircase, leads the aspirant to the high pinnacled Raja Yoga" (Pancham Singh, trans.,

from the ground floor to the upper floor of a house. Concentrating on breathing is one way of reaching the beyond within. Drawing his senses inward, like the tortoise gathering its limbs beneath its shell,[315] and keeping consciousness devoid of all conceptions, the yogi devotes himself to an inner experience: he listens to the rhythm of his breath, which in turn sings and resounds.

At first, it is like the ringing of the dainty bells worn around the ankles when one dances in honor of the gods; they set the tone as one's naked feet drum the earth; it sounds like the hand bell tolled to mark the beginning of prayer, to dispel everything unholy and disturbing from the devotional space; it sounds like Vishnu's conch in his battle against the daimons of the I:[316] lust, wrath, and blind infatuation.[317] And it sounds like a lute playing on the nerve cords, the channels of the breath of life, stirred by the world-effecting divine power, mother of the world, who playfully goes to work within us and everywhere throughout Maya. Like Penelope,[318] however, she may undo the world's fabric again; then it is as if this divine power—which is identical to what manifests in us as the Kundalini,[319] that

---

Haṭha Yoga Pradīpikā [Varanasi: Indica Books, 2012]): i.e., to the "royal yoga," referring to Patanjali's systematization (see below, n. 341).

[315] The metaphor of the turtle withdrawing its limbs into its shell is an ancient topos of Hindu literature. Moreover, the turtle symbolizes the human being as a middle term of the triad between sky (the superior and rounded part of the caress) and earth (the inferior and flat part of the caress). See, for instance, *Bhagavadgita* 2.58: "And when this one wholly withdraws / all his senses from their objects, / as a tortoise draws in its limbs, / his wisdom is well-established" (Flood and Martin, *Bhagavad Gita*, cit., 21).

[316] The conch shell (*shankha*), whose sound terrifies the demons, is one of the primary attributes of Vishnu, a divinity mentioned already in the Vedas, who belongs to the *trimurti* that includes Brahma and Shiva. Traditionally, Brahma acts as the creative principle, Vishnu the preserver, and Shiva the destroyer. The other symbols carried by Vishnu in his four arms are a disk, a mace, and a lotus flower. See the entry "Vishnu" in Jacobsen et al., *Brill's Encyclopedia of Hinduism*, cit., vol. 1: *Regions, Pilgrimage, Deities* (2009), 787–800.

[317] Cf. the Indian scholar and mystic Chanakya (4th–3rd century BCE): "There is no disease (so destructive) as lust; no enemy like infatuation; no fire like wrath; and no happiness like spiritual knowledge" (Chanakya, *Sri Chanakya Niti-Shastra: The Political Ethics of Chanakya Pandit*, trans. by Miles Davis [Patita Pavana dasa)]), ch. 5.12, available at https://sanskritdocuments.org/english/chaaNakyaNiti.pdf (accessed March 28, 2024).

[318] The daughter of the nymph Periboea and Spartan king Icarius, Penelope was the wife of the Greek hero Odysseus, ruler of Ithaca, in Homer's *Odyssey*. Awaiting his return from the Trojan War for twenty years, she unpicked her day's work at the loom each night to keep her suitors at bay, having promised to choose between them only after finishing a shroud for her father-in-law. She is regarded as model of patience and fidelity.

[319] Sanskrit female term deriving from *kundela* (wound/rolled up, in the form of a spiral) attested in the Advaita philosophy and in other yogic traditions. Manifesting the universal, divine, feminine energy known as *shakti*, the Kundalini corresponds to "generative strength." It is represented as a sleeping snake, to indicate its quiescent state, and is located

snake-like force of life—clanged our mind and our breath of life together like two cymbals. In its dance, this power is at work within both the world and the I, and it worships the otherworldly, supraworldly highest divine being. It sounds like a reed flute, with whose tune Krishna, the incarnation of the almighty god Vishnu, delighted shepherdesses, herds, and birds as a young shepherd. The spine is like a great tree, beneath which our senses lie, listening and turned inward; there, too, lie the women, the "vein of highest lust"; *sushumna*,[320] the innermost nerve channel of the spinal cord, is the flute, while the mind and breath are the two hands holding it. The six lotus-shaped centers that lie along the Sushumna, one placed on the other like knots, each bears within itself a coiled snake, a Kundalini (their life-force). These six snake queens dance to the imploring sound of the flute, and their husband, the snake king, whose head is located in the uppermost lotus chakra of the skull while the end of his tail lies in the

---

at the base of the spine, above the first chakra's plexus (see next note), coiled three and a half times around it. The Kundalini and chakras are vividly described in Tantric texts, translations of and commentaries on which were diffused in particular by Arthur Avalon: in, for example, *The Serpent Power: Being the Ṣaṭ-cakra-nirūpana and Pādukā-pañcaka; Two Works on Laya-yoga* (Mineola, NY: Dover, 1974 [1918]). Additionally, and in a style more accessible than Avalon's scholarly compendiousness, works by the controversial theosophist Charles Webster Leadbeater, such as *The Inner Life* (1910) and *The Chakras: A Monograph* (1927), popularized the chakra system, being for decades among the few Tantra sources available to Westerners: see Kurt Leland, *Rainbow Body: A History of the Western Chakra System from Blavatsky to Brennan* (Lake Worth, FL: Ibis Press, 2016), esp. 197, 202. For recent references to Kundalini, see James Mallinson and Mark Singleton, eds, *Roots of Yoga* (London: Penguin 2017), 178–80; for a translation of old Sanskrit textual passages on Kundalini, ibid., 213–18.

[320] *Sushumna* (Skr.: "kind") is the central of the three principal *nadis* (Skr.: "channel, flow") including the side channels *ida* and *pingala* (according to Tantric sources, there are seventy-two thousand channels for the flow of consciousness). Situated within the spine, the *sushumna* (also referred to as the inner flute, the middle pillar, the path of Kundalini, etc., and mentioned in the *Maitri Upanishad*, 6.21) is an energy pathway beginning at the perineum and moving up through the body to the crown of the head that connects the body's energy centers (chakras), creating a flow of energy, or a pranic current, by which awakened sexual energy, or kundalini, travels up. The *sushumna* rises in the *muladhara* chakra, the pelvic plexus, and from there continues to flow upward in the central channel, while *ida* passes to the left and *pingala* to the right side. The three *nadis* reunite in the *manipura* chakra, the solar plexus, then meet between the eyebrows, in the pineal gland (*ajna*, or *agya*) chakra and, finally, in the *svadhisthana* chakra. "This energetic pathway is not generally recognized as an anatomical reality by Western science. However, the channel does follow through the body actual physical and neurological pathways that connect the endocrine glands" (Donna Farhi, *Yoga Mind, Body and Spirit: A Return to Wholeness* [New York: Holt, 2000], 207). In addition to the major chakras, another twenty-one minor, subsidiary centers are postulated.

depths of the abdomen, dances with them.[321] Furthermore, this breath sounds like a kettledrum beaten in the battle waged by our spark of life against the I's passions. It sounds like the hand-drum beaten, as he dances, by Shiva—the supraworldly one from the beyond, who manifests himself as the inner guide of our small world body. It roars like the thundercloud standing midway on the horizon at the juncture of our eyebrows; flashes of a higher light shoot forth from this cloud while its rains extinguish the blaze that we habitually use to cling to the world and to individuation, fate, birth, and death.

It is a matter of hearing all these sounds and grasping their meaning, because they are all part of the great primordial sound AUM, with which the inner beyond, which resides in everything, manifests itself in the realm of sound.[322] Consciousness, which can focus entirely on this sound, "melts" therein and reaches its other world within: "if the consciousness of the yogi gives itself wholly to this sound, then it forgets everything exterior

---

[321] As elsewhere (see, e.g., *Indische Sphären*, cit., 2nd edn, 108, 112 and fig. 8; cf. *Yoga und Maya* [Munich: Oldenburg, 1940], 107, fig. 10), and in line with most Yoga and Tantra schools, Zimmer lists six energy centers through which Kundalini energy ascends to *sahasrara*, the thousand-petaled lotus flower, otherwise called the crown, or parietal, chakra. The latter is not in fact always regarded as a chakra, as it is deemed to belong to a higher order, being—in Avalon's words—"'beyond' all categories of dual experience as well as the 'Supreme' of all categories—the 'Limit' of all definitions" (Arthur Avalon, *The World as Power* [1922] [Delhi: New Age Books, 2013], 360). Avalon's influential book included the translation of the *Sadcakranirupana* or "Description of the Six Centers," which belonged to a text composed in 1577. As observed by Gordon White, the six-center tradition of the Tantra system originated from the eleventh-century *Kubjikamata-tantra*, a text belonging to the Kaula, a Tantra tradition focused on the worship of Shiva and Shakti. The goddess Kubjika is connected with, and considered a personification of, the Kundalini (White, *Alchemical Body*, cit., 5 and 73). See also below, n. 345.

[322] "AUM [. . .] is a sound-symbol of the whole of consciousness-existence, and at the same time its willing affirmation" (Zimmer, *Myths and Symbols*, cit., 154). Moreover, "A the waking state, U the dream, M deep sleep, and the Silence, *Turiya*, 'The Fourth': all four together comprise the totality of this manifestation of Atman-Brahman as syllable" (Zimmer, *Philosophies of India*, cit., 377; see also 372–78). According to Eliade, "OM incarnates the mystical essence of the entire cosmos. It is nothing less than the theophany itself, reduced to the state of a phoneme" (Mircea Eliade, *Yoga: Immortality and Freedom* [1954], trans. by Willard R. Trask [Bollingen Series LVI] [Princeton, NJ: Princeton University Press, 1958], 51 n. 7); see also Feuerstein, *Deeper Dimension*, cit., ch. 65, where he describes the OM as "the oldest mantra, or sound or numinous power, known to the sages of India," adding that this "'sound' or vibration, which pervades the entire universe ans is audible to yogins in higher states of consciousness. In the Western hermetic tradition, this is known as 'the music of the spheres'" (304 and 305); and Giuseppe Tucci, *The Theory and the Practice of the Mandala, with Special Reference to the Modern Psychology of the Subconscious*, trans. by Alan Houghton Brodrick (Mineola, NY: Dover, 2001 [orig. Ital. edn 1949]), 122–23. Cf. Jung, *Psychology of Kundalini Yoga*, cit., 73.

and finds peace together with that sound. Whoever practices this yoga overcomes the sphere of manifold qualitative differentiation, of the tangibly developed I and of the world body; he leaves behind all activity and melts into the pure ether of undifferentiated spirit devoid of quality." He inwardly dissolves like a cloud in the ethereal blue of the firmament.

Another exercise is to listen to the chanting of the breath within. Rhythmically flowing inward and outward, it resounds "ham–sa, ham–sa," or indeed "sa–'ham, sa–'ham." "Ham–sa" is how breath speaks its name, because "hamsa" means the "wild swan" (which corresponds to our German *Ganser* [gander]).[323] The breath of life is a wild swan that incessantly flies up and down inside us if one does not quiet it by arresting its breathing. It is the tangible manifestation of our spark of life, to which our present individuation adheres like all previous ones. But at the same time, it announces its secret: "sa 'ham, sa 'ham"—whereby "'ham" or "aham" is "I," and "sa" is "he"; thus, says the breath, "I am He," "I am He." I, individuation, am the supra-individual, which disguises itself as its appearance; I, the this-wordly, am the other-worldly; I, the creature, am in effect the immortal divine. From sunrise to sunset the human being breathes 21,600 times,[324] which is how often the swan moves through all the lotus centers within the cage of the body. It lives in the eight-petaled heart-lotus as a principle of life. Depending on the particular lotus leaf on which it happens to be, our feeling is determined as either lust or wrath, or fatigue, or hunger, and so forth. But if the swan is situated at the middle

---

[323] In Hinduism, *hamsa* means "wild goose" (usually translated simply as "swan"). Its flight represents the wish of the soul to free itself from the perpetual cycle of existence. From the cosmic egg volatile Brahma was born, often shown mounted on the *hamsa*. Cf. Zimmer, *Myths and Symbols*, cit., 48–49. According to Feuerstein, "the hamsa also stands for psyche (*jiva*), which lives through the breath" (*Deeper Dimension*, cit., 314; see also below, nn. 326 and 333).

[324] According to the *Gheranda Samhita*, one of the three classical texts of Hatha yoga, probably composed in Bengal around 1700 CE, "[t]hroughout a day and a night there are twenty-one thousand and six hundred such respirations (that is 15 respirations per minute). Every living being (*jiva*) performs this *japa* unconsciously, but constantly. This is called Ajapā Gāyatrī" (*Gheranda Sanhita: A Treatise on Hatha Yoga Translated from the Original Sanskrit by Sri Chandra Vasu* [Bombay: Bombay Theosophical Publication Fund, Tookaram Tatya, 1895], 5.83, 45; see also next note). A recent translation into English is James Mallinson's *The Gheranda Samhita: The Original Sanskrit and an English Translation* [Woodstock, NY: YogaVidya.com, 2004]). Such enumeration is part of the complex architecture of Hindu numerology, which works out innumerable relationships in the micro- and macrocosms; for instance, concerning breathing, allowing an average of four seconds for each breath, then 24 hours × 60 minutes = 1440 minutes; 1440 minutes × 60 seconds = 86,400 seconds; and 86,400 divided by four (seconds for one breath) = 21,600 cosmic breaths (10,800 each of solar and lunar breaths).

of the lotus, then we yearn for what lies beyond the world and everything within it that draws us here and there, inspiring liking and disliking. Its sound, "ham-sa" "saˊham," is the unspoken whispering, which incessantly resounds, without our involvement, on the threshold to our inner beyond.[325] Its movement, its sound, is the innermost Maya, the disguising of our supra-individual, immortal, divine essence as its opposite, the masking of the "brahman" as "jiva,"[326] the "one who lives," which takes effect within us as the principle of life and individuation. To melt this sound into silence is to become aware of brahman. "Whoever seeks the upper kingdom (the omnipotence) in yoga should simply concentrate on this sound; he should abandon strenuous thinking, because this sound, like a noose or gazelle trap, works upon the 'mental disposition' or 'mood,' just as the gazelle trap is used to hunt and slay that animal." The snake-like, inconstant sense is hypnotized by this sound like a snake by the imploring sound of a flute; as our consciousness vibrates with the sound, it dissolves with that sound, "like fire with the wood upon which it burns, consuming itself when it has consumed the wood."

The same technical principle is at work here, just as with most yoga exercises. Our mood or mental disposition is open to the impressions without, and the stirrings within. Under the spell of involuntary participation in the exterior and interior, it is "scattered," drawn variously here and there and involuntarily fixed on the things and moods approaching it. Single-mindedness[327] concentrates the disposition "in a point," and with it holds fast to what has been fixated, makes the disposition adhere to the

---

[325] This refers to the "*ajapa*-mantra" (cf. Zimmer's "Yoga und Maya," cit., 387; *Indische Sphären*, cit., 118 [2nd edn, 114]). The *ajapa*-mantra—also known as Ajapa Gayatri, Hamsa mantra, or Hamsa Gayatri—indicates the practice of *japa* (Skr.: "meditative repetition") without the mental conscious effort usually needed to repeat the mantra: in this way, the practice of constant remembrance and awareness becomes automatic and breath offers itself as prayer. Regarding the So'ham (which is considered a mantra, the *ajapa*-mantra, in Tantrism and in Kriya yoga), see Zimmer, *Myths and Symbols*, cit., 50–51, and *Indische Sphären*, cit., 117 [2nd edn, 114]). The Gayatri Mantra, first recorded in the *Rigveda* (3.62.10) and still recited by millions of Hindus during their morning ablutions, is considered "the receptacle of India's most ancient wisdom that subsequently led to Hinduism" (see Feuerstein, *Deeper Dimension*, cit., 300–303).

[326] *Jiva* is the individual Self or the soul, an innate force which inherits the physical body. Avalon defines it as the result of Brahman's "original reflection [. . .] on the mirror of its own Maya," and as "the Supreme 'Personality,' the first Reflex, the individual Self" that in turn "is tending to liberate himself"; while "*Brahman* is the absolute alogical Real, that is, Reality, not as conceived by Mind, but as it is in itself beyond (in the sense that it is exclusive of) all relations" (*World as Power*, cit, 307).

[327] Elsewhere, Zimmer adds the term "concentration" ("Yoga und Maya," cit., 387; *Indische Sphären*, cit., 118 [2nd edn, 114]).

fixated, and holds fast to this fixation until the counterparts—the fixated and the fixating—coalesce into one. This is "simplification," "unification,"[328] or "coalescence."[329] In the daily cult of Tantra yoga, this can occur by way of an inwardly visualized image of God. Whoever is initiated into this cult thus becomes a deity, which he has formed out of his own intangible inner substance. His devout concentration, his devout disposition, melts into the deity, and this in turn melts into the formless pure being of the inner depth, from which its form has built itself up. Instead of such a divine figure, there stands the sound of breathing, this manifestation of the jiva. The sound has absorbed consciousness, and disappears with it into the unconscious. Thus, the jiva returns to its true hidden nature of brahman.[330]

According to one myth, what the swan sings, as the breath of God, is the same song that resounds within us. Every evening, God takes back into himself the world that he has created, with its withered abundance of forms. Then he bears it, slumbering, in his body. Just as a dream of the exterior world enacts itself within us, thus the world continues to be at play within the God in ideal perfection. The song of the eternal swan sings, and the breath of God harps, about its unfolding as a dream: "I assume many forms and swim slowly in the great ocean when the moon and sun have faded." This is the nocturnal state of the world, when it has once

---

[328] In the other redactions Zimmer adds, in parenthesis, "samadhi" ("Yoga und Maya," cit., 387; *Indische Sphären*, cit., 118 [2nd edn, 115]).

[329] In the other redactions Zimmer adds, in parenthesis, "laya" ("Yoga und Maya," cit., 387; *Indische Sphären*, cit., 118 [2nd edn, 115]).

[330] Brahman is the absolute, universal principle which oversees terrestrial evolution and also corresponds to individual essence, the Atman. According to Zimmer, "from Vedic times to the present day" Brahman is "the most important single concept of Hindu religion and philosophy" (*Philosophies of India*, cit., 75; cf. its homonymous chapter, 74–83). Also, the neuter noun *brahman* is "beyond the differentiating qualifications of sex, beyond all limiting, individualizing characteristics whatsoever. It is the all-containing transcendent source of every possible virtue and form. Out of Brahman, the Absolute, proceed the energies of Nature, to produce our world of individuated forms, the swarming world of our empirical experience, which is characterized by limitations, polarities, antagonisms and cooperation" (Zimmer, *Myths and Symbols*, cit., 123). See too the note on the same page: "Brahman (neuter) and Brahmā (masculine) are not to be confused with each other. The former refers to the transcendent and immanent Absolute; the latter is an anthropomorphic personification of the Creator-Demiurge. Brahman is properly a metaphysical term, Brahmā mythological"; Coomaraswamy adds in a note (ibid.) that Brahman is the principle of all differentiations, "just as in Genesis the 'image of God' is reflected in a creation 'male *and* female.' In general, masculine gender implies activity and procession, female gender passivity and recession, the neuter static or absolute condition."

again become the waters of the unconscious.[331] "I am the Lord and the swan. I brought forth this world from within myself and I dwell in the circling passage of time."

This same swan has very likely levitated through many of our deeper dreams. For India this is a great symbol. It is probably the highest form of the soul bird. It expresses the metaphysical-transcendental aspects of the soul, that final dimension lying deeper than everything that resembles form within us, everything in our unconscious that is pregnant with form and generates significance. The swan denotes that part of our essence which is superior to the I and the world, unentangled and untouchable in its apparent enmeshment with and devotion to the spellbinding vortex of existence. The swan pursues its course above the mirror of the water of life, dipping its neck into that depth. But it is not bound to that depth, because it is elated. It ascends from the lower tide to the crystal upper ocean and rows there even more freely, in greater accord with itself than in heavier waters. It migrates wherever it pleases. The homeless wild swan that pursues its course over great distances without trace, descending at will on lake or bay, in India signifies the unfettered deepest principle within us, which playfully lends itself to embodiments here and there, time and again, for the duration of a lifetime.

Why are we so particularly enraptured, time and again, by Leda's encounter with the swan, as if this image contained more than Europa's adventure or Danae's devotion—that is, more than an hour of profound love and the grace of a great god who possesses knowledge of a young woman's dream? Is not Leda, just as Psyche, haunted by and yet devoted to the divine depth within her, which only visits her briefly—an untenable and untouchable depth that wounds and dissolves her?[332] In India,

---

[331] This passage will be replaced in the other redactions by "the nocturnal state of the world, in which the world has once again melted into the primordial waters of the unconscious, the shapeless body of the world's essence" ("Yoga und Maya," cit., 388; *Indische Sphären*, cit., 119 [2nd edn, 115]).

[332] In Greek myth, the king of the gods Zeus (Roman Jupiter) becomes a swan to seduce Leda, who gave birth to the twins Castor and Pollux (the Dioscuri) and Helen and Clytemnestra. (In another myth, Castor and Clytemnestra are fathered by Leda's husband Tyndareus.) The Zeus–Leda encounter, prominent in Greek and Roman art, appeared too in a much-copied lost painting attributed to Leonardo da Vinci. Zimmer also refers to the Phoenician princess Europa (from whose name that of the continent derives), whose rape by Zeus in form of a bull, recounted in Ovid's *Metamorphoses*, is another recurring motif in art; as is Danae's rape (or "seduction") by Zeus in form of golden rain, which resulted in the birth of Perseus; and the myth of Psyche (narrated in Apuleius's *The Golden Ass*; see below, n. 352): a beautiful princess whose yearning for her (at first) invisible divine lover Eros (Roman Cupid) brought her many tribulations, not least from his jealous mother the

the swan is the sign of Brahma, as well as both his mount and his symbol in the animal kingdom. The homeless ascetic, who is no longer at home anywhere on earth, but instead soars freely and aspires to dissolve everything within him that takes form in brahman's intangible ethereal space, from which it has condensed like clouds, is referred to as the "swan" and the "highest swan" (parama hamsa).[333]

The breath, the spark of life voiced as "I am He," always resounds within us, but only the yogi hears and understands it. At one point in the myth, Krishna, the incarnation of the almighty god Vishnu, struggles in the guise of a shepherd boy with the snake daimon. To the horror of the shepherdesses, it seems that he is overwhelmed in the water by the daimon, encoiled and paralyzed by the snake, and rendered unconscious by its poisonous breath. Then his half-brother[334]—who is also an incarnation of Vishnu—calls out to him from the shore, "Divine Lord of the gods, what are you doing, bringing forth this human being within yourself? Do you not know your other essence? You are the navel of the world, the creator, destroyer, and custodian of all worlds; you are the world! You have shown that you are human and that you can play boys' games, so therefore defeat the daimon!" Thus, Krishna is reminded that a smile divides the chalice of his lips; he slaps the windings that bind him and easily frees his body; he bends the snake's head under his feet and breaks into a stamping dance until the snake-daimon pleads for mercy.[335] The two faces of our

---

goddess Aphrodite (Roman Venus), before the pair were at long last allowed by Zeus to marry, and she became the goddess of the soul. On the rape in ancient Greek culture, see Eva Cantarella, "Dangling Virgins: Myth, Ritual and the Place of Women in Ancient Greece," in Susan Rubin Suleiman, ed., The Female Body in Western Culture: Semiotic Perspectives = Poetics Today 6, nos 1–2 (1985): 91–101.

[333] Parama means "supreme, transcendent"; the honorific title paramahamsa or paramhamsa is awarded to supreme teachers who have attained enlightenment (for example, Ramakrishna). Hamsa is also a denomination of "Vedantic monks [. . .] because they are wanderers, like the great birds that migrate from the jungles of the south to the lakes of the Himalayas, at home in the lofty sky as well on the water-surfaces of the earthly plane" (Zimmer, Philosophies of India, cit., 262). Wild swans or ganders "are at home in the trackless lofty sky as well as in the waters of the lakes of the land, just as saints are at home in the formless sphere devoid of attributes as well as in the garb of the human individual, seemingly moving among us in the phenomenal sphere of bondage" (ibid., 461 n. 233).

[334] In Indische Sphären, cit., Zimmer adds, "Rama" (121 [2nd edn, 117]). On Rama, see Heinrich Zimmer, Maya: Der indische Mythos (Stuttgart: Deutsche Verlags-Anstalt, 1936), 3.1, 2, 3 and 4.1.

[335] The story of Krishna and the demon-snake Kaliya is narrated in the Bhagavata Purana, Book 10, ch. 6: see Edwin F. Bryant, trans. and ed., Krishna: The Beautiful Legend of God; Śrīmad Bhāgavata Purāṇa, Book X, with Chapters 1, 6 and 29–31 from Book XI (London: Penguin, 2003).

being and of all beings manifest themselves in the symbol of the two
divine-human brothers, who are parts of one and the same divinity, in the
self-forlornness of the human god-child in its human role, in the windings of the snake, in the watery grave, and in the invincible clarity and
certainty of his brotherly second I: we are jiva and brahman in one, spellbound creatures and abandoned, and at the same time immortally free.

An ancient Indian parable speaks about two birds sitting on the same
tree.[336] One is eating a sweet berry, the other calmly watches. These two
birds are one and the same bird seen under two aspects. But yoga is the
way to experience their unity and to grasp the world-forlornness of the
bird as our third attitude, which leaves us untouched even as the bird's
nature impels it to nibble the sweet berry.[337]

There are always optical means of capturing the beyond within. The
yogi concentrates on the two-petaled lotus between the eyebrows, where at
first he sees the pure ethereal universe of the heart. With deepening concentration, five ethereal spaces supplant each other; these are, in the most subtle aggregate states, the five elements—earth, water, fire, air, and ether—
from which everything both subtle and coarsely material in the world is
formed. The first of these ethereal spaces is like a dark wood at night, the
second glows like an apocalyptic fire, the next one glints in many kinds of
light, the fourth shines like a thousand suns, and the last lacks all nameable
qualities. Within them, the clear waves of a higher light break open, like the
pure waves of the milky ocean of the cosmic life essence and the milky potion of immortality.[338] They shimmer like bolts of lightning, like glow-
worms and self-luminous gemstones in the night. Under a myriad of rays
the sun of knowledge rises, to which the inner darkness of the surrounding

---

[336] The story is narrated in *Rigveda* 1.164, 20–22. It also appears in the *Mundaka
Upanishad* (3.1.1) and in the *Shvetashvatara Upanishad* (4.6).

[337] Cf. Zimmer, *Indische Sphären*, cit.: "The two birds of the old symbol merged into a
single animal over the course of the symbol's history; Jean Paul [referring probably to the
German Romantic author (1763–1825)] at one point speaks of a 'double-headed eagle appearing in fables, which stoops one head to devour its prey, while looking around and keeping watch with the other'" (121 [2nd edn, 117]). The metaphor of the paired birds suggests
Samkya's fundamental dualism between spirit or Self (*purusha*), which is innate, uncreated,
and unconditioned, and nature or substance (*prakrti*), which is manifest and conditioned.
Already Shankara (788–820), founder of Advaita-Vedanta, saw in this account an allegory
for the polarity between the individual Self (*jiva*) and the god Ishvara. In *The Psychology of
Kundalini Yoga*, cit., Jung mentions it to exemplify the psychic state in which "consciousness is removed to a sphere of objectlessness" (83): such an advanced stage in the individuation process can be likened to the "feeling of deliverance" pursued by yoga, by fostering a
detachment of consciousness "from the objects and from the ego" and its "freeing from the
tamas and rajas" (82–82; see above [I], 44. Cf. the final chapter of Jung's "Commentary on
The Secret of the Golden Flower" [1929/1957], CW 13).

[338] That is, the *amrita*.

world and I yields. The swan who pronounces the words "I am He," and who always lives in the opposition between jiva and brahman, gains the strength to move with complete freedom from lotus to lotus in the light of this sun. All activity of the senses ceases and thinking (manas), which is like the moon (candramas), vanishes in the ocean of bliss. A pair of birds, which according to the Indian conception are separated every night and yearn to find each other in the morning,[339] meet during this sunrise—united at long last: it is the mediated knowledge about the unity of jiva and brahman, as taught by the doctrine, and the immediate knowledge, the experience that it is really so. Both become one, and a knowledge of everything, an indescribable experience, is present. All darkness dissolves into light.

Now the yogi is "himself an enlightening light." This sun of knowledge is the pure observing eye beholding his entire interior and exterior world. This light, indescribable, is the deepest darkness for those turned toward the world. It is the manifestation of brahman; in him light vanishes, like the glow of a lamp in bright daylight. In this state one is like the purest heavenly ether, intangible, colorless, formless, and beyond all pronounceable quality; it is as if one were in a deep sleep, beyond the formed world of waking and dream, in utmost bliss, beyond words. This is the highest simplification,[340] in

---

[339] *Cakravaka* birds are believed to be separated at night and reunited with the day. They thus become a stock example in Sanskrit poetry for depicting lovesickness. In Advaita-Vedanta, which is possibly the reference here, the *cakravaka* bird longing for the sun becomes a metaphor for the devotee longing to be reunited with God. Cf. the entry "Cakravaka" in R. N. Saletore, *Encyclopaedia of Indian Culture*, 5 vols (New Delhi: Sterling Publishers, 1981–85), vol. 1 (1981), 277; and Cinzia Pieruccini, "The *Cakravāka* Birds: History of a Poetic Motif," in Giuliano Boccali, Cinzia Pieruccini, and Jaroslav Vacek, eds, *Pandanus '01: Research in Indian Classic Literature* (Prague: Charles University Faculty of Arts, Institute of Indian Studies, 2002), 85–105.

[340] To which Zimmer adds in *Indische Sphären*, cit., in parenthesis, "samadhi" (123 [2nd edn, 119]). Feuerstein translates this (Skr. *sam*, "union," and *adhi*, "lord") as "ecstasy" and recalls the two "fundamental types" of *samadhi*, a "lower" than "union with the Lord" and a "higher," whereby only the latter means "a total absence of contents in consciousness. [. . . .] The meditative experience of the deity of a particular *yantra* belongs to the 'lower' mode of *samadhi*. It is considered an invaluable preparation for the ultimate realization of the universal Self" (*Deeper Dimension*, cit., 325). Jung was influenced by Hauer's translation of *samadhi* as "enfolding, i.e., introversion" and defines it as "contemplation in the passive sense," adding, "therefore the goal of yoga is so-called rapture," or "ecstasy" as the in Greek *ekstasis* (*Psychology of Yoga and Meditation*, cit., 215 and 213). He continues, "In fact this state is not unconscious, although certain texts compel us to assume that it is a question of unconsciousness, because consciousness is so disconnected from objectivity that it is virtually empty. But one must be conscious of something, otherwise one cannot be conscious at all"—a topos of Jung's understanding of the supreme condition of Eastern orthopraxy which led him for instance to regard, as affirmed in his next sentence, that Buddhist state of emptiness called *shunyata*, or "the absolute void" as a "*contradictio in adjecto*" (213). At this juncture—and by contrast—one might recall the famous distinction between ecstasis, or extasis, and "enstasis," the latter referring to an inward status consequent to a

which Hatha yoga elevates itself to Raja yoga,[341] to the goal of the Vedanta, the "fourth state" of brahman, beyond waking and dream and deep sleep.

Symbolically speaking, the most important such yoga process is "to move like a snake."[342] Just as the "infinite" world-snake in Indian cosmology, as a formulable symbol of the primal water of life in the depths, bears on its head the great cosmos with its axis, that is, the world-mountain Sumeru,[343] the Kundalini snake lies at the base of the body and carries its

---

supreme (conscious) concentration, popularized by Eliade in, e.g., *Yoga: Immortality and Freedom*, cit.: "Contemplation makes enstasis possible; enstasis, in turn, permits a deeper penetration into reality, by provoking (or facilitating) a new contemplation, a new yogic 'state.' This passing from 'knowledge' to 'state' must be constantly borne in mind, for, in our opinion, it is the characteristic feature of *samādhi* (as it is, indeed, of all Indian 'meditation')" (82). Eliade also points out that "from time immemorial India has known the many and various trances and ecstasies obtained from intoxicants, narcotics, and all the other elementary means of emptying consciousness; but any degree of methodological conscience will show us that we have no right to put *samādhi* among these countless varieties of spiritual escape. Liberation is not assimilable with the 'deep sleep' of prenatal existence, even if the recovery of totality through undifferentiated enstasis seems to resemble the bliss of the human being's fetal preconsciousness. [. . .] [T]he yogin works on all levels of consciousness and of the subconscious, for the purpose of opening the way to transconsciousness (knowledge-possession of the Self, the *purusha*)" (ibid., 99; see also 170–73).

[341] Raja yoga ("the royal path," *raja* being equivalent to *rex*), also called Ashtanga yoga, is considered one of the four paths to salvation (together with Bhakti yoga, Jnana yoga and Karma yoga), and among them its highest form. Raja yoga focuses on self-mastery through concentration, via mental discipline, in contrast to Hatha yoga's focus on physical discipline. It is a comparatively late development, and was first introduced to a wider Western audience at the 1893 World Parliament of Religions in Chicago by Swami Vivekananda, with reference to Patanjali's yoga system as the quintessential "classical yoga." Thanks to this scholarly Hindu monk, Raja yoga soon became very popular in the West, for "its message was what many in cultic milieus worldwide had been waiting for: a flexible set of teachings that would meet their craving for exotic but nevertheless accessible and ideologically familiar forms of practical spirituality" (see Elizabeth De Michelis, *A History of Modern Yoga: Patanjali and Western Esotericism* [London: Continuum, 2006 (2004)], 150 and passim).

[342] In the other redactions, Zimmer replaces these sentences with the following: "Among the yoga exercises, the 'stirring of the life energy' (Shakti-cālana) is unsurpassed in terms of its complexity and meaningfulness. Here the technique of controlling, concentrating, and guiding the breath faces its most complex task" ("Yoga und Maya," cit., 389; *Indische Sphären*, cit., 123 [2nd edn, 119]). Both texts further expand on the dynamics of subtle energy transformation. Zimmer explains that "the sensory-motor life energy, *prana*, which is stored in the body as in a pot," is activated thanks to special respiratory exercises through an "act of violence" that can awaken "the serpent of the life energy" sleeping in the *muladhara* ("in the deepest lotus center of the body") and lead it to ascend the spinal column (and the chakras) "as a column of mercury does in a thermometer" ("Yoga und Maya," cit., 389; *Indische Sphären*, cit., 123–24 [2nd edn, 119]).

[343] Great Meru (Sumeru) is the sacred world-mountain in Hindu, Buddhist, and Jain cosmologies, located in the northern Himalayas. It represents the center of all physical, metaphysical, and spiritual universes, whose summit is the paradise of the gods. According to

microcosm in its axis, the spine. It lies asleep in the deepest lotus chakra, known as *muladhara* the "root place" in the genitals.[344] Above it towers the spine with its innermost channel, the Sushumna, "the vein of purest lust," which carries five more lotus-shaped chakras located in the intestines, the heart, the throat, the center of the eyebrows, and the [top of the] skull. Surrounded by a thousand-petaled lotus, the latter center is the uppermost of these five.[345] The lowermost lotus or "root place" bears earth signs and centers these elements in our microcosm. The four lying above it focus the ever-finer and increasingly evanescent elements—water, fire, air, and ether. The undeveloped, undifferentiated divine world-matter has evolved into each of these, one after the other, through incremental self-transformation and condensation, in the cosmogony of both our body and the world. At the same time, these lotus flowers, which the yogi sees before his mind's eye, are the centers of our five sensory powers linked to the qualities of the five elements: sound and hearing to the ether, touch and feeling to the air, light and seeing to fire, taste and savoring to water,

---

Eliade, "in a number of cultures we do in fact hear of such mountains, real or mythical, situated at the center of the world; examples are Meru in India, Haraberezaiti in Iran, the mythical 'Mount of the Lands' in Mesopotamia, Gerizim in Palestine—which, moreover, was called the 'navel of the earth.' Since the sacred mountain is an *axis mundi* connecting earth with heaven, it in a sense touches heaven and hence marks the highest point in the world; consequently the territory that surrounds it, and that constitutes 'our world', is held to be the highest among countries. This is stated in Hebrew tradition; Palestine, being the highest land, was not submerged by the Flood" (Mircea Eliade, *The Sacred and the Profane: The Nature of Religion*, trans. by Willard R. Trask. [Orlando, FL: Harcourt, 1987 (1957)], 38).

[344] *Muladhara* is the name of the first chakra (*cakra* [Skr.]: "wheel" or "round object"), and means roughly "foundation, root support," situated in the perineum. Psychologically, Jung considers it "the earth upon which we stand, [. . .] our conscious world, because here we are, standing upon this earth, and here are the four corners of this earth. We are in the earth mandala. And whatever we say of *muladhara* is true of this world. It is a place where mankind is a victim of impulses, instincts, unconsciousness, or *participation mystique*, where we are in a dark and unconscious place" (*Psychology of Kundalini Yoga*, cit., 14–15; see also 23, 60ff., and passim).

[345] Reference to *sahasrara*, also called the "crown" chakra. As the apex of Kundalini energy, it constitutes the primordial unity of energy, revealing divine knowledge (as described, for example, in the *Gheranda Samhita*). This reunification is symbolized by the sexual union of Shakti with Shiva. *Sahasrara* is also compared to Sumeru's peak. It should be noted that Zimmer's list of the chakras here follows the Buddhist Tantra tradition in considering (only) five chakras (for the first and second, *muladhara* and *svadhisthana*, located below the navel, are fused, as are the sixth and the seventh, *vishuddha* and *ajna*, located in the head) and associating them with the five classical elements, following the Upanishads. Likewise Vajrayana (Tibetan) Buddhism singles out only five centers, located respectively at the forehead/crown, the throat, the heart, the navel, and the genital/anal region (Feuerstein, *Encyclopedia of Yoga and Tantra*, cit., 84).

and smell and smelling to the earth. The thousand-petaled lotus located at the zenith of our microcosm represents the supraworldly, otherworldly divine, into which the full diversity of all sensations and of the world-formation, consisting of all the elements, conclusively merges. The step-ladder of these centers from top to bottom frames the step-by-step passage of cosmogony from the undeveloped to the developed in increasing density and weight.[346]

It remains for the yogi to awaken the life-snake of the depths through inner pneumatic pressure and to force it upward through the innermost channel of the spine. This upward movement of the Kundalini is the process of step-by-step fusion and backward transformation of the five elemental developmental stages of the world-matter from denser to finer, until they merge in the intangible supraworldliness of the skull's thousand-petaled lotus—in preworldly undifferentiated fullness. The uppermost lotus—a thousand petals for innumerably many—exists beyond finite categories of size, lying beyond their reach, while the lower chakras consisting of four, six, ten, twelve, sixteen, and two petals are individually distinct and tangible. They have different colors, while the highest is colorless. This whiteness in its fullness contains all the colors in the lower chakras. All petals in the lower five lotus chakras singly play host to the fifty signs of the Indian alphabet, symbolizing their differentiated essentiality. On petals of the highest lotus, however, these signs recur in infinite abundance. Thus that chakra lies beyond all individuation, yet contains all possibilities. It is said to dwell beyond the egg of Brahma, that is, beyond the human microcosm and the macrocosm of the world. Thus the

---

[346] To this account of Kundalini energy's ascent, in "Yoga und Maya," and more extensively in *Indische Sphären*, Zimmer adds a comparison of Hindu and Western embryology. The first posits that intrauterine human life develops from the head, later condensing the life-force in the sexual and excretory zones (therefore "the Lothos muladhara, in which it [the life-force] sleeps, bears the sign of the earth, and the zone of evacuation bears the earth-like: the excrement. But as an animal life-force, the serpent is seated at the sexual zone" [*Indische Sphären*, cit., 124 (2nd edn, 120)]). Western embryology, however, teaches that "the fructified egg" falls into "this lowest zone, that of evacuation, and the highest, that of the head, and thereby constitutes the first polarity of embryonic development, which arises from the undifferentiated togetherness of fructifying union" (ibid., 124–25 [2nd edn, 120]). Accordingly, Zimmer infers that "the path imposed upon Kundalini by the yogi, so that the unfolding may ascend from its vegetative slumber at the base of the body to enter its counterpole of unfoldedness at its highest point, as it were leads back to that embryonic pre-existence, to the earliest stage of the cosmogony of our body, before our microcosm drove apart the intimate coalescence of the polar zones" ("Yoga und Maya," cit., 389–90; *Indische Sphären*, cit., 125 [2nd edn, 120]).

Kundalini snake, the force of life, ultimately cancels out both itself and the world unfolded by its upward ascent.

According to Indian thinking, this upward, stepladder-like path taken by the progressively developing elements into their otherworldly fullness corresponds to the natural process of human death. This becomes clear from the *Bardo Thodol*, a Tibetan-Buddhist book of the dead.[347] This treatise describes the agony experienced by the dying person as a sequence of sensations: first, a terrible feeling of pressure is said to occur, as if the earthly body were sinking in water; thereafter, the earthly (which corresponds to the lotus located at the "root place") dissolves in that water (in the sphere of the next lotus), and the dying person experiences a stiffening sensation of cold. Now the dying person is fully submerged in cold water, but at its height this sensation of cold (understandably) turns into a sensation of blazing heat, and thus the detachment of the jiva from the body has reached the zone of fire located in the heart lotus. This terrible heat turns into a feeling of wanting to be scattered as a thousand atoms. Here, the jiva reaches the air sphere of the throat lotus. When this feeling subsides, the dying person has attained the ethereal sphere. This corresponds to the fifth lotus, located between the eyebrows.[348] Only he who has been redeemed during his lifetime, thanks to yoga, will rise spontaneously above this fine-material sphere. Precisely this element of that "intermediate state" (*bardo*) ethereally envelops the spark of life when it leaves the body, until it enters a new one.

Thus, the ascent of the Kundalini through the lotus centers of the body is at the same time a willfully practiced anticipation of natural death. This

---

[347] Attributed to the Buddhist mystic Padma Sambhava, the *Bardo Thodol* was probably composed in the fourteenth century. Only in 1927 did the first edition in English appear, being one of the first key texts of Tibetan and Vajrayana literature made available in a European language, translated by Walter Y. Evans-Wentz in collaboration with his teacher, the lama Kazi Dawa-Samdup, with the fortunate title *The Tibetan Book of the Dead, or, The After-Death Experiences on the Bardo Plane, according to Lāma Kazi Dawa-Samdup's English Rendering* (with a foreword by Sir John Woodroffe) (London: Humphrey Milford, 1927). Instrumental in disseminating knowledge of Tibetan Buddhism in the West, and exerting an influence upon such authors as Aldous Huxley and Timothy Leary, the book contains the teachings for the dead during the *bardo* state, the forty-nine days between death and rebirth. It was republished with Jung's commentary in 1935 (Oxford: Clarendon Press); see Jung's "Psychological Commentary on *The Tibetan Book of the Dead*" [1935/1953], CW 11; and Donald Lopez, *The Tibetan Book of the Dead: A Biography* (Princeton, NJ: Princeton University Press, 2011). See too William McGuire, "Jung, Evans-Wentz and Various Other Gurus," *Journal of Analytical Psychology* 48, no. 4 (2003): 433–45.

[348] Reference to the *ajna* chakra.

ascent gives the adept a new, immortal life already in the flesh. He is reborn into this immortal life when the Kundalini returns to this world by crossing the highest point of the thousand-petaled lotus at the threshold that separates the hereafter and this world, and then gradually develops the adept's microcosm anew down to the lowest lotus. Then, the adept is granted a consciousness. While this is an everyday consciousness, once again, it is at the same time transformed by crossing the threshold to the beyond and back again.

With this insight into the meaning of Kundalini passage, this exercise moves closer to what is known in the West. In *The Magic Flute*,[349] the two men in armor who open to Tamino and Pamina the path of initiation that is meant to transform them sing the following words:

> Der, welcher wandert diese Strasse / voll beschwerden / Wird rein durch Wasser, Feur, Luft und Erden; / Wenn er des Todes Schrecken ueber- / winden kann, / Schwingt er sich aus der Erde / himmel an; / Erleuchtet wird er dann imstande sein, / sich den Mysterien der Isis ganz zu weihn.

> (Whoever walks this sorely trying path / will be purified by water, fire, air, and earth. / If he can surmount the terror of death / he vaults from earth to heaven; illumined, / he will be fit to devote himself wholly / to the mysteries of Isis.)[350]

Knowledge about the mysteries of Isis possessed by initiates and lodge brothers in the eighteenth century[351] goes back to the well-known last

---

[349] *The Magic Flute* was Mozart's last, and surely most enigmatic, opera (libretto by J. E. Schikaneder), first performed a few months before the composer's death at the age of thirty-five in 1791. It is commonly regarded as a dramatic representation of an initiation into Freemasonry (which Mozart joined in 1784). A principal source for both the opera and Masonic borrowings from Egyptian ritual was the Abbé Terrasson's novel *Sethos histoire, ou Vie tirée des monuments anecdotes de l'ancienne Egypte, traduit d'un manuscrit grec* (1731).

[350] *The Magic Flute*, 2.28 (English translation by Lea Frey). These words of warning from the guards to Tamino in *stylus gravis* invoke death and creation, echoing a passage from the *Sethos* novel (see preceding note): "defeating the horrors of death, he will escape from the bosom of the earth, he will see the light once again, and he will be entitled to ready his soul for the revelation of the mysteries of the great goddess Isis" (cf. Jules Speller, *Mozarts Zauberflöte: Eine kritische Auseinandersetzung um ihre Deutung* [Oldenburg: Igel, 1998], 224–225). In contrast to Terrasson's *Bildungsroman* and the contemporary orthodox Freemasonry which prohibited the initiation of women, in his opera Mozart also has a woman, Pamina, undergo initiation.

[351] It is commonly accepted by scholars that Masonic symbolism and ceremonies borrowed crucial aspects from Egyptian mysteries (as well as from Hebrew tradition, on account

chapter of Apuleius's *The Golden Ass*.[352] Its description of the path that initiates of Isis must follow corresponds exactly to the ascent of the Kundalini.[353] Isis, this goddess of late Antiquity who gathers into herself all great goddesses and mothers of the Near East and the Mediterranean

---

of Israel's enslavement by Egypt). Most commentators thus follow suit from the Greeks, notably Herodotus, who derived the Pythagorean, Eleusinian, Orphic, Gnostic, Kabalistic, and Zoroastrian mysteries from Egypt. Reginald Witt, who considers Freemasonry as well as Rosicrucianism to be the successors of the ancient mysteries, points out that the entry of Mozart's Tamino and Papageno into the temple of Isis for their initiation "must be thought of from two standpoints. They typify modern Freemasonry. They also follow in the steps of their ancient brethren, the Cabiri, and Isis who went in search of the slain Osiris, or met the same god, no longer a corpse but resurrected as the god Ptah, The Lord of Life, opening and closing the day, creating 'with his heart and tongue,' superintending architects and masons at the building of temples and pyramids, ordaining that the house should be established firmly, and holding hat the *djed* column as his emblem of stability" (R. E. Witt, *Isis in the Graeco-Roman World* [Ithaca, NY: Cornell University Press, 1971], 157).

[352] The *Metamorphoseon libri XI* by the Numidian philosopher and writer Lucius Apuleius Madaurensis (ca. 125–180 CE), called *Asinus aureus* (*The Golden Ass*) by Augustine (*The City of God* 18.18), is the only surviving complete Latin novel and a crucial record of the initiation practiced in the temples of Isis during the imperial age. The protagonist Lucius, wishing to learn magic in Thessaly, is transformed into a donkey, and in this form encounters all manner of episodes in the *comédie humaine*. Apuleius marks out and simultaneously conceals, in a philosophical yet satirical style, a mystagogic path to wisdom during an era of syncretism, warning against dabbling in the magic arts. In the last chapter, eating roses during the worship procession for Isis, Lucius resumes his human form and beholds the goddess as the Great Mother in all her known forms, a prefiguration of the eternal feminine already testified by Plutarch (*De Iside et Osiride* 372e). Subsequently he is initiated by stages into the mysteries of Isis and Osiris. Apuleius's novel is also rich in digressions and tales, one of the most celebrated being the myth of Eros and Psyche (see above, n. 332): for a Jungian perspective on this, see Erich Neumann, *Amor and Psyche: The Psychic Development of the Feminine; A Commentary on the Tale by Apuleius*, trans. by Ralph Manheim (Princeton, NJ: Princeton University Press, 1971 [1956]); Marie-Louise von Franz, *The Golden Ass of Apuleius: The Liberation of the Feminine in Man*, revised edn (Boston, MA: Shambhala, 2001 [1970]).

[353] Zimmer's analysis appears to be in line with Jung's views in *The Psychology of Kundalini Yoga*, cit., on the "worldwide and ancient symbolism" of rebirth as attested in, for example, Christian baptism, the Isis mysteries, and ancient Egyptian funerary symbolism (31). By correlating the phases of awakened Kundalini energy with the initiatory itinerary in Apuleius's book, Jung ties the apex of the Isis ceremony to the third (*manipura*) chakra, "the fullness of jewels," as "the fire center, really the place where the sun rises," and where "the first light comes after the baptism" (30–31, and cf. Appendix 2: "the analogy to Kundalini is the sun serpent, which later in Christian mythology is identified with Christ. The twelve disciples are thought of as stations of the annual cycle supported by the zodiac serpent. All these are symbols for the change of the creative power. In *muladhara* is the night sun, and beneath the diaphragm the sunrise. The upper centers starting from *anahata* symbolize the change from midday up to sunset. The day of sun is the Kundalini passage—ascent and descent—evolution and involution with spiritual signs.") Jung also links Kundalini to the dragon which appears in an episode of the *Hypnerotomachia Poliphili* (21; see also below [MT], 251ff., and esp. n. 546. Shamdasani suggests (*Psychology of Kundalini Yoga*, cit., 22 n. 41) that Jung's linkage

basin,[354] as the different aspects and names of the all-embracing world mother, represents the highest divine force (*shakti*). She is the creative power of the world, the world mother, Kundalini, the feminine ruler of the secret of life, who allows herself to be worshiped in the form and name of all Indian goddesses and gods.

What Apuleius says about Isis holds true equally for Shakti-Kundalini:[355] "In the hands of Isis lies the life of every human being"—that is, Kundalini is the creative force that develops the microcosm of our living body. And "in her hands lie the keys to the realm of the shadows"—that is, Kundalini is the guide to the beyond of the thousand-petaled lotus. "In

---

of Kundalini with the anima may stem in part from Avalon's description of Kundalini as the "Inner Woman" in *The Serpent Power*, cit. Nonethless, to Westerners, Jung says, "the anima always first appears so grotesque and banal that it is difficult to recognize the Shakti in it" (ibid., 86).The connection between Isis worship and Kundalini energy has been also pointed out, for instance, by Dion Fortune (pseudonym of Violet Mary Firth [1890–1946]), a British psychologist, spiritualist, and occultist. (On this, see Gordan Djurdjevic, *India and the Occult: The Influence of South Asian Spirituality on Modern Western Occultism* [New York: Palgrave Macmillian, 2014], ch. 4).

[354] The cult of Isis, Egyptian goddess of fertility, motherhood, magic, and "the goddess with a thousand names" (entry "Isis," in Lindsay Jones, ed., *Encyclopedia of Religion*, 2nd edn, 15 vols [Detroit, MI: Thomson Gale, 2005], 7:4557), sister and wife of Osiris and mother of Horus, spread through the Roman Empire following the rise of the Ptolemaic dynasty. Testimonies are found in Greece, the Aegean islands, Asia Minor, northern Africa, Sicily, Sardinia, Spain, southern Italy, and Germany. Isis was celebrated in two solemn festivals: the *Navigium Isidis* (Vessel of Isis) on March 5 (Apuleius, *Metamorphoseon* 11.8–17), and the *Inventio* (or *Heuresis*) *Osiridis* (Discovery of Osiris), from October 28 to November 3. Isis-worship found its center around the second century CE in Rome, reaching its apogee under Caracalla (188–217), then started to decay after the advent of Constantine, and waned after Theodosius, who prohibited all non-Christian cults (Edict of Thessalonica, 380); the last temple of Isis was closed in 536 by Justinian and transformed into a Christian church. Greeks equated Isis with Demeter (Herodotus, *Histories* 2.42ff.). Both early Christianity and the cult of the Virgin Mary bear her imprint; Tertullian cites the trinity of Osiris, Isis, and Horus as a basis for the Trinity of Christian dogma: see T. R. Glover, *The Conflict of Religions in the Early Roman Empire* (New York: Cooper Square, 1975 [1909], 23. Apuleius uses cult phrases that characterize Isis as the mother of god (*Metamorphoseon* 11.25), and it is noteworthy that an Egyptian theologian, Cyril of Alexandria, led the Council of Ephesus (431) to decree that the Virgin Mary, as mother of Jesus, has two natures, divine and human, and therefore is *theotokos* (God-bearer, or birth-giver of God): see Isolde Penz, *Wege zum Göttlichen: Die Sehnsucht nach dem Einssein mit dem Göttlichen in Mythos, Gnosos, Logos und im Evalgelium nach Johannes* (Berlin: LIT, 2006), 93.

[355] The following citations from Apuleius (*Metamorphoseon* 11.23) are from Robert Graves's version, *The Transformations of Lucius, Otherwise Known as The Golden Ass*, trans. by Robert Graves (Harmondsworth: Penguin, 1950), 241ff. Cf. Jan Assmann's translation and commentary on this katabasis in his "Death and Initiation in the Funerary Religion of Ancient Egypt," in W. K. Simpson, ed., *Religion and Philosophy in Ancient Egypt* (Yale Egyptological Studies 3) (New Haven, CT: Yale University Press, 1989), 135–59 at 152.

her mysteries, the devotion to a voluntarily chosen death and the reattainment of life would be celebrated and conceived by the mercy of the gods. Through her almighty power the initiates would be reborn and restored to a new life"—that is, to the overcoming of transience and death by way of a sacramental process, which effects mysterious transformation. What Apuleius says about this process corresponds to the course of the Kundalini process: "I approached *the very gates of [life and] death* and set one foot on the threshold of Proserpine"—that is, the realm of death is entered, the undoing of the natural I is carried out. Here, Persephone would correspond to the divine force and world mother as the crone-like, skeletal Mahakali, the goddess of death. "Yet [I] was permitted to return, *rapt through all the elements*"—what else do Kundalini and her initiates do?— "At midnight [literally: "in the time of the deepest midnight (Zur Zeit der tiefesten Mitternacht)] I saw the sun shining as if it were noon"—the highest brahman breaks out of the darkness of Maya—"I entered the presence of the gods of the underworld and the gods of the upper world—this is an ancient Egyptian formula for 'all gods'—"[and I] stood near and worshiped them"—just as the yogi finds all the gods located in the lotus centers of his microcosm, and worships each at his appointed shrine.

The initiate of Isis symbolically experiences his return from the otherworld of death as the nocturnal passage of the sun: from its death in the west it returns by way of an underworld journey at night through the twelve hours to its birth in a new ascent. The initiate is dressed in twelve different robes, one after another, corresponding to the twelve hours of the underworld passage, in which the lower gods and the dead hail the sun, which passes by them "at midnight and shining as if it were noon"—after which the initiate is presented to the gathered community of Isis, "as when a statue is unveiled, dressed like the sun." He stands on a wooden pulpit beneath a portrait of Isis which apparently depicts her holding her child Horus, in whom Osiris, her husband and brother, has been reborn.

All this amounts to a process which passes before the initiate during his initiation, a show replete with chambers and images, costumes, formulae, symbols and gestures, with passages through brightness and darkness—it is a sacramental action, performed from without so that the *mystes* will be transformed within. It mimes the sun in order to partake of its essence: that is, being immortal while forever destined to die.[356]

---

[356] The initiation ritual of the Isis mysteries described by Apuleius has been connected with the Egyptian funerary religion and cult of death, while in the appearance of the sun at midnight there seems to be a similarity to the hymns to the sun-god included in chapter 15

The yogi practices this relationship not in the halls and corridors of a temple, but instead in the world of his body. On his passage to transformation, the yogi uses his own body as a cultic instrument (yantra,[357] or mandala[358]).

In other devotional exercises, the initiate creates a circular, concentric image (mandala) by way of introspection. Using his imagination, he places himself at the center of this structure, at the self-canceling point or drop

---

of the ancient Egyptian funerary text *Book of the Dead* (fifteenth to fourth century BCE) (Friedrich Junge, "Isis und die ägyptischen Mysterien," in Wolfhart Westendorf, ed., *Aspekte der spätägyptischen Religion* [Wiesbaden: Harrassowitz, 1979], 93–115 at 105; Assmann, "Death and Initiation," cit., 152 and 154). Moreover, "a Mithraic overtone" has been pointed out, principally in Apuleius's emphasis on the sun, in his portrayal of the initiatory experience as "both infernal and celestial," and—last but not least—in his giving the name "Mithras" to the priest of Isis (Roger Beck, *Beck on Mithraism: Collected Works with New Essays* [Aldershot: Ashgate, 2004], 121ff.).

[357] The homology between the human organs and the cosmic rhythms was well known in India and found its culmination in the Tantric traditions (see Eliade, *Yoga: Immortality and Freedom*, cit., 227ff.). The *yantra* (Skr.: "loom," or "machine") is the linear paradigm for the *mandala* in its geometrical, essential schema, usually adopted in Hinduism. The term is composed of *tra*, a Sanskrit suffix forming nouns denoting "instruments or tools" and the verb *yam*, "curb, subdue, rule, control." It literally means "object serving to hold," "instrument," "engine," and is composed of a series of triangles framed in a square with four 'doors' (ibid., 219). As Zimmer elsewhere points out, "yantra denotes, primarily, any kind of machine-machine, that is to say, in a pre-industrial, pre-technical sense; a dam to collect water for irrigation, a catapult to hurl stones against a fort—any mechanism built to yield energy for some definite purpose of man's will. In Hindu devotional tradition, 'yantra' is the general term for instrument of worship, namely, idols, pictures, or geometrical diagrams" (Zimmer, *Myths and Symbols*, cit., 141). Again, "This intricate linear composition is conceived and designed as a support to a meditation—more precisely, to a concentrated visualization and intimate inner experience of the polar play and logic-shattering paradox of eternity and time [. . .]. This composition summarizes in a single moment the whole sense of the Hindu world of myth and symbol" (ibid., 140). See also Heinrich Zimmer, *Artistic Form and Yoga in the Sacred Images of India*, trans. and ed. by Gerald Chapple and James B. Lawson with J. M. McNight (Princeton, NJ: Princeton University Press, 1984), 28 and passim; Tucci, *Theory and the Practice of the Mandala*, cit., 60 and passim.

[358] The *mandala* (Skr.: "circle"), often in fact square, filled with god-images and geometrical figures, is a complex arrangement of patterns or pictures used in Hindu and Buddhist Tantrism to represent the cosmos (or the totality of psychic and super-psychic forces) in a metaphysical and symbolic perspective. It is a microcosmic representation of the universe as well, according to orientalist Giuseppe Tucci: a "psycho-cosmogram" aiming at reintegrating and renewing the integrity of consciousness (*Theory and the Practice of the Mandala*, cit., vii and passim). Cf. Eliade's chapter 6, "Yoga and Tantrism," in *Yoga: Immortality and Freedom*, cit., where he states that "like the yantra, the mandala is at once an image of the universe and a theophany—the cosmic creation being, of course, a manifestation of the divinity. But the mandala also serves as a 'receptacle' for the gods" (220).

(*bindu*) of the transcendental.[359] He allows this to flow out of its self-abrogated, intangible fullness, consisting of all conceivable forms, into the concentrically shaped diversity of the phenomenal world, which he lets stream at random from this center, only to take it back again—into itself—as the sphere of the transcendental. Kundalini yogi is such a mandala for himself, or rather his body is like the circular segment of such a mandala: its center point, the transcendental, which eludes human grasp, is the thousand-petaled lotus of the other world, lying beyond the brahman's egg of the body.[360] But the concentric layers into which this "drop" develops are designated by the remaining five lotus centers. The "root place" lies on the periphery of the microcosm's development, and the Sushumna is a radius running through all concentric sphere-segments that connects this periphery with the center point. Practice of the Kundalini process takes place along this radius: this willed end of the world, the deliberate dying to, and being subsequently reborn as, world and body.

This is an experience of how we come into being, and of how we build ourselves up every day into who we are. One reaches behind the secret of the world process, to how the transcendental One (the supraworldly, auspicious Shiva, the Supreme God) becomes the manifoldly ephemeral, the world, due to the infinite agility of his creative force, Shakti, who is the Kundalini within us. One experiences, enlighteningly, the unity of opposites, Shiva and Shakti: transcendence, resting in itself, and the immanence

---

[359] The *bindu* (Skr.: "point," "dot") is the centerpoint of the yantra and mandala, representing the unity of human and divine dimensions; see Tucci: "the drop or particle, that is to say the egg created by the mingling of the male seed (*sukla*) and the female ovum (*rakta*)" (*Theory and Practice of the Mandala*, cit., 122); and Avalon: "Bindu or Sakti is neither male or female, but partakes of the characteristics of both. In its Advaita, or non-dual nature, is a dual characteristic, a polarity [. . .], which may be expressed by the terms 'positive' and 'negative', and which is denominated in the Tantra by Shiva and Shakti" (Arthur Avalon, ed., *Principles of Tantra: Tantra-Tattva of Sriyukta Siva Chandra Vidyarnava Bhattacharya*, Part 2 [Madras: Ganesh, 1978 (1916)], 41). Feuerestein defines the *bindu* as "that point in space and time where any object comes into manifestation. [It] stands between manifestation and the unmanifest, between actuality and potentiality. It is the creative matrix, the primary structure, from which issues the whole cosmos in its multiformity. This is true both of the physical world and the psychological universe, macrocosm and microcosm" (*Deeper Dimension*, cit., 325–26). Cf. Jung: "Bindu corresponds to the self-generated lingam around which the serpent lies. A shell surrounds both; this shell is Maya" (*Psychology of Kundalini Yoga*, cit., 76).

[360] Reference to the *sahasrara*, also regarded as the seventh primary chakra. On the importance of the body and cosmophysiology in Tantrism, see Eliade, *Yoga: Immortality and Freedom*, cit., 227ff. He observes, "Hatha Yoga and Tantra transubstantiated the body by giving it macranthropic dimensions and assimilating it to the various 'mystical bodies' (sonorous, architectonic, iconographic, etc.)" (236). Cf. Tucci, *Theory and Practice of the Mandala*, cit., ch. 5.

of the world that comes into play from and within itself, are two aspects of a *single* existence.

> illumined,
> he will be fit to devote himself wholly
> to the mysteries of Isis.

How does this translate into Hindi? The secret of Isis is Kundalini. Where would it not be? It is itself the evident secret, which she unfolds from the Maya of the world and the I, and then dissolves on her return to the supraworldly, auspicious Shiva of the highest lotus. Her secret is that she is one with him, never separated nor different, locked in her husband's infinite embrace, and that He seems to be in both this world and the next one. Only the yogi, who has illuminated himself, can "devote himself wholly" to Shiva. For him, the force that binds all beings, the spell of involuntary participation in the transience of the world and I, has lost its power. He stands in the midst of life, devoted to it as the manifold revelation of divine force which is his innermost life, neither overcome by his worldly and bodily Maya nor spellbound in bottomless calm.

The Berlin Seminar of C. G. Jung

Notes of the Seminar Given by
Dr. C. G. Jung in Berlin between
26 June and 1 July 1933

Transcriber's Note [original manuscript]

This report has been reproduced as a manuscript, and is reserved exclusively for use by the seminar participants. Its distribution or publication, in part or in whole, be it only the lending or borrowing of this copy, requires the explicit permission of Dr. Jung.

Transcriber's Note to Jung's Seminar

The grammar and spelling of the original manuscript (e.g., the use of "ss" instead of "ß," as customary in Switzerland) have been preserved in most instances. Obvious errors (e.g., dass instead of das) or typos, however, have been silently corrected. Text spaced out in the manuscript has been italicized. Remarks which appear in square brackets, as well as a few explanatory footnotes, have been added by the transcriber and integrated by the editor.

Ernst Falzeder

Lecture 1

MONDAY, 26 JUNE 1933

LADIES AND GENTLEMEN,
Attendance is far more numerous than I would have expected. It will thus be somewhat difficult to hold a seminar in the customary manner. Usually, I ask participants questions. In view of the large number of participants, this is evidently more difficult. Nevertheless, I shall try to ask you questions to establish which aspects of the matter at hand cause you particular difficulties. We shall be taking a rather thorny and difficult path, and I shall not assume that you will be able to readily follow each and every step. Consequently, it will be fruitful for the seminar as a whole if you ask questions. So, should you have any questions that require no immediate answer, could you perhaps submit these to me in writing at the beginning of the second hour or tomorrow evening at the beginning of the first hour?

Yesterday, Professor Zimmer drew an excellent picture of matters. One felt transported to the dark shadows cast by old mango trees set against a brilliant white foreground. Under these trees sit wise or enlightened men, from whose heads magnificent figures rise, full of sparkling gemstones, like the figures in the world of Amitaba.[361] While it flatters him, Professor

---

[361] Sanskrit name of one of the five Dhyani Buddhas, mainly associated with longevity. Jung devoted his "The Psychology of Eastern Meditation" [1948] to the *Amitayurdhyana-sutra* (Treatise on Amitabha meditation), a sutra whose original was never found, considered apocryphal and translated into Chinese between 424 and 442 CE, and published in Max Müller, ed., the *Sacred Books of the East*, 50 vols (Oxford: Oxford University Press, 1879–1910), vol. 49 (1894), Part 2, 159–201. Jung points out that Amitabha is "the Buddha of the *setting sun* of immeasurable light," as well as "the protector of our present world-period" (*CW* 11, § 912), and "the dispenser of the water of immortality" (§ 916). The first attribution of infinite life to the Buddha arose in the Mahayana sect; the Buddha Amitabha, according to most scholars originated either in Buddhism, Hindu mythology, or the sun worship of Zoroastrism (see Guang Xin, *The Concept of the Buddha: Its Evolution from Early Buddhism to the "Trikāya" Theory* [London: Routledge, 2005], 168 and passim).

Zimmer's introduction to my seminar is most unpleasant for me, since I must take the same path in the opposite direction. While he has spoken about the highest perfections, I can speak only of beginnings. And while he has illustrated how over thousands of years of culture certain ideas have reached their ultimate flowering or perfection, I must consider paltry beginnings and downright impoverishment: namely, how these same things begin with us. It is precisely this that helps us establish a sense of just how much we lack in this respect. With us, the entire inner side of the human being, that is, that whole inner path, is as it were uncharted territory. It is not even new territory as yet, since it has not even been properly discovered. We have no more than vague inklings, as you will gather from the material that I shall present during this seminar. I shall refrain from cherry-picking, but instead I present the matter as a whole, in all its savoriness and unsavoriness.[362]

Therefore, the material that I have chosen for this seminar concerns a case that occurs on a daily basis, and none too rarely either, particularly nowadays. I have not included any pathological material, based on the assumption that my audience has no particular medical interest. I have chosen a normal, intelligent, and educated average person who consulted

---

[362] Jung's stance is consistent with his psychological interpretation of the symbolism of the chakras, with the first chakra, *muladhara*, corresponding, in terms of Western psychology, to the conscious world, "a place where mankind is victim of impulses, instincts, unconsciousness, of *participation mystique*, where we are in a dark and unconscious place," and the next chakra, *svadhisthana* being "the unconscious, symbolized by the sea, and in the sea is a huge leviathan which threatens one with annihilation" (C. G. Jung, *The Psychology of Kundalini Yoga: Notes of the Seminar Given in 1932*, ed. by Sonu Shamdasani [Bollingen Series XCIX] [Princeton, NJ: Princeton University Press, 1996], 15). In this hermeneutic framework, Jung remarked that the Hindus "have the unconscious above, we have it below" (16) and then explained, "We would put *muladhara* above because this is our conscious world, and the next chakra would be underneath—that is our feeling, because we really begin above. It is all exchanged; we begin in our conscious world, so we can say our *muladhara* might be not down below in the belly but up in the head" (17). Again, he pointed out that "in adapting that system to ourselves, we must realize where we stand before we can assimilate such a thing [. . .]. We do not go up to the unconscious, we go down—it is a *katabasis*"; after which, he refers to the common custom of locating the ancient mystery cults underground—from the Mithraic cults to the Christian crypt (18). Subsequently he expanded on this, stressing the necessity for the West to "begin with [. . .] the practical and the concrete," to dive down into—and assimilate—the unconscious, rather than "explain the world in terms of *sahasrara*," and risk identifying with brahman (65–66). "Only after having reached this standpoint—only after having touched the baptismal waters of *svadhisthana*—can we realize that our conscious culture, despite all its height, is still in *muladhara*. And as long as the ego is identified with consciousness, it is caught up in this world, the world of the *muladhara* chakra [. . .]. Observed from that angle we ascend when we go into the unconscious, because it frees us from everyday consciousness" (66–67).

me in my practice. Thus it would be quite incorrect to call him a patient. Should I nevertheless use this term from time to time, I'd ask you to correct it accordingly. The gentleman concerned is not sick, but, as I have mentioned, a normal person. I have brought along several of his dreams, notably from the beginning of the whole sequence, in order to show you what the path that leads to inner self-development looks like at its inception. How we begin, what we do, how we must take pains to start out along the path which in the East has already become a main highway with magnificent bridges. Here, with us, it's a dark wilderness, which begins in the most banal realms of daily life and winds its way only slowly out of our everyday interests and preoccupations, and leads only gradually into those depths where we slowly begin to understand that Eastern symbols are in actual fact connected with something that can be experienced. We cannot readily make this out. If you consider these vast conceptions about which Professor Zimmer has spoken, you must admit that they are very far removed and that we are all hopelessly ignorant in this respect. It is inconceivable for us that a Westerner has ever had such an experience. We also realize that Westerners commonly consider this some kind of nonsense, a phantasm of the East, the luxuriant, exuberant fantasy of a tropical climate. But that is not the case. These things are all based on actual inner experiences, while the question of how we might somehow approach these experiences can be answered only with difficulty. You will soon gather that the way in which the inner path or inner development is conveyed to us corresponds not in the least to our expectations. Rather, they come to us in a way that one cannot possibly call typical. It happens in the most individual manner, and therefore at first it goes unrecognized. This explains why it has taken us so long finally to realize that the unconscious events occurring within us have a *direction*, and that they move toward a certain *goal*. We have therefore assumed for so long that the unconscious must have a certain goal, and that we are the lords and masters who know exactly what everything is aimed at, and that everything the unconscious does is at bottom nothing else than just what ought to be. Thus we have come around again to so-called normal human life, for instance, by assuming that precisely this and nothing else needs to happen; that is to say, that the patient need only marry and all will be well, or that one makes the right adjustment to life and other such carryings-on. Now obviously there are very many people who are not married and have neuroses, just as there are very many who are not adjusted; and we'd like to think that doctors will strive to set these people more or less right so that they can more or less go on living as everyone else does. Naturally, this far from answers *the question about the soul*. It is merely the prerequisite for the fact that one must help people back onto their feet, so that they can eat properly, sleep properly, and walk properly. Needless to say, that's the basic condition. But if you are confronted with a case like the one I shall present here, then these issues carry no weight at all; instead, it's a matter *de luxe*. These are all luxury problems, you will say, which don't occur in normal life. Laying claim to anything psychic or spiritual, shall we say, that's sheer intellectual or aesthetic luxury, or indeed superstition or the like. If, then, I describe this atmosphere in such detail, I do so simply to prepare you for the state of mind in the gentleman concerned.

The case concerns a man aged forty-seven, who leads a normal married life and has several children. He is quite affluent, owns a very successful business, or rather has retired from active entrepreneurship and now leads the life of a gentleman.[363] For someone involved in trade and commerce, he is actually highly educated. While he has no formal university education, he is very well read, has a very good high-school education, and he is, as mentioned, not at all neurotic. If I call him not at all neurotic, you must of course make small allowances. For if you examine anyone in slightly greater detail, then of course you will discover some kind of leak. We'd be nothing less than Pharisees if we assumed that perhaps we haven't sprung a leak somewhere, even if this were simply a matter of possessing certain peculiarities.

The first peculiarity of the gentleman concerned is his extremely correct behavior. He is overly proper. This prompts me to ask you: what crosses your mind when you meet that frequent type, who is too correct? What do you think is the matter with such a person? Does extraordinary correctness epitomize what it means to be human?

ANSWER: No.

I hear dissenting voices. But what, then, do you think he lacks? Why does someone behave so correctly?

ANSWER: Because he is insecure.

Insecure. Perhaps. But more could be involved. Might there be reasons for this insecurity that one combats with the antidote of correctness?

ANSWER: At bottom he's not so correct.

He's not so correct, or could be. You may safely assume this in the case of extremely correct behavior. It's always a compensation for something

---

[363] "gentleman": in English in the protocol.

that is possibly very incorrect. You mentioned insecurity. That's quite true. His behavior is not only very correct but he is also very shy, in fact peculiarily shy, so that he'd knock not only on the door before stepping into a room, but also on the doorjamb, and that's a bit too much. What would you think if someone even knocked on the doorpost?

ANSWER: He wants no surprises.

You mean that he doesn't want to surprise me.

ANSWER: That less so.

But he's not being taken by surprise. He might be surprised if he comes to see me. But what you would chiefly assume?

ANSWER: He is apologizing for having been born.

For the fact that he's actually present. He reduces himself to the minimum, and might that not mean, in other words, that he is compensating for something?

ANSWERS:—Inferiority.—Impudence.

He is compensating not his inferior, but his superior value. In actual fact, he feels very important. But he dare not acknowledge this. He is filled with a sense of unnatural importance. To begin with, we can therefore assume that this is someone who feels terribly important, in fact illegitimately so, which is quite stupid in human relationships. It's a sign of mild idiocy if people consider themselves very important. Because in fact he is by no means idiotic; I know him well enough to say so. Instead, he possesses extraordinary common sense. He does not belong to the common category of imbeciles,[364] but we can safely say that he is full of unnatural importance.

---

[364] "imbecile" (from Latin adjective *imbecillus* "weak, weak-minded," substantivized only from the early nineteenth century), is a medical term associated with *Debilität* (debility) or *degenerativer Schwachsinn* (degenerative mental weakness) which emerged with political, educational, and legal connotations in the late nineteenth century alongside "idiot" and "feeble-minded," indicating a pathological intellectual deficiency: see Mark Jackson, *The Borderline of Imbecility: Medicine, Society and the Fabrication of the Feeble Mind in Late Victorian and Edwardian England* (Manchester: Manchester University Press, 2000), esp. 33ff. The construction of the feeble-minded "was closely linked to the emergence of increasingly sophisticated classifications of mental deficiency. Significantly, however, both the construction of a new category of deficiency and the pathologisation of that category were parallel products of a range of interwoven social anxieties" (34). Subsequently, especially with Henry Herbert Goddard's taxonomy (Goddard, *The Criminal Imbecile: An Analysis of Three Remarkable Murder Cases* [New York: The Macmillan Company, 1915]), the categories of "moral imbecile" and "moron" (from the Greek *moros*, "foolish, stupid") acquired greater prominence, to refer to a higher grade of "feeblemindedness" linked with

What circumstances do you think allow one to be filled with such importance that one cannot recognize oneself? He cannot recognize his importance, because he doesn't possess it in actual fact. Instead, it possesses him. He does not have his importance, but it has him, and somewhat unwillingly. He'd rather not have it and therefore he apologizes for his existence. He asks himself if it may enter the room, because it strikes him as somewhat outrageous that such importance, for which he'd prefer not to be responsible, comes into the room with him. What circumstances do you think have caused him to feel like this?

ANSWER: Because he has been forced into this situation.

He's been drawn into quite a few situations. He is the president of various large stock corporations and so forth. He has suffered no hardship in that respect. It's something quite different. Think of something quite out of the ordinary.

ANSWER: Christina Alberta's father.[365]

Well, how might he come into the picture here?

ANSWER: I think that this is the type of person in question.

Yes, it might well be that type. I never met him while he was in the mental hospital, nor do I know whether that was how he behaved. He

---

allegedly hereditary phenomena such as crime, pauperism, and addiction. See also Henri F. Ellenberger, "Psychiatry from Ancient to Modern Times," in Silvano Arieti, ed., *The American Handbook of Psychiatry*, vol. 1: *The Foundations of Psychiatry*, 2nd edn (New York: Basic Books, 1974), 3–27; David Wright and Anne Digby, eds, *From Idiocy to Mental Deficiency: Historical Perspectives on People with Learning Disabilities* (London: Routledge, 1996).

[365] The novel *Christina Alberta's Father* (1925) is by H. G. Wells (1866–1966), who with Jules Verne pioneered science fiction. (Wells's *The Time Machine* [1895] has entered several streams of popular culture; *The War of the Worlds* [1897], which inaugurated sci-fi catastrophism, was for Jung a fine example of inner transformation.) The heroine of *Christina Alberta's Father* "is constantly under the surveillance of a supreme moral authority, which tells her with remorseless precision [. . .] what she is doing and for what motives. Wells calls this authority a 'Court of Conscience'" which personifies the Animus (Jung, *The Relations between the I and the Unconscious* [1928], CW 7, § 332; see also § 270). In *DAS* the novel is described as "a very good example" of the fact that "a man will admit every kind of sinful thinking but no feeling, and a woman cannot admit thoughts" (because her thinking is "inexorable") (95). It originated in a conversation between Jung and Wells (see E. A. Bennett, *What Jung Really Said*, 2nd edn [New York: Schocken Books, 1983], 93). In the 1928 "Vienna: Three Versions of a Press Conference," Jung describes the father character's pent-up megalomania: he is a retired laundryman convinced that he re-embodies Babylon's Sargon I, "reincarnation of the King of Kings," and is the coming ruler of the world (William McGuire & R.F.C. Hull, eds, C. G. *Jung Speaking: Interviews and Encounters* [Bollingen Series XCVII] [Princeton, NJ: Princeton University Press, 1977], 42). Jung's joke below—"I never met him while he was in the mental hospital"—alludes to this figure.

might have, but I don't think so. Christina Alberta's father, after all, did have a strong pathology. In our present case, however, it's not a question of pathology.

ANSWER: He is worth more than he himself knows.

Now that's an idea. Inside, this man is actually more than he himself knows, and so he apologizes for constantly assuming that he is more than he seems to be. This causes him quite some discomfort. By no means can he explain how he could be more than he believes himself to be. He is forty-seven years old, the president of various stock corporations, and very wealthy. From experience, he knows that he is able to do things, and that he is an important person. Yet now he feels that this importance is standing behind him, incessantly, and that he cannot get a hold on it. Somehow, this is very unpleasant for him.

Another peculiarity of his is that he's not only correct, but also anxious and correct, so that the following incident happened to him. He was driving a little too fast in his motor car, and a police officer stopped and warned him. As a result, he suffered a stomach cramp and felt as though he had to vomit. This experience made quite an impression on him. Such incidents happened to him from time to time over the years. What would you conclude from this?

ANSWER: He has an inner fear.

Fear? Of his inside? Of his stomach?

ANSWER: No, that he has had to swallow this.

Yes, one could see it that way. He felt sick because he had to swallow a minor police warning. In light of the vague inklings of his importance, one could, for instance, assume that he had the following train of thought: How can something like this have happened to me, to me who is so correct? One could grasp it thus. But in reality he feels very sick. He's on the verge of vomiting. We must assume this to be a fact. But what makes him sick?

ANSWER: He cannot stomach the admonishment.

Yes, that would almost be the same. He could not stomach what the situation was compelling him to swallow. He could, after all, have given the police officer his opinion on the matter.

ANSWER: Because that would have agitated him too much.

But then would I need to know why this incident agitates him so much?

ANSWER: He believes he knows himself quite well.

He believes he knows himself, and therefore he is very surprised by this reaction. This is very important. He is someone who is very conscious of himself. As I discovered later on, he had done a great deal to become conscious of himself in every respect. Whenever you meet a correct person, you will always find that he at least tries to be conscious of himself to the greatest possible extent. Therefore he is always self-aware. In English, one says that he is "self-conscious."[366] This particular man was self-conscious in the highest degree. And then came this particular reaction, like a bolt out of the blue. It surprised him beyond all measure. Even though the incident had occurred several months before, more or less the first thing he mentioned was that it had somehow deeply shaken him. That is, that something so curiously unconscious could happen to someone like him who was, after all, so conscious of himself. And precisely that made him sick. You see, if a person feels sick under such circumstances, then they'll always feel sick about themselves. That's because one has that particular attitude that makes one feel sick. One could indeed vomit, after all, and that is precisely what the police officer evoked, namely, that all this correctness could ultimately cause him to be sick. In the meantime, this mysterious reaction occurred—that is, that he very often reacted with his stomach if his correctness was subject to some particular excitement or agitation. Our first consultation began as follows: "Doctor, I must apologize for coming to see you. You see, I am not sick. But a few months ago, I had this stomach cramp. But this doesn't require treatment. In actual fact, I don't know why I have come to see you. You only treat sick people, don't you?" I replied, "Yes, but occasionally healthy people consult me, too." "I'm very relieved. I thought I would not be allowed to consult you."

"What exactly do you want," I said, "why this preamble?"

He replied, "You know, I have read a great deal. Since the nature of life and the world is becoming increasingly unclear to me, instead of clearer, I thought that perhaps Doctor Jung might know a thing or two. I have no worldview, you see. Religion does not mean anything to me. I am forty-seven years old, in good health, and I have a professional life. It is quite interesting, but also not all that important, such as the question, well, what happens to us, where are we going, where is the world going, what is the meaning of life, and what sense does everything make?"

Then he asked me, "Well, can you tell me what the point of life is?"

---

[366] "self-conscious": in English in the protocol.

Naturally, I am surprised and reply, "What sense does your life make to you? I would need to be the good Lord himself to answer this question. I know what meaning my life can have, but yours?"

"Where can I raise this question? Who can tell me?"

I reply, "You'll be knocking on many doors in vain. For nowhere in the outside world is it written what the meaning of life for this particular gentleman might be. But what I can tell you is that if you are serious about finding out, then let us begin with you, because you have someone inside who could supply an answer. You have a very particular personality inside you. This personality is not you. You are forty-seven years old. Naturally, that's a pretty short period of time. Of that period, you are reasonably well aware of, shall we say, twenty-seven years. You can attribute the rest to the unconscious. In these twenty-seven years, you have experienced only the smallest part of life, of course, and therefore you're in absolutely no position to form any idea whatsoever about life. However, within you there is a person who is perhaps millions of years old: namely, your unconscious, which has experienced this situation millions of times and has stored all these human lives and their sediments accordingly.[367] Thus, if we find ourselves in a situation that we find strange and incomprehensible, we can safely assume that this very situation has occurred already at least thousands or millions of times in the life of mankind. If we possessed historical traditions, moreover, then we could consult the ancient books about what human beings thought or did at the time. But since we possess such traditions only in the smallest part, we must turn to the unconscious to discover whether it perhaps knows something about these matters and whether it has an answer. We must turn to the Old Man who is a million years old and possesses experience that reaches back across

---

[367] "The Old Man," "the Great Man," or "the 2,000,000-year-old man" is one of Jung's metaphoric personifications for the collective unconscious (see, e.g., "The 2,000,000-Year-Old Man" [1936] and "A Talk with Students at the Institute," in McGuire and Hull, *C. G. Jung Speaking*, cit., 88–90 and 359–64 respectively). Cf. a 1948 letter of Jung's to Father Victor White, in which he recounts of a dream of the Man: "While I stood before the bed of the Old Man, I thought and felt: *Indignus sum Domine* [I am not worthy, Lord]. I know him very well: He was my 'guru' more than thirty years ago, a real ghostly guru [. . .]. It has been since confirmed to be by an old Hindu. You see, something has taken me out of Europe and the Occident and has opened for me the gates of the East as well, so that I should understand something of the *human* mind" (C. G. Jung, *C. G. Jung: Letters*, ed. by Gerhard Adler with Aniela Jaffé, trans. by R.F.C. Hull, 2 vols [Bollingen Series XCV] [Princeton, NJ: Princeton University Press, 1973], 1:491 [January 30, 1948]); Jung's reference here to his (inner) "guru" relates more specifically the Philemon figure in the *Liber Novus*: see also *MDR*, 182ff.

all human ages, and who thus harbors a range of experience that surpasses human experience as such in an unimaginable way."

This idea struck him as extraordinarily peculiar and paradoxical, which I quite understand. For it is actually quite a marvelous idea if properly entertained. But what else could I have said? If there is anything I wanted to do with him, I had to establish a hypothesis of some kind.

In reply, he asked me, "What are we going to do now? How do we draw nearer to this wealth of experience?"

I said, "Well, we need some kind of message or some kind of communication channel with this Old Man. For instance, those could be products that reach us from him. For instance, dreams. Dreams are something that we do not make. Dreams happen to us. There we are passive. They quite simply take place.[368] Dreams are simply nature—undistorted, impartial nature. We are merely the recipients of these messages, and if we turn to dreams, then we are doing more or less what human beings have done always and everywhere for countless millennia. Only recently, for about the last hundred and fifty years, have there been people in cities whose minds were somewhat uprooted and who imagined that there was nothing one could do with dreams. These are so-called educated people or intellectuals, while everyone else, ordinary people, keep a dream book. Every maid has a dream book. When I was in Africa, the Swahili boys kept a dream book which they consulted every morning.[369] Equally so the primitives. You know from the reports by other researchers how very often

---

[368] Cf. the Children's Dreams Seminar: "This is the secret of dreams—that we do not dream, but that we *are* dreamt. We are the object of the dream, not its maker. The French say: 'Faire un rêve.' This is wrong. The dream is dreamed to us. We are the objects. We simply find ourselves put into a situation. If a fatal destiny is awaiting us, we are already seized by what will lead us to this destiny in the dream, in the same way it will overcome us in reality" (C. G. Jung, *Children's Dreams: Notes from the Seminar Given in 1936–1940 by C. G. Jung*, ed. by Lorenz Jung and Maria Meyer-Grass, trans. by Ernst Falzeder with Tony Woolfson [Philemon Series] [Princeton, NJ: Princeton University Press, 2008], 159).

[369] Reference to Jung's North African "Bungishu Psychological Expedition" to Mount Elgon, between Kenya and Uganda in 1925–26 (see *MDR*, ch. 9, part 3 and, on Jung's Somali and Swahili assistants interested in dream interpretation, 265; see also Blake W. Burleson, *Jung in Africa* [London: Continuum, 2005], 52–54). Burleson's detailed study lists among Jung's most crucial lessons from Africa the pivotal relevance of dreams—that is, "big" dreams; his understanding of Fate; and "the idea that part of the human psyche is not essentially human but animal" (224–26). The Elgonyi were at that time a tribe numbering approximately five thousand, considered by the colonial administration a "dying race" (221). *Swahili*, or *suaheli*, meaning "edge" or "coast," commonly referred to by native speakers as Kiswahili, names the language and culture developed among east-coast Africans and Arabs from the tenth century, which spread into Kenya, Tanzania, Angola, and Somalia from the nineteenth century when Arab merchants began to interact with local populations. Although Swahili belongs to the Bantu language family, it has borrowed many words from Arabic and is the most widely spoken language in Sub-Saharan Africa,

dream books have played a role. Vierenhuys,[370] for instance, who explored the Nile and Congo, writes that he was very often forced to stop on marches because his porters, who had been engaged in continuous dream analysis all morning, had arrived at an unfavorable outcome, and said, 'The dream was not good,' upon which they immediately set down their load. Then, the whole caravan was forced to sit down and wait until others had dreamed other dreams the following night, which might allow the journey to go on or not. That is, among primitives the dream is a very well known, even recognized function, and often a social one. If a man has a dream, then he summons the tribe to gather. The men sit down in a palaver circle and he narrates his dream, which afterwards is more or less interpreted by the expert, the medicine man, the chief. Thus, it's only the intellectual upper class which, for approximately the last hundred and fifty years, no longer attaches any psychological meaning to dreams. But there are other, very essential things to which the intellectual upper class has not attached any particular significance in the last hundred and fifty years. The dream is thus in good company. In such matters, which concern the meaning of life, I find it appropriate to adhere to historical data. What human history has considered important from time immemorial, that is valid truth, not what is babbled today. That is all nonsense, and falls away. But what mankind has asserted from time immemorial, that is psychological truth. We must only find new forms for it. Everything else lacks instinct and is uprooted." I continued talking along much the same lines as I have been here, and he then said what such people very often say:

"But I have no dreams!"

To which I replied, "You will have a dream tonight."

As these matters tend to go, people who have no dreams have no need for any. Because there is nothing they would do with them. But if one needs dreams, then they will happen. And if no dreams happen, then the situation is not yet clear. Now if this man had returned the next time and said, "I have no dream," then I would have answered, "If there is more on your chest, then tell me what matters. And when you have spoken, when the table has been cleared, then the dreams will come." Believe me, this happens like clockwork.

---

nowadays used by over seventy million Africans, and, along with English, the official language of the East African community.

[370] Incorrect transcription, possibly intending the Dutch explorer, ethnologist, and natural scientist Anton Willem Nieuwenhuis (1864–1953), who did not travel in Africa, however, but explored central Borneo extensively in the 1890s, and was professor of geography and ethnology at Leiden University and editor of the *Internationales Archiv für Ethnographie*.

He came to our next consultation with a dream. I must obviously mention that his life, his intimate life with his wife and family, was not quite clear, as was usually the case. Marriage is one of life's most difficult problems, and this issue is never either resolved or clarified. That's what makes us entertain such ideal notions about it, to constantly soothe ourselves, just as we stroke the neck of an agitated horse. Marriage is one of the most difficult chapters. Hence, the French quite rightly say, "Il n'y a qu'une tragédie; celle de l'alcove."[371] Both he and his wife consider their relationship under threat. A strange alienation of affection between them has occurred. Notwithstanding goodwill on both sides, and despite the greatest possible effort, the unconscious has drawn them both away from the relationship; he had enough decency not to assume this had occurred on account of his wife, but instead he assumed that matters were not as self-evident on his side as they should have been. But he's a very upright person, one capable of shouldering the blame. For the moment, I shall not divulge any details; as you will gather, the issues will become clear from the dreams.

Let me narrate the first dream that he brought along. In this kind of treatment, the first dream is always an event. It provides an indication, and I'm always very eager to know what the Old Man has to say about the situation. Customarily, analysis acquaints us with a very complicated and difficult situation. One readily devises a whole series of good little remedies to advise the patient. He wants this, as becomes very apparent. One wants to take the swiftest route to a solution, and one's disposition is generous enough to exercise one's imagination: perhaps this might work, or perhaps that. Over the years I have learned to push all of that aside. I know nothing. I cannot know anything. No one can know. If someone claims he knows, then he quite simply belongs to a certain category of imbeciles. One can know nothing in this respect. For here destiny speaks, and destiny is always individual, and it's as sure as death that we cannot know. Therefore, I take no pains to seek to know anything. Thus, we have

---

[371] "There is but one tragedy, namely, that of the bedchamber," likely modeled on the sentence from Leo Tolstoy (1828–1910), "Man survives earthquakes, epidemics, the horrors of disease, all the agonies of the soul, but for all time his most tormenting tragedy has been, is and will be—the tragedy of the bedroom," quoted in D. H. Lawrence's unfinished novel *Mr. Noon* (D. H. Lawrence, *Mr. Noon*, ed. by Lindeth Vasey [Cambridge: Cambridge University Press, 1987, 190]). Tolstoy's marriage, which inspired the idealized portrait of Levin and Kitty in *Anna Karenina* (1878), though notoriously complicated, lasted almost fifty years, until his death. His wife Sofja Andreevna Tolstaya, née Behrs (1844–1919), known as Sonja, sixteen years his junior, had a deep admiration for her husband's genius and served him devotedly as an amanuensis, repeatedly copying out, for instance, the drafts of his magnum opus *War and Peace* (1865–69).

a purer and unprejudiced disposition, allowing us better to hear the voice that speaks impartially and always puts its finger on the right spot. All of us have our best analyst within ourselves, and therefore I try to bring forth this inner voice in everyone I treat, which then replaces me. I can then leave. The voice does everything in my place.

The first dream begins thus:

> *I hear that one of my youngest sister's children is sick. Then, my brother-in-law (that is, my youngest sister's husband) asks me out to the theater and dinner with him. It occurs to me that I have actually already eaten, but nevertheless decide to join him. We enter a great hall, in whose midst stands a long table, upon which the cloth has been laid. Several rows of benches have been arranged on all four sides of the hall, as in an amphitheater, with their backs facing the table. Thus, my back faces the dinner table, not the hall. We take our seats and I ask my brother-in-law why his wife has not joined us; I answer my own question, thus: probably because the child is sick. I inquire about its health, and my brother-in-law replies that it is feeling much better and now has a fever of only one or two degrees. All of a sudden we are inside my brother-in-law's house; I greet the child, a girl aged about two (which does not correspond to reality).[372] The child does not look good. I am told that it refuses to repeat my wife's name. I utter her name—Aunt Marie—several times. It amuses me not to pronounce the 'a' at the end of the name Maria, but instead to yawn, in spite of the protests of those standing around me.*

So, he cannot pronounce his wife's name properly in the dream, but instead he must yawn. This immediately affords you a little insight into his marital difficulties. Now let us consider this dream without prejudice. If you want to understand dreams, then you should never try to be clever, but as stupid as possible. You must grasp the art of being as simpleminded as that. You must not forget that we are dealing with a state of mind that should be judged by those means that he [the dreamer] happens to have. For instance, if you are treating the dream of an ordinary, uneducated person, then you must think this dream just as these people think it, as it occurs to them. You'd be utterly mistaken, of course, to think it as it would occur to you. That's a glaring error, which has always been committed in the assessment of primitive psychology. We have attempted to grasp with our discursive thinking how primitives think. Their thinking is completely

---

[372] Cf. the addition in the almost verbatim dream exposition in *DAS*: "There is no such girl in reality, but there was a boy of two" (7).

different from ours. It's more primitive. If we want to understand them, then we must be able to follow their particular manner of thinking. If you set to work on a dream, then you must immediately abandon all your presuppositions and seek to climb down to the dreamer's naivety. Thus one should not strive for theoretical knowledge from the outset, but at first throw out all theory. We are always advised first to speak about such a dream with the person concerned: allow them to talk about it, without any constraints. Some dreams are so difficult that I refrain from going into detail, but rather encourage the person to tell me about it, which as a rule permits the sense of the dream to emerge. Often thus, we can at least form some initial idea of the dream, such as where the drama is set. The dream, after all, is somehow related to the drama. It is also characterized by drama's structure. It opens with the dramatis personae,[373] followed by the location and the time of day. Place, action, and time. Subsequently, we observe how a problem is touched upon at the beginning of the dream. That's the exposition. Thereafter follow the entanglement, the peripeteia, and ultimately very often a lysis, or result.[374] But this happens only in the case of fairly complete dreams.

In our present case, the action occurred several years ago, about five or six years ago to be precise, when the gentleman concerned was conducting business affairs in an exotic country; as I told you, he later retired from his business. What would you conclude, on your own, from the fact that place and time are set about five or six years in the past? What, then, given that we are dealing neither with the present nor with the future, but with events that occurred so many years ago?

---

[373] Lat.: "masks of the drama," meaning characters represented on stage by the actors in both classical and contemporary theater. Jung consistently applied the term "persona" to the complex of attitudes towards the external world, or the "outer attitude, the outward face," which is complementary to the anima (and animus). See *Psychological Types* [1921], *CW* 6, § 803ff. and passim. Cf. *The Relations Between the I and the Unconscious* [1928], *CW* 7, § 243ff., and *The Structure of the Unconscious* [1916], *CW* 7, § 296ff.

[374] Elsewhere too Jung points out the resemblance of the dream's structure to the plot phases of classical Greek drama—broadly recalling the canonical model set out in the earliest surviving work of dramatic theory, Aristotle's *Poetics* (ca. 330 BCE), which essentially identifies the following phases: "exposition," developing by *desis* (complication) to the *metabasis* (point of transition), involving—what Aristotle regards as the more satisfactory "complex" type of plot—*anagnorisis* (recognition) accompanying *peripeteia* (reversal of circumstances) (ch. 11); and proceeding from this climax to resolution or catastrophe (meaning "overturning" in Greek) by *lysis* (or *lusis*: loosening/unraveling) (ch. 18). See Jung's "On the Nature of Dreams" [1948], *CW* 8, § 561–64, and cf. *DAS*: "Each dream is like a short drama. At the beginning is a sort of exposition, giving a statement of things as they are, just as is shown very beautifully in the Greek drama. First there is a demonstration of the situation from which things start; then comes the entanglement or development, and at the end the catastrophe or solution" (241).

Figure 3.

ANSWER: The problem lies in the past, so to speak.

Yes, indeed, and it relates to his previous situation. It's neither here nor now. The characters whom we encounter are his youngest sister and her husband, as well as a child, who in reality does not exist, as I told you, and who cannot pronounce his wife's name. To begin with, we come across something that has attracted the dreamer's attention first of all, and I must draw a picture for you to illustrate this [see Fig. 3].

This drawing shows the layout of the hall. It's a square or quadrangle, in the midst of which, at this point here, stands a somewhat longish table that has been set for dinner. Then we have several rows of benches, which are arranged to face the walls, that is, with their backs to the dinner table, their seats toward the wall, so that the people seated there have their backs toward the laid table. You must bear this in mind. Let's take a short break now and resume later.

* * *

LADIES AND GENTLEMEN,

Let us continue our analysis of the dream. We now come to a section that is somewhat boring. It consists of a psychological inventory, that is, the assembling of the context. If we want to treat a dream, then we must first ascertain where the words occurring in the dream and its images come from; in other words, what context they occupy in the mind of the dreamer. In this respect, we proceed exactly as if we were dealing with a text that is too difficult to decipher, for instance. If there are gaps, or if some words elude us, either because their usage is unusual or because their meaning completely escapes us, then we may have to look elsewhere to gather whether the same words recur or how certain words are used in other texts, in order to establish their exact meaning in our text. We adopt this method rather than that of free association, for instance, which is, of course, never free association, but always bound. Instead, we apply ordinary common sense, which we also adopt whenever we are seeking to determine what we cannot understand. Thus we perform no tricks, but proceed utterly naively and allow the patient to make his comments. At the outset, when they have no idea how to grasp a dream, I allow people to speak directly. Later on, I have them prepare a dream in a straightforward manner. I ask them to fold a sheet of paper in half. On one side, they note down the individual parts of the dream: for instance, I dream that I am in a railroad carriage, I am traveling to so and so, the ticket inspector comes along, inquires about me, and so on. On the other side, I write down what comes to my mind, what I think about the railroad, what I can establish about the railroad in general. Now there are dreams where one can determine nothing at all, dreams that are apparently so stupid and so empty of associative material that patients simply refuse to fill in the latter half of the sheet because nothing occurs to them. In that case, one must encourage them to go on talking. This difficulty will not concern us in the present case. Our man brought along ample material from the outset, as long as we remained in familiar territory. But, as you will see, we soon went astray and lost sight of the highway upon which he still had associations.

My dreamer at first addressed the subject of this curious hall. What made a particular impression on him was its layout, which he described at length.

From the door one must ascend some steps like those leading up to a rostrum; both sides of these stairs were flanked by benches, which stood facing the walls. I saw people sitting down on the benches, and also that the meal, which had been laid out on the table in the middle, should not yet be eaten. I did not notice the strange layout of the hall in the dream, which was unrelated to its continuation, since, upon sitting down, I immediately raised my question, and seconds later I was no longer in the hall.[375]

So it is a fleeting, momentary discernment of this hall. Thereafter comes another situation.

The strange layout of the hall reminded me of a hall where Basque pelota was played.

Basque pelota, known as *pelota vasca* in Spanish, is played in Basque regions, as the name already suggests, and also all over the Levant. It's a variety of tennis.[376] The seats are arranged in rows, as in an amphitheater, almost reaching the middle of the space, while the court occupies the remaining area. He had watched Basque pelota only on certain very rare occasions. He went on,

We sat down and I asked my brother-in-law why his wife had not come along.

---

[375] The corresponding portion of the dream account in *DAS* reads, "We arrive in a large room, with a long dining-table in the center already spread; and on the four sides of the big room are rows of benches or seats like an amphitheater, but with their backs to the table—the reverse way. We sit down, and I ask my brother-in-law why his wife has not come. Then I think it is probably because the child is ill, and ask how she is. He says she is much better, only a little fever now. Then I am at the home of my brother-in-law" (7).

[376] In *DAS* the analysand "remembered having seen a room like that [the one in his dream] in an Algerian town, where they were playing *jeu de paume*, a kind of *pelota basque*" (21–22). *Jeu de paume* (originally spelled *jeu de paulme*; Fr.: "palm game"), is the precursor of a variety of ball games such as *pelote basque*, *pelote valencienne*, *balle pelote*, the *jeu de balle au tambourin*, and most racquet sports including tennis. Originally the ball was thrown by hand over a taut rope at midfield. The two main forms are the *court paume* and the *longue paume*. The Spanish term *pelota*—cf. *pilote* in Basque and Catalan—comes from Latin *pila*, and has successively designated various types of games whose origins are lost in antiquity. In Book 7 of the *Odyssey*, Homer may allude to such a game when Nausicaa plays with a ball in dance rhythm; and analogous games were to be found among Native Americans, and also in Byzantium, Egypt, Iraq, and Japan. In the West, modern ball games and tennis find their origins in the variants of *ludum pilae* (or *pile*) originating in cloisters and ecclesiastic venues. Practiced since the thirteenth century, the pre-racquet version of tennis also spread through France and Italy (where it is called *pallacorda*) rapidly becoming the most popular game and the first professional sport of the modern era, accompanying the progressive decline of the most successful early modern European sport, the knightly tournament. In Italy the game *pallone*, which Goethe saw being played by four noblemen in Verona in 1786, and *calcio*, became extremely popular—and remain so (as does *bocce*: see also below, n. 392). Nowadays several varieties of *pelota* permeate Basque society: see Paddy Woodworth, *The Basque Country: A Cultural History* (Oxford: Signal, 2007), 66 and passim; also Olatz González

The dreamer says that as a rule he doesn't go out in the evenings without his wife, nor does his brother-in-law. He might as well have answered his own question, insofar as,

> While I was asking, the probable reason for her absence occurred to me and so I didn't wait for an answer, since I wanted to show that I had not forgotten about the sickness.

The issue here is the sickness from which his sister's child had suffered in real life. The child died of dysentery. It was a boy. In the dream, he thus assumes that the woman, his sister, had not come along for dinner because of the child's sickness. He adds that his wife never took part in any pleasant pastimes if there was something wrong with the children or if she believed that they were not properly attended to in her absence. Then he takes up the sentence in which he asks about the child's condition; in reply, he discovers that it's feeling much better, and has only a slight fever, one or two lines on the thermometer. Then he remarks that the deceased boy, who had died of dysentery, had always had a fever.

> Whenever our children are not feeling quite as they should, the thermometer is produced on the spot.

Then, when he arrives at his brother-in-law's house in the dream, he says,

> My father lived in the house (where my brother-in-law now lives) and he left it to my sister. It's only a hundred paces from my house, and so we often get together. The house and shutters are painted in the same gray color. For my taste, this lends everything a somewhat monotonous appearance I'd like the shutters to be painted in a different color to liven things up.

When he makes a remark like that, one involuntarily pricks up one's ears. Why? One must immediately take note of these connections.

Abrisketa, *Pelota vasca: Una ritual, una estética* (Bilbao: Muelle de Uribitarte, 2005). The International Federation of Basque Pelota was established in 1929 in Buenos Aires. Today Basque *pelota* is played not only in Basque-speaking regions and neighboring areas in Spain and France, but also in Belgium, Greece, the Netherlands, northern Italy, in Latin America (for instance, Argentina, Chile, Peru, Cuba, Mexico, and Uruguay) and in other countries such as India. For further details see Gabriel Garcia i Frasquet and Frederich Llopis i Bauset, *Vocabulari del joc de pilota* (Valencia: Generalitat Valenciana, Conselleria de Cultura, Educació i Ciència, 1991). For a detailed historical survey of the jeu de paume and ball games, see John McClelland, *Body and Mind: Sport in Europe from the Roman Empire to the Renaissance* (London: Routledge, 2007), especially 77ff. and 122ff.

ANSWER: It's actually an aside.

One could call it a cursory remark that bears no relevance. It's not even worth mentioning that the house is painted gray. Obviously, however, it matters to him. Why?

ANSWER: Because of the monotony.

The monotony, that's the point. Because the dream ends, by way of a lysis, with him yawning. Officially. So if his context affords him the possibility, he can therefore take up the subject of monotony. Please bear in mind that the house occupied by his brother-in-law and his youngest sister has been left to his sister. Besides, the fact that it's painted gray leads us immediately to make an assumption, doesn't it? What function would his brother-in-law have?

ANSWER: He means himself.

Yes, of course. His brother-in-law stands in an analogous relationship to himself. He is his parallel. His brother-in-law's marriage also parallels his own. They live in a desperately boring house, which immediately casts a certain light on the situation. Then he continues. He greets the sick child, a girl about one or two years old. Then he observes that his brother-in-law's family only has a boy, who is now two and a half years old.

I always liked to spend time with this boy.

He has two other sisters, who both have female children. The girls are about seven years old, and he also has spent a lot of time with them.

I find little girls much funnier and more affectionate than boys.

Well, there you have an association, the parallel between that family and his own.

Let me ask you a question that I must account for first. Whenever precise time indications are made in a dream, the dreamer will as a rule overlook these, unless he is very shrewd. Establishing such time indications, however, is of the utmost importance for interpretation, because they lead us to quite specific events. So if the child is purportedly three years old or so, I routinely ask the dreamer, "What happened three years ago?"

And so I asked him, "What happened about two years ago?" He did not ponder this for long, but said quite readily,

I began studying the occult.

Already at that time, the problem of the meaning of life had held a strong attraction for him, and he began his search; at first he started where most people start looking. He has two particularities. One is his psychic desire, one could say, and the other is his extremely sensitive body. He had observed that from time to time his stomach reacted strangely to certain situations. Consequently, he inferred both correctly and logically that this reaction was related to his eating habits. The stomach, after all, has to do with eating. Accordingly, he began studying different dietary regimes for his physical welfare; he has fallen for Bircher and vegetarianism.[377] Subsequently, it occurred to him that a similar regime must exist for the mind. Well, there are several regimes, as you know, comprising all kinds of notions; for the most simpleminded, Christian Science is a very good regime,[378] or anthroposophy for that matter.[379] They require no effort, as is well known, but instead everything, all concepts, are delivered

---

[377] The Bircher diet is based on the theories of the Swiss physician and nutritionist Maximilian Oskar Bircher-Benner (1867–1939), who was inspired by Christian and Eastern values (he was an admirer of Gandhi and Raimon Rolland yet he was also initially favorable toward National Socialism). Bircher muesli, his famous cereal recipe, was part of a regimen of raw foods (particularly fruit and vegetables). The Bircher-Benner Klinik in Zurich, known as the Sanatorium Lebendiger Kraft (Life energy sanatorium), hosted such illustrious figures as Rainer Maria Rilke, Wilhelm Furtwängler, and Thomas Mann (who took inspiration from the clinic for his novel *The Magic Mountain* of 1924). The clinic ran on a strict schedule based on gymnastics, sunbathing, hydrotherapy, and outdoor exercise. Founded in 1904, it closed in 1994, but has recently reopened as a medical center in Braunwald.

[378] Christian Science, or the "Church of Christ Scientist," was founded by American religious leader and reformer Mary Baker Eddy (1821–1910) in Boston in 1879. She aimed to restore primitive Christianity, believing in an opposition between the material and spiritual worlds, the first being the product of mental fallacy, and the second, sustained by God, being an absolute good guaranteeing recovery and health. Eddy worked out her doctrine following sudden recovery from a fracture due to a fall: she became convinced that illness basically derives from moral weakness. Faith and prayer therefore became the cure for every physical and moral evil. The movement peaked during the 1920s, building a solid following in Switzerland.

[379] Anthroposophy is a spiritual movement founded in 1912 by the Austrian philosopher, teacher, writer, and esotericist Rudolf Steiner (1861–1925), who was largely inspired by Eastern and German Romantic thought, notably Hegel and Goethe (four volumes of whose scientific writings Steiner edited). Steiner diverged from the Theosophical Society (of whose German Section he was the first secretary) and developed his own system which, besides sharing crucial aspects with theosophical anthropology and cosmology, strongly emphasizes education and practical applications. Harmonizing body, soul, and spirit, Anthroposophy promotes a multidisciplinary development beginning in childhood, which aims at spiritual growth and higher creative faculties. Artistic activity—in drama, architecture, music, picture, sculpture, and dance—bridges faith and science. Anthroposophy also embraces homeopathy, eurhythmics, and biodynamic agriculture, and still maintains its main center in Dornach, near Basel. For an overwiew of Steiner's vast output, see Rudolf Steiner, *Anthroposophical Leading Thoughts*, trans. by George Adams and Mary Adams (Forest Row, East Sussex: Rudolf Steiner Press, 1973); see also the scholarly study by Helmut Zander, *Rudolf Steiners Ideen zwischen Esoterik, Weleda, Demeter und Waldorfpädagogik* (Paderborn: Ferdinand Schöningh, 2019). A critical approach is to be found in Peter Staudenmaier, "Race and Redemption: Racial and Ethnic Evolution in Rudolf Steiner's Anthroposophy," *Nova Religio* 11, no. 3 (2008): 4–36. For a valuable assessment of similarities and differences between anthroposophy and analytical psychology, see Gerhard Wehr, *Jung and Steiner: Birth of A New Psychology* (New York: Anthroposophic Press, 2002).

to one's doorstep, cut and dried. The same is true for magic of any kind, just as for spiritualism. His interests ran along these lines, and thus he engaged in peculiar studies at the time.

"Recently," he said, "I have moved away from such studies, in fact more because a certain disrepute clings to them than from lack of interest."

Thus, you can see that respectability grips him to the marrow, so much so that he does not forget his self-consciousness,[380] even when he is reading a book. It's not quite correct, you see.

"Two years ago," he said, "I read H. Dennis Bradley's *Towards the Stars*."[381] Some of you will know this book. It is spiritualistic, and quite briskly written, which in particular appealed to him.

"I gave the book to my sister after her little boy's death. She wrote to tell me that it had offered her solace. That summer, I also read *The Seeress of Prévorst*."[382]

---

[380] "self-consciousness": in English in the protocol.

[381] *DAS* informs us that the patient's little nephew apparently died while he was reading this book (39). Herbert Dennis Bradley (1878–1934) was an idealist philosopher, and himself a medium, who wrote in support of psychical phenomena, spiritualism, and channeling. His debut novel *Towards the Stars* (London: T. Werner Laurie, 1924)—whose title invokes the official motto of the Royal Air Force, in which Bradley had served: "Per ardua ad astra" (Through adversity to the stars)—belongs to a trilogy including *The Wisdom of the Gods* (1925) and *After* (1931), in which a series of séances with the American medium George Valiantine are recounted, and his mediumship championed. One might recall Jung's long-lasting interest in, and exploration of, psychic phenomena: for instance, until at least 1931 he used to join (often along with Eugen Bleuler) séances with the German-Swiss esotericist, graphologist, and medium Oskar Schlag (see Oskar R. Schlag, *Von alten und neuen Mysterien (Die Lehren des A.)*, ed. by Antoine Faivre and Erhart Kahle [Stäfa: Rothenhäusler Verlag, 1995]); on Jung's exploration of psychic phenomena, see Sonu Shamdasani, "'S.W.' and C. G. Jung: From Mediumship to Analytical Psychology," in Ehler Voss, ed., *Mediality on Trial: Testing and Contesting Trance and other Media Techniques* (Oldenburg: De Gruyter, 2020), 121–42.

[382] Physician and poet Justinus Kerner's (1786–1862) *The Seeress of Prevorst* [1829], trans. by Catherine Crowe (London: J. C. Moore, 1845) met with great success in the nineteenth century, provoking intense debate on the psychophysical phenomenology of the book's subject, the German mystic and visionary Friederike Hauffe (1801–1829). Jung was acquainted with this "first history of a case of somnanbulism psychologically observed" (*DAS*, 39) at least from 1897 and—as McGuire points out (ibid., n. 5)—had discussed it already in *The Zofingia Lectures* (*CW* supp. vol. A, § 93ff.). He was impressed by Hauffe's visionary and mythopoetic abilities, which were akin to those of his cousin Hélène Preiswerk, on which he wrote his dissertation (see *On the Psychology and Pathology of So-Called Occult Phenomena* [1902], *CW* 1). A lengthy analysis is dedicated to the Seeress of Prevorst in C. G. Jung, *The History of Modern Psychology: Lectures Delivered at ETH Zurich, Volume 1: 1933–1934*, ed. and trans. by Ernst Falzeder (Princeton, NJ: Princeton University Press, 2019), 39ff.

He wanted to commission a pychological study of this book, and he would have remunerated the author, because he would have liked to have seen the psychology of this "seeress" of Prévost [Prevorst] published. But the gentleman concerned had been somewhat pathological, and opined that studying this book could damage him. So he [the analysand] discarded the plan.[383]

So the child does not look well. He says:

Bad complexion, old, distorted features, just as the deceased boy had exhibited during his last sickness. I am told that the child does not want to utter my wife's name. My wife is very popular with her nephews and nieces. Her name is among the first that crosses their lips. Lately, I received a letter from my sister, my younger sister, in which she reports how her little boy had sung a melody he had composed himself, namely, that Aunt Maria is "a good boy." He obviously seemed to be identifying with his aunt already.

This instance demonstrates—emphatically—that his wife is utterly different from how the dream represents her! He reiterates his wife's name several times to the boy, and it occurs to him that he had always made sure, both with his own children and his sister's, that they had to pronounce correctly those words that they mispronounced. He had always pronounced any given word correctly, because as a result of his correctness he obviously makes sure that nothing incorrect happens.

Let us now turn to the sick child. He mentions that his sister's first child, a boy, had died. He had shared the concern for the child during the sickness, and its death had shaken him deeply. Moreover, it was his godchild. His youngest sister had always been his favorite sister, even though he had pestered her a lot when they were children. She is eleven years younger than him. Her first boy had suffered greatly from bowel disorders, to which he ultimately fell victim. Therefore, pathological disorders experienced by the second boy caused the parents and relatives much anxiety. "Last summer my sister turned to Christian Science, since when little seems to be wrong with her child; whether this is accidental, or due to more agreeable treatment, remains to be seen."[384]

---

[383] *DAS* reports that the "gentleman" was "a certain doctor" whom Jung knew and who, he noted, was "not a psychological light; if he had attempted to write this analytical study, it would have been poor stuff. A good thing he desisted!" (40).

[384] Cf. *DAS*: "That the death of the first child had the effect of making his sister take to Christian Science is a fact belonging to the sister" (10). In any case, the analysand's approach to Christian Science "has also to do with the female character in his own psychology; it is decidedly a hint. The female factor underwent a certain conversion," evidenced by

About his brother-in-law he says, "We have been friends since my first professional appointment. We often went to the opera at the time. He is very musical and taught me the little notion of music that I have. He came to the company through me. He's now a director, but I was disappointed by how long it took him to learn the ropes. Like his brother, my other brother-in-law, he deals with people more easily than I do."

As we know, this brother-in-law asks him out to the theater and for dinner. "I cannot recall ever going to the theater with him since my wedding. Other than in the company of our wives. I cannot remember ever being asked out for dinner other than to his house."

At this moment, he remembers that in fact he has already had supper. Nothing comes to his mind at this point. He is empty. Nevertheless, he joins his brother-in-law. He observes that he does not like going out in the evenings. That requires a special occasion, which either appeals to his special interest or which he believes will be of interest to his wife. "However, unless I take the initiative, she also prefers to go to bed early."

"We enter a large hall," he says. It reminds him of a ballroom or banquet hall; while built in that style, it has never been used for banquets. Then he recalls the banquets that he has attended during the meetings or assemblies of the various stock corporations. The table, laid with a white tablecloth, ran almost from one end of the hall to the other. "My attention was distracted almost immediately, however, since we had to look for our seats, although not at the long table, but instead on the benches facing the wall."

He then turned to his yawning at the letter "a." Yawning, he told me, was an expression of boredom, and he found the protests of those standing around him quite justified. One should not exhibit such bad habits in front of children, because unlike adults they are unable to tell the difference between jest and earnestness. So that's the material that he added.

Can you already provide a satisfactory interpretation of this dream with the help of this material? I believe that would be somewhat difficult. So let's work our way through this dream a bit further with the assistance of this material.

The first message that we receive from the unconscious begins with his observation that one of his youngest sister's children is sick. We have heard

---

his growing and rather confused interest in "philosophy, occultism, theosophy, and all sorts of funny things; he was too level-headed to be much affected by them, though he had a mystical streak" (ibid.). Again, "[h]e shows an entirely female character in his mystical and philosophical interests. So we know that the child is a child of that female factor in him" (ibid.).

that his youngest sister is his favorite sister, that she is married to a man who holds a subordinate position to his own in the company, who is not taken quite seriously because he is evidently struggling to learn the ropes, and whose career has thus not advanced as swiftly as he [our dreamer] had actually hoped it would. So his brother-in-law is evidently of a somewhat inferior caliber.

Now the child is not a real child, but a fantasy child. The real child was a boy and has died. The sick child in the dream is a little girl. Thus, the dream has introduced into the situation a new idea, the fact that there is a sick child, and that this child is female. The characters appearing in his dream are, it is obvious, figures within his *drame intérieur*,[385] his inner spectacle, or the dramatic, conflict taking place within him.

Let us now try somehow to determine these characters, in order perhaps to establish whether or not they are typical. Naturally, this attempt is purely hypothetical, nothing but a working hypothesis. His brother-in-law, for instance, evidently represents him, our dreamer, and his brother-in-law's marriage represents what also concerns him. In other words, his brother-in-law somehow occupies his position, and thus a lesser personality has taken his place. He does not appear in this character with his actually existing values, but instead he is represented by a lesser figure. What would you call such a lesser figure? That is, please assume, for a moment, that you also had such an inferior figure, a figure that is neither as good nor as clever nor as proficient as you are, but that everything is a bit dampened. What would you quite naturally call such a figure?

---

[385] Allusion to *L'Hérédo: Essai sur le drame intérieur* (Paris: Nouvelle librairie nationale, 1917) by novelist and playwright Léon Daudet (1867–1942). Jung regarded this novel as a brilliant exemplification of the capacity of the imagination to enter and reawaken forefatherly, primeval memories. For instance, in "Concerning Rebirth" [(1939) 1940/1950], *CW* 9.1, he refers to that "confused but ingenious book *L'Hérédo*" which "supposes that, in the structure of the personality, there are ancestral elements which under certain conditions may suddenly come to the fore. The individual is then precipitately thrust into an ancestral role" (§ 224). Jung summarizes Daudet's description of possessional states brought on by ancestral souls commanding the I through "autofécondation intérieure" or "self-fertilization" in *Visions: Notes of the Seminar Given in 1930–1934*, ed. by Claire Douglas, 2 vols (Bollingen Series XCIX) (Princeton, NJ: Princeton University Press, 1997), 2:1266. Conceiving the individual itself as "a sort of conglomeration of ancestral lives," their constellation brought to mind reincarnation and migration of souls (ibid.). Elsewhere he explains the collective unconscious as a "functioning of the inherited brain structure," which in turn consists of "the structural deposits or equivalents of psychic activities which were repeated innumerable times in the life of our ancestors" ("Analytical Psychology and Education" [1946], *CW* 17, § 207). Daudet's subsequent *Le Monde des images* (1919) and *Le Rêve éveillé: Étude sur la profondeur de l'esprit* (1926) dealt with interior fantasies termed "person-images."

ANSWER: A shadow.

Indeed, a shadow, isn't that so? A shadow figure, a slightly negative figure, which is dependent upon him like the shadow that follows our every step, trotting alongside us, ever-present, and yet forever empty. Thus, a personification as it were of his inferior attributes. Well, what are his brother-in-law's positive attributes? Have you noticed anything in this respect?

ANSWER: Music.

Yes, indeed, music is very important. His brother-in-law has taught him the little knowledge of music that he has. This is one advantage that his brother-in-law possesses. What does this mean? What would characterize this man, for instance? Which function?

ANSWER: Well, feeling, naturally.

Correct. His brother-in-law represents feeling, his inferior function.[386] Yes, this shadow is evidently the personification of the feeling person. For music is the art of feeling, and wherever music is concerned, the feeling function is concerned. The dreamer is neither a musical nor a feeling person, but instead he is quite evidently a matter-of-fact person, in fact a decidedly matter-of-fact person, that is, a thinking person. Now you will be familiar with the contradiction that exists between these two functions. When you are thinking, then you do not really want the feeling. Because feeling always somehow contradicts thinking, because it confers upon things a certain value. If you are feeling, then you are often well advised not to think at the same time. Because that can be most unpleasant for you. No one knows this better than men. Consequently, a certain natural opposition exists between these two functions. Therefore, it is quite natural that his shadow represents the feeling function in him. At this moment, his shadow is not alone, but is married to a female figure, namely his youngest sister, who is eleven years his junior. We must now account for this figure. Where does she belong? How can we accommodate her within the male soul?

As you know, we have the hypothesis of the so-called "anima," that is to say, the feminine side set alongside the masculine, which has been called into serious doubt. But if you take an unbiased view of male psychology, then you'll discover the feminine without great difficulty. It's precisely with the most masculine man that you will, as you know, find a great

---

[386] According to Jung's typology, in someone whose first function is thinking, feeling is the inferior function (cf. *DAS*, 269ff.; and see *Psychological Types* [1921], CW 6, § 595–600 and § 763–64).

tenderness of feeling, which can extend as far as intolerable sentimentality, which is what one reproaches women for, who are as a rule much less sentimental than men. This allows you to account for the efficacy of this figure in all the accusations or criticism that the masculine man levels at women—for instance, moods. But the real man has moods moods galore. And if you put all these moods into a retort, heat them, and then distill the "refined spirit" from it and afford this spirit the opportunity to become a small homunculus, a sweet little female develops. This is the extract of all the emotional reactions of the strong masculine man. The result is a female figure. Whenever the masculine man is possessed by a certain mood, which happens at least once a week, he is thus actually possessed by femininity. In the Middle Ages, as you know, this was felt so plainly that if a person showed an affectionate emotion, one assumed that *she* was a witch, in that one simply projected this type of anima. From experience, we know that when such an image is projected, a fascination is ever-present. This image exists as an *a priori* archetypal structure. Which is why you can ask so many men what their anima looks like. They know exactly what this anima is supposed to look like; they can even say "there she is" when they meet her in the street. Whenever a woman appears who corresponds to this type, the man knows this exactly: this is the figure onto whom one could project this, that is, that unknown ideal or the more or less ideal image of womanhood.

It is the feminine in the soul of the man that appears to our dreamer in the guise of his younger sister. From the fact that this anima is eleven years younger than him, we could conclude that he has a very fatherly disposition. In actual fact, that is the case. He exhibits a decidedly protective behavior toward women. For instance, if a woman experiences any sort of mischance, then he finds this particularly agreeable. Then he draws out his manly protection and bestows upon the woman the necessary safeguard. He also fathers his feelings. He knows exactly what to do and he has always tried to do the right thing for his emotional life, such as educate his children properly and so forth. These educational intentions occur very frequently in men who have such a protective attitude toward their emotional life. They want the whole world to benefit from this disposition, in that this femininity is inclined to spread over the whole world and is projected everywhere and onto everything. So, let us make the hypothesis that this female is what one calls anima.

Now let us attend to the fact that his sister's child is sick. That would mean that the child of his anima and brother-in-law is sick. What does this mean? What would you say? What is this child?

ANSWER: His marriage.

No. His marriage is not two years old.

ANSWER: His attempts to gain a worldview.

His attempts at a worldview are two years old. They have played a considerable role and are the child of his soul. The soul's child is a try at a conception of the world, or at developing the contents of one. For that is where his soul was searching. He does not know, in actual fact, what he is seeking, a worldview or a religion, something that could put an end to this unsatisfactory state of affairs, in order to rid himself of all difficulties. Whatever it may be, this desire abides in him. It concerns his feeling, which is represented in the dream by this complex fantasy, that is, by the shadow, by the image of the soul, and by a child that they have together. In other words, his feeling has begotten with the female in him these attempts at formulating a concept of the world, which the unconscious now calls sick. How can we grasp this? How shall I explain it? In what way is this [feeling] sick if it carries out such occult studies? What is sick about this?

ANSWER: The studies are unsatisfactory.

ANSWER: They cannot help him.

Yes, there are many things that offer us no help, but which we would not necessarily call sick.

ANSWER: This kind of worldview does not correspond to his nature.

Indeed, not only does it not correspond to his nature, but . . .

ANSWER: It is even harmful.

Yes, like an illness. The illness is a foreign agent that enters our body and poisons it. His attempts at a worldview are perceived similarly by the unconscious, in that his studies of the occult are not only futile, but also represent an illness, precisely the illness of his psychic exertions. In other words, this is something that must be cured no matter the circumstances, or the outcome will be bad. Basically, the parallel with his sister's boy, who actually died, always poses a threat.

Let us suggest that what we might call a hypothesis affords us a positive indication from the Old Man, who, in other words, is saying, "What you have created with your yearning for a worldview is sick; it makes you ill, it's not right; indeed it's even abnormal."

This is the exposition of the dream. The drama begins thus, and now we can turn to the entanglement, that is, the attempts made by the Old Man to work on his [the dreamer's] problem, or to resolve it, or to somehow treat the question. Now the action begins. He goes along with his brother-in-law, or rather the latter asks him to join him at the theater and for dinner. As you can see, the inferior brother-in-law has never done this before. Thus, something new happens, something that he has never done with his brother-in-law, namely, to go to the theater and out for dinner. We must ask ourselves: what does it mean if the shadow invites us to go to the theater? What is a theater? What happens there? What does he know about the theater?

ANSWER: A performance.

Yes, indeed, a performance. A play is performed at the theater, and what is this play about?

ANSWER: His life.

About his life? That's not necessary. Instead, human situations are represented in images, and images are presented. Where are images, notably psychic ones, presented?

ANSWER: In dreams.

Yes, in dreams, for instance, and where else?

ANSWER: In fantasy.

Indeed. Thus we could say, for instance, that in the dream his brother-in-law invites him to enter the dream, that is, to go to the theater, the place where images are presented. Or it could mean that he prompts him to look at the images, the images of life. We can translate this without further ado, and so I translated this as indicating that he should go look at images which, according to the Old Man, are fantasy images. He should have fantasies. Well, why should he have them? What do you think? To what end? What good does this do him, now, when he has reached the end of his consciousness?

ANSWER: To show him that he is at the end.

Yes, of course, that his consciousness has reached its end. Here, the Old Man advises him to go to the theater and look at the images. If we formulate the matter thus, it reminds us directly of the report of Socrates's daimon. For instance, of the daimon's words, "Socrates, you should make music." Whereupon Socrates, with his bottomless rationalism, went and bought himself a flute.[387] The unconscious spoke to him—and we can

---

[387] From Plato's *Phaedo* (60e). Socrates understood his recurrent dream promptings to compose music as an encouragement to keep on practicing philosophy, in his view the most sublime form of music (μυσική in fact implies τέχνη, "the art of the Muses," and thus is not limited to musical performance). In *DAS* Jung vividly paraphrases Socrates's reception of

translate this as freshly today as though it had been put on ice. He was a thinker, a rationalist. But it tells him: You should use your feeling once in a while. Thus speaks the unconscious. Moreover, it invites him out for supper. He should have supper with his shadow, go to the theater, look at images, and then have supper on top of it all. Which already concerns the layout of the hall [see Fig. 3, p. 141]. What does it mean to have supper with someone?

ANSWER: Pleasure.

ANSWER: It is a sign of community.

Yes, indeed. If we sit down at table together, from time immemorial this has meant community. If you sit down for dinner with the members of a tribe, then that means community, connection, unity. One shares peace. One shares the same table or the same meal. Therefore, the meal also signifies community, which reaches as far as the customs of our Christian church, where supper, the Last Supper, is the most profound and strongest expression of Christian community. It means, thus, that he should have *communio* with his shadow.[388] What does it mean to have communion with one's shadow?

ANSWER: He should have supper with his inferior function.

Yes, he should get together with his inferiority, sit down at the same table with it. What arises from this?

ANSWER: A compensation.

He has been busy doing this in his life.

---

the prompting: "Of course the daemon meant: 'Do practice more feeling, don't be so damned rational all day long.' This could be applied to the patient very suitably," being as he is "very intellectual and dry" (11). Cf., as recalled by McGuire's note (ibid.), Marie-Louise von Franz, "The Dream of Socrates," *Spring* (1954), republished in her *Dreams: A Study of the Dreams of Jung, Descartes, Socrates, and Other Historical Figures* (Boston, MA: Shambhala, 1991); and Jung's "Foreword" to the *I Ching* [1950], CW 11, § 995. Later in DAS Jung adds, "And another time, [Socrates heard] 'Take the street to the left,' and by listening to the voice of his daemon [. . .] avoided a great heard of pigs that were rushing down the street he had been on. I was consulted recently by a woman who hears such a voice; she is just nicely mad, sort of home-like. She has a voice that speaks from down below, in her belly, and gives excellent advice; she comes to be cured of her voice yet wants to keep it. It is the voice of the shadow, of course [. . .]. And that woman is bewildered, she doesn't know whether it is the voice of God or the devil. One ought to be afraid and yet it should not be taken entirely seriously" (58–59).

[388] Lat. *communio* and the verb *communicare* (cf. Gk. κοινωνεω [koinoneo]; It. *comunicare*]) originally referred to sharing, joining, or having something in common with someone. *Communio* derives from both *cum-moenus*, shared possession of a defense or wall, and *cum-munus*, a shared commission or task, and translates Gk. κοινωνία (koinonia), "community," covering broader common initiatives.

ANSWER: A unity.

He is not up for unity.

ANSWERS:—A community.—To get acquainted with the dark aspects.

He bears a grudge against his shadow, and would rather have nothing to do with it. He would rather not have this shadow at all. There is a conflict here, and he is not at all in agreement. He is supposed to do something that he finds most abhorrent. He will need to make a tremendous effort to come together and sit down at the same table with his inferior I, with his brother, who is darker, weaker, and more inferior. As we know, nobody can bring himself to accept this: namely, that things are not as nice as we wish them to be. Actually doing this once and looking things in the eye is a great task. There, matters become unpleasant.[389]

Now comes the fundamental question: will he evade this challenge or not? Will he be able to look at himself, that is, look his own darkness in the eye?[390] That is the touchstone.

---

[389] Cf. DAS: "Communion means eating a complex, originally a sacrificial animal, the totem animal, the representation of the basic instincts of that particular clan. You eat your unconscious or ancestors and so add strength to yourself. Eating the totem animal, the instincts, eating the images, means to assimilate, to integrate them" (12). Jung's linkage of the Christian term to the shadow seems to to re-establish a connection not with the Self, but with the earthly, forgotten roots of the individual's primitive soul. On the subject of the Christian communion, from a comparative perspective, see "Transformation Symbolism in the Mass," [1942/1954] CW 11, § 338ff. It is interesting to recall here a passage in Jung's Zarathustra Seminar commenting on the origins, transformations, and transhistorical filiations of the symbol ("A symbol is never an invention. It *happens* to man") and its primeval ritual performance: "Did the early Christians think that behind the idea of the holy communion lay that of cannibalism? We have no evidence for it, but of course it is so: that is the very primitive way of partaking in the life of the one you have conquered. When the Red Indians eat the brain or the heart of the killed enemy, that is communion, but none of the Fathers of the Church ever thought of explaining the holy communion in such a way. Yet if their holy communion had not contained the old idea of cannibalism it would not have lived, would have no roots. All roots are dark" (C. G. Jung, *Nietzsche's Zarathustra: Notes of the Seminar given in 1934–1939*, ed. by James L. Jarrett, 2 vols [Bollingen Series XCIX] [Princeton, NJ: Princeton University Press, 1988], vol. 2, May 18, 1939, 1251).

[390] This phrase, echoing somewhat Matthew 6:22–23 ("The eye is the lamp of the body. If your eyes are good, your whole body will be full of light. But if your eyes are bad, your whole body will be full of darkness. If then the light within you is darkness, how great is that darkness!"), seems to recall a passage of Jung's interview for Radio Berlin of June 26, 1933 (see "An Interview on Radio Berlin," in McGuire and Hull, C. G. *Jung Speaking*, cit., 59–66 at 64), quoted above in the Introduction (21–22) See also below, 174 and n. 416.

Lecture 2

TUESDAY, 27 JUNE 1933

LADIES AND GENTLEMEN,
Let us begin. You will remember that yesterday we managed to see what
the theater and dinner could mean for the dreamer. Let us now make our
way further through the dream. The idea to which the dinner has given
rise is obviously no more than an assumption, just as everything else is a
matter of conjecture. Naturally, we set to work on such material experi-
mentally at first, and must always await certain confirmation before we
can be be sure of things. Thus, we can never say with certainty: this means
this and that means that, at least not before we have obtained evidence
from the further course of events and from more material; that is, from
further parallels that might point in the same direction. You'll recall that
after one instance in which he [the dreamer] mentions going out with his
brother-in-law, a small reservation occurs to him—namely, that he has already eaten supper. He nevertheless decides to go along. Let us now assume that our assumption in regard to the theater and the supper more
or less applies, in which case we could perhaps imagine what kind of reservation occurs to him at this moment. Would you have any idea what
this brief reservation, which he actually pushes aside quite readily, could
mean?

ANSWER: An unconscious warning not to sacrifice himself too soon.
   What for?

ANSWER: For the communion.
   I wouldn't consider that to be a sacrifice, but more or less self-evident
honesty. Wouldn't you agree?

ANSWER: Yes.
   So what kind of reservation might he have? Have you no further ideas
in this respect? "I had already had supper" refers to a past event, and also

to the idea of communion. By implication, this suggests that he has heard this before. Well, where has he heard this?

ANSWER: This is his marriage.

Well, yes, indeed we can see this in his marriage, for instance. Have you further thoughts about this? The idea of the theater, which refers to fantasy activity, occurs beforehand. He tells himself: I have had fantasies here before. He has also had communion before, and he has been with his shadow. In other words, he has known and seen this previously, and therefore it's not actually necessary that he swallow this yet again. Now this is actually true. In his opinion, he has already done this. It's exactly as if he had experienced nothing new. This is quite an ordinary mechanism, which you will often come across: namely, that when we visit our psychology as it is, we always entertain the idea, "Yes, we already know this, it's nothing new." For we all know ourselves, after all.

In a manner of speaking, we all adopt the very naive point of view that we know ourselves quite well, and that it's superfluous for anyone to mention anything new. Therefore, we are so deeply convinced that we could never find out anything about ourselves from within, and especially in no way that might lead beyond ourselves, but at the very best only back to ourselves. This view is supported in the strongest terms by certain theories.

Well, after he has decided nevertheless to make an attempt, he enters a great hall, amid which stands this long table, together with his shadow. Let me draw this picture again to refresh our memories [see Fig. 3, p. 141]. It is a quadrangular hall. Benches flank the walls, here, which I shall merely hint at, and here in the middle, running across almost the entire space, is the dinner table. Thus, this would be a kind of graduated seating, as in an amphitheater. You'll recall that he associates this hall with Basque pelota. Laid for dinner, this is where the table stands. Is there anything strange about this?

ANSWER: Yes, it is much too large.

Indeed, much too large for two persons. That's it. And there is no mention of a large banquet about to take place. But there seems to be a large table that has been laid for very many people. The hall also indicates this, in that a large hall is ready to accommodate a larger number of people, as is the case with Basque pelota. For not only players are in attendance, but also a large number of spectators. Thus, it's a kind of theater. Basque pelota is also used as theater, as it were. There are spectators, and they are enjoying the game.

Let us pursue this idea of pelota in greater detail. If a dream takes up such a matter even only indirectly, then it's always worth pondering. You have already seen that the technique, or the method, if I may call it that, that I am following is a kind of meditation method.[391] I consider the dream or the relevant part of the dream, and listen to what the dreamer tells me about it. I attempt to see matters as he sees them, in order to perhaps support his fantasy in this respect. So I queried the dreamer about this game, and he told me about it. He did not play pelota himself. As far as he knew, it's a kind of tennis, where a ball is hit with a racket. It's a ball game. Astonishingly, however, a table laid for many people stands in this room. Thus, if the whole group gathered for the game also have supper at the same time, what would that mean? What are your thoughts?

ANSWER: The ball game is also a communion, a community.

Yes, such games, in particular national sports, are plainly forms of community. You can see, for instance, that *bocce* is played in every small Italian town or village. Everyone comes together, the whole male population; politics are discussed, people see each other and discuss matters. *Bocce* is played, too.[392] The same holds true for pelota. It is a communion, a social event, conducted with much devotion, a gathering of men.

---

[391] Cf. Jung's revised version of his Eranos lecture in 1934: "The therapeutic method of complex psychology consists on the one hand in making as fully conscious as possible the constellated unconscious contents, and on the other hand in synthesizing them with consciousness through the act of recognition." However, since civilized man "possesses a high degree of dissociability and makes continual use of it in order to avoid every possible task," recognition alone is not enough. "As the archetypes, like all numinous contents, are relatively autonomous, they cannot be integrated simply by rational means, but require a dialectical procedure, a real coming to terms with them, often conducted by the patient in dialogue form, so that, without knowing it, he puts into effect the alchemical definition of the *meditatio*: 'an inner colloquy with one's good angel [*colloquium cum suo angelo bono*]'" ("Archetypes of the Collective Unconscious" [1934/1954], CW 9.1, § 84, § 85; his note refers to Ruland's *Lexicon alchemiae* (1612), s.v. "Meditatio").

[392] Although generally declining in popularity, the game of bowls is still played today, the rules varying, particularly in Italy (*bocce*, from Lat. *bottia*, "ball") and in France (*boules*). Its origins are ancient (Egyptian artifacts from 3500–4000 BCE testify to a similar game), and was a principal pastime of Roman soldiers, who spread it throughout the empire. In the Middle Ages it was popular in among all classes, even clerics and noblewomen. In England its heyday came during the reign of Elizabeth I. Nonetheless, many attempts were made to quash the game. Earlier in the sixteenth century in England Henry VIII had prohibited it, and the doges of Venice proclaimed an edict warning strongly against "il pericolo grande delle balle" (the great danger of balls). But these were the last fading anathemas against a custom by then diffused through most of western Europe. While the American Puritans banned the game in 1685, in Britain Charles II legalized it and even drew up a kind of rule book. In 1753 in Bologna the game was officialized and regulated, thanks to

This fits quite well with having supper together, which also symbolizes communion. This symbol lends particularly strong emphasis to the idea of communion: that is, of social connection or human connection. Thus, we are evidently on the right track if we assume that this dream actually concerns the idea of communion. Consequently, it might perhaps be worth contemplating this communion somewhat further. Our dreamer is not an unsocial person. He does what is normal in this respect. In which case we must really ask ourselves: why does the dream insist on the idea of communion? Why should he be guided to the gathering of men, to the ball game, to a joint supper? What do you think? It's as if he had completely missed out on such moments.

ANSWER: The sense of community then overcomes the feeling of inferiority.[393]

ANSWER: Well, he did mention that he goes out only with his wife.

Yes, he did indeed mention that when he does go out, he does so mostly with his wife. So there is something that speaks against him experiencing

---

Raffaele Bisteghi's *Gioco delle bocchie* (1753). See Mario Pagnoni, *The Joy of Bocce*, 3rd edn (Bloomington, IN: Authorhouse, 2010).

[393] Community feeling (*Gemeinschaftsgefühl*) and the inferiority feeling (*Minderwertig-keitsgefühl*) are key concepts in the "Individual Psychology" of the Austrian psychiatrist, psychoanalyst and social psychologist Alfred Adler (1870–1937). After his split from Freud in 1912, Adler developed his system in *The Neurotic Constitution* (1912) and *Healing and Education* (1913), while *The Science of Living* (1929) and *The Meaning of Life* (1933) provided a more philosophical outlook on the interrelation between "social feeling" or "sense of community" and family dynamics and the power drive. Adler regarded community feeling (like the inferiority feeling) as an innate impulse which needs to be consciously developed as a criterion for individual health, since, contrary to selfish, narcissistic tendencies, it can transform the power drive into an impetus toward responsibility and completeness within society: see Gisela Eife, *The Development of Alfred Adler's Individual Psychology: Theory of Personality, Psychopathology, Psychotherapy (1912–1937)* (Göttingen: Vandenhoeck & Ruprecht, 2019), 31 and passim. Conversely, Adler regarded the inferiority (and the connected superiority) feeling as an obstacle to a healthy *Gemeinschaftsgefühl*: "All individuals have a sense of inferiority and a striving for success and superiority which makes up the very life of the psyche. The reason all individuals do not have complexes is that their sense of inferiority and superiority is harnessed by a psychological mechanism into socially useful channels. The springs of this mechanism are social interest, courage, and social-mindedness, or the logic of common sense" (Alfred Adler, *The Science of Living* [Eastford, CT: Martino Fine Books, 2011 (1929)], 215–16). Adler referred to a healthy striving attitude as the "striving for completeness" (Adler, "Advantages and Disadvantages of the Inferiority Feeling" [1933], in H. L. Ansbacher and R. R. Ansbacher, eds, *Superiority and Social Interest: A Collection of Alfred Adler's Later Writings* [Evanston, IL: Northwestern University Press, 1964], 51–52 and passim). In 1934, following the closure of his Vienna clinics by the Austrofascists, he emigrated to the USA. His concepts (in particular *Gemeinschaftsgefühl*) were widely (mis)used for political ends by the Nazis, but also by radical Marxists.

a sense of boundless communion. Because he is quite evidently restricted by this dinner table. Let us indeed hope that the sense of community overcomes the feeling of inferiority. In that way a favorable turn would set in. But ordinarily that does not happen. There are countless people who feel exceptionally inferior, precisely in the company of others, even if they have a sense of community. But as a rule this sense of community is negative. Therefore, I believe that it is very important for our dreamer that he feels restricted by his wife in social company, and in fact that she somehow thwarts communion. Have you perhaps observed anything in this dream that might indicate that the communion is somehow incomplete?

ANSWER: That benches have their backs turned toward the dinner table.

We'll return to this issue later. It is obviously a clear expression of the fact that there is no communion. If people sit with their backs toward each other, then we can safely assume that there is no communion, although they have gathered for the purpose of community. Here we have a negative communion, but I would like to know if you have observed anything in the dream or in the material that might explain his inferior sense of community, which must be compensated in this way.

ANSWER: That he finds himself in his brother-in-law's house immediately afterwards.

Indeed, the hall is not situated in his brother-in-law's house. This comes only later on. We are in fact inside this particular public building.

ANSWER: He leaves immediately thereafter.

ANSWER: This points to a lack of masculinity.

In what sense? Because these are men whose fellow he is meant to be. Or rather "should" not be. This would indeed be a place frequented chiefly by men. But that does not prove that he would have no communion. No, it's something different. The shadow accentuates his inferior I, his alter ego. For he must go there with it. That is, he must hold communion with the Self in him at that place,[394] with the ulterior motive that, if he holds communion with the dark I there, then a communion with that factor has been prepared, with a community in the first place—that is, primarily a supper in the company of men. It's as if his dissociation from the dark I posed an

---

[394] Interestingly, the Self is named only here in the seminar, and notably in connection with the shadow (see their definitions in the appendix to *Psychological Types* [1921], CW 6). Jung designates it "the totality of Man, the sum total of his conscious and unconscious contents" ("Psychology and Religion," [(1937) 1938/1940], CW 11, § 140) and regarded the mandala as its typical symbol (ibid., § 156, and cf. *Aion* [1951], CW 9.2, ch. 4). Cf. above [I], 37ff., and n. 116; [ZL], nn. 304 and 306; and below, n. 525.

obstacle, *as if one found no relationship with the neighbor without becoming conscious of one's dark I*. Could you imagine something like that?

In our conscious, differentiated I, you see, we are all separated from one another. We make believe to others, and we deceive them. We place our best sides center stage, in the desire to be taken for whom we imagine we are, or what we wish to be like. Wherever possible, we prefer not to show our inferior side. Consequently the human, all too human,[395] darkness of course gets pushed into the background, and as a result this darkness, which prevails between everyone, is like a small abyss, which must first be crossed or bridged. Placing emphasis on our differentiated personality has a separating effect. In our differentiated functions we're in fact independent, and we don't actually need the other. Someone who possesses differentiated thinking, for instance, needs no authorization to be able to think. He can do this by and for himself. He is independent of others. In his inferior, undeveloped function he cannot declare himself independent of others. *The real connection among people never exists in strength, but in weakness*. That is where we are really dependent. Therefore, we also dislike being dependent. A man must make a tremendous effort to somehow declare himself dependent, since his pride demands that he not be dependent. Therefore, he seeks to keep those sides of his personality that make him seem dependent as secret as possible. He will find his relationship with others only when he can admit his own weakness; thus only when he disagrees less with his "darkness" and holds communion with his alter ego. In doing so, he will be better able to find a connection with others than if he adopts a pose and plays a role on the basis of his differentiated function. Obviously, that posture will never help him establish a real human relationship, but at most a position of power or opposition, a confrontation, which is, however, never a human relationship.

Accordingly, the benches in the hall are arranged so that everyone turns his back on the others. So, there is no concentration on the spectacle, because everyone is looking outward by turning their backs. Now, the essence of such a dream naturally is not that one looks out the window, but instead looks inward, to where what really matters is happening: namely, that faces are turned toward each other, that a concentration occurs, as it were, a convergence of gazes, because that is the essence of the process occurring here. If this is the meal, then naturally this is a ritual being performed, one that expresses communion. If this is the game of pelota, then it symbolizes the spirit of community.

---

[395] Allusion to Nietzsche's *Human, All Too Human: A Book for Free Spirits* (1878).

I was wondering whether you have ever heard of rites that have chosen precisely the symbol of playing a ball game as an expression of communion? Have people gathered for this purpose—regardless of ordinary, worldly, local events at which people gather to engage in a tennis tournament or a game of football or the like, or, as mentioned, to play *bocce*— that is, where it would be a ritual that would require explanation? Do you know of such a symbol, a religious symbol?

ANSWER: In India, for instance, in honor of the gods.[396]

Yes, many ball games are played there to honor the gods. But what about Europe, I wonder. Can you think of an example?

ANSWER: The *jeu de paume*.[397]

That's a secular game, which would correspond to a game of tennis. But I mean a religious game.

QUESTION: Involving balls?

Yes, indeed. This is something quite unknown, because it was suppressed in the sixteenth century by a papal decree. It occurred in the Catholic Church. The clergy would play pelota in the churches, in spring, naturally. I have brought along some extracts from documents that give us a sense of this strange practice. In those days, these ball games were played inside churches. They were part of the rituals customary during Carnival [preceding Lent], which were also suppressed considerably earlier by papal decree, because all kinds of mischief took place. For instance, the clergy, who were supposed to celebrate Mass, in actual fact staged a masquerade, drank themselves silly, danced in the streets, and put on a lot of tomfoolery, causing a great uproar.[398]

---

[396] Probable reference to the game mentioned in the *Mahabharata* as *iti-danda* or *gulli-danda/gilli danda*, played with a ball and two sticks and similar to present-day cricket. See also Renate Syed, "Anmerkungen zum Ballspiel in Indien," *Berliner Indologische Studien* 13–14 (2000): 231–52.

[397] See above, n. 376, and below, n. 509.

[398] Cf. *DAS*, 26. Jung deals with such customs, which are usually grouped as expressions of the so-called *libertas decembrica* ("December freedom," a phrase from Horace, *Satires* 2.7.4), and include the *festus puerorum*, the election of an *episcopus puerorum*, and the *festum asinorum* (or *asinaria festa*), and still reveal "the spirit of the trickster in his original form," in "On the Psychology of the Trickster-Figure" [1954], CW 9.1, § 460–63. See also Jung's quoted source (ibid., § 460 n. 4), Charles du Fresne du Cange, *Glossarium ad scriptores mediae et infimae graecitatis* (Niort: L. Favre, 1883–87 [1688]), entries "Kalendae," "pelota," and "festum asinorum"; and the detailed study by Max Harris, *Sacred Folly: A New History of the Feast of Fools* (Ithaca, NY: Cornell University Press, 2011), in particular Part 2, chs 6, 7, and 8, and Part 5, chs 20 and 21.

But the ball game survived much longer. It went like this: the clergy stood in a circle and threw the ball to each other in a ritual fashion. I have not been able to find out how exactly the game was played, or which particular forms it involved. The archbishop, for instance, was said to have taken the ball and thrown it to each of the monks one by one. The game was played in France until the sixteenth century, until it was banned, as I said before.

There are other, highly peculiar, parallel customs: for instance, the so-called "burial of Halleluja." "Halleluja" was a female who was buried ritually, also in spring. I mention this because Halleluja was also represented as a ball. The choir boys had to gather in the sacristy on a certain day, the Saturday before Septuagesima Sunday, and they would form a ball out of earth. The ball was placed upon a bier, together with holy water, a frond, incense, and various other items; this lump of earth was then carried into the church where plaintive chants were sung over it. Then it was interred with great pomp in the central courtyard of the cloister. Assuming that it is a cloister, then the lump of earth representing Halleluja was buried here, in the middle.[399] The idea that this creature was female has

---

[399] Jung is paraphrasing a passage in the statutes of the church of Toul in Lorraine "redacted in 1497 by Nicolas le Sane, Licentiate of Law and Canon of the cathedral," consulted by the French priest and scholar Abbé Jean Lebeuf (1687–1760) and quoted by him in a "Letter Written from Burgundy to M. de S. R. about Some Curious Peculiarities of two MSS.—the one of Toul and the other of Sens," in *Mercure de France*, December 1726, 2656–73, as cited by G.R.S. Mead, "Ceremonial Game-Playing and Dancing in Mediaeval Churches," *The Quest: A Quarterly Review* (October 1912), 91–123; repr. in Mead, ed., *The Sacred Dance in Christendom* (*The Quest* Reprint Series 2) (London: J. M. Watkins, 1926), orig. pagination, at 110–11. Jung was greatly indebted to the erudite theosophist and translator George Robert Stowe Mead (1863–1933), and owned several of his books, including *A Mithraic Ritual* (1907) (which was crucial for Jung's discussion of the "Solar Phallus Man"), the voluminous second edition of his *Fragments of a Faith Forgotten* (1906 [1900]), and the entire run of his journal *The Quest*. For a contextual study of Mead's pivotal role in the study of Gnosticism and Hermetism, see Clare Goodrick-Clarke and Nicholas Goodrick-Clarke, eds, *G.R.S. Mead and the Gnostic Quest* (Berkeley, CA: North Atlantic Books, 2005). *DAS* provides other details borrowed from Mead: "In the Middle Ages Alleluia was believed to be [. . .] an unknown woman who was buried at Eastertide, so she would be a sort of queen of the past year [. . .]. The Latin text which refers to it prescribes what to do at the burial" (34). Then Jung quotes almost verbatim Mead's rendering of a "curious ordinance" discovered by Lebeuf (ibid.; Mead, "Ceremonial Game-Playing," cit., 111–12). In the following, I quote Jung's report, inserting in square brackets excerpts from Mead's translation quoting Lebeuf's letter, to indicate variations: "On the Saturday preceding Septuagesima Sunday [That is, sixty-three days before Easter], at nones, the choir boys are to assemble in [the great vestry, in] festival attire and [there to] arrange for the burying [or burial] of Alleluia; and after the last benediction [or Benedicamus] they are to go in procession with [crosses,] torches, holy water, and incense, carrying a clod of earth on a bier through the cloister, [cf. Mead: "carrying a clod of earth as at a funeral, and are to proceed across the choir, and go to the cloister"], wailing [(*ululantes*)], to the place where

most probably to do with the fact that these excellent Latinists considered "Halleluja" to be female rather than male, on account of the final "a."[400] By that time, knowledge of Latin among the lower clergy had become nearly extinct.

Let me mention some extracts from other texts. The last traces of pelota in the Church can be found in Auxerre in the sixteenth century. Regrettably, the text is too meager.[401] All it mentions is that the tables were laid

---

Alleluia is to be buried; there to sprinkle water and grain on the clod of earth, to swing the incense, and to return the same way. [And after one of them has sprinkled the water and a second censed (the grave), they return by the same way. Such is the custom from of old.]" Mead refers to Lebeuf's conjecture that "in this ceremony, which took place between nones and vespers, with the full sanction and approval of the Chapter, the boys must have carried a sort of bier on which was the representation of the deceased Alleluia who was laid in a grave in the cloister" (*DAS*, 34; quoting Mead, "Ceremonial Game-Playing," cit., 112; see also Mead's next section, "The Whipping of Alleluia," 112–13). Then Jung goes on, "The clod of earth is the ball, and the ball is the sun, which is renewed through sacrifice at Easter time. Alleluia was simply Mother Earth, a feminine potency made to suffer death, burial, and resurrection, and supposed to be responsible for the new sun. The American Pueblo Indians assume that they support the sun by their ritual, and this is the same thing, death, burial, and resurrection. My Indian friend Mountain Lake in a letter to me said: 'If the white man keeps on interfering with our religion, in ten years they will see something!'— the sun would not rise again" (*DAS*, 34 and n. 5; cf. Jung's letter to Antonio Mirabal, October 21, 1932, in Jung, *C. G. Jung: Letters*, cit., 1:101. One may recall that Jung instituted a knotted-napkin version of the Alleluia game at the Zurich Psychological Club (Barbara Hannah, *Jung: His Life and Work: A Biographical Memoir* [New York: Putnam's Sons, 1976], 194–95).

[400] According to Mead, the original Hebraic term for Hallelujah (*Hallelu Jah*: "Praise Jah") "had supplied the basis for a number of Latin substantival, adjectival and verbal forms," and "Alleluia or Alleluja had been personified as a feminine potency, and not only personified but made to suffer death, burial and resurrection" (Mead, "Sacred Game-Playing," cit., 111; cf. *DAS*, 34). Like *Amen* and *Hosanna*, this Hebrew mystical interjection was not translated in Greek or Latin scripture because of its high mystery ("propter mysterium"), for all three are parts of the Angels' hymn (see Henri de Lubac, *Éxégèse médiévale: Les quatres senses de l'Écriture*, Part 2 [Paris : Aubier, 1961]). Saint Jerome (*Epist.* 137, to Marcellus) analyzes the word into *hallelu*, second-person imperative *lode*, meaning praise, and *Iuh*, God or the Lord, rendered in Latin as *Iubilate Deo, Exultate Deo*, to underscore exaltation. In psalmody this word signals that the whole psalm must be raised as a doxology to God; Dante has the righteous (*i Giusti*) sing it in Heaven.

[401] (This last sentence was deleted in Jung's copy of the Seminar.) The reference is to the Latin manuscripts preserved in the archive of the cathedral of Auxerre, compiled shortly after the customs they describe were practiced and reported in the original by the Abbé Jean Lebeuf, "Lettre [. . .] touchant une ancienne danse ecclésiastique, abolie par l'arrest du parlement," *Mercure de France*, May 1726, 911–25 at 921–22 (see Mead's translation in *The Sacred Dance*, cit., 96ff.). Lebeuf gives an account of a liturgical variant of the *lusus pilae*, which was celebrated for almost two hundred years on Easter Sunday after the Mass, taking place on floor maze in the nave of the cathedral of Saint-Etienne at Auxerre (and very probably in other French churches such as Chartres, Sens, and perhaps Reims and Amiens), and which possibly involved around a hundred canons. (The first mention of the dance, included in a 1396 ruling of the cathedral chapter, assigned it to Easter Monday, but

for dinner, and the newly appointed canon had to take the ball, known as *pelota*, after which he intoned the antiphon with the people singing ancient hymns in honor of Easter; the canon danced with the ball in his hands, rhythmically, and everyone else began dancing as well. Then the ball was thrown from one dancer to the next. Once this ceremony had been completed, the participants rushed to dinner; all the officiants took their seats on benches arranged in the chancel, and they were served a moderate helping of bread and white wine—and I'm translating here—while a homily was read aloud from the pulpit. During this ceremony, the newly appointed canon stood in the middle holding the ball. These games were held on behalf of Saint Stephen, at around two o'clock in the afternoon. So that time was somehow important. Subsequently, the dean reversed the *amice*[402]—that is, a white shoulder cloth made of linen, originally worn over the head and later over the shoulders—and began manipulating the ball, but in what way I do not know. So goes this development dating from the sixteenth century.[403]

---

by 1471 at the latest the date had shifted to Easter Sunday: see Mead, "Ceremonial Game-Playing," cit., 94). Thus from Auxerre, a formerly important monastic center where the earliest conciliar decrees against dancing in church were issued (in the years 573 and 603), comes "the fullest description of the dance of the maze, indeed the most complete account of any ecclesiastical dance in the premodern world" (Craig Wright, *The Maze and the Warrior: Symbolism in Architecture, Theology, and Music* [Cambridge, MA: Harvard University Press, 2001], 139). These ecclesiastical dances have been widely discussed: see, e.g., Louis Gougaud, "La danse dans les églises," *Revue d'histoire ecclésiastique*, 15, no. 1 (1914): 229–45 (for Auxerre, 235–36); the seminal work by Wright, *Maze and the Warrior*, cit.; Penelope Reed Doob, "The Auxerre Labyrinth Dance," in *The Myriad Faces of Dance: Proceedings of the Eighth Annual Conference of the Society of Dance History Scholars (U.S.)* (SDHS, 1985), 132–42; Penelope Reed Doob, *The Idea of the Labyrinth, from Classical Antiquity through the Middle Ages* (Ithaca, NY: Cornell University Press, 1990). Traces of these ceremonies involving a ball are documented as having been performed approximately up to the sixteenth century, when various clerical voices were raised—as declared in capitular decree of 1517—against "the potential for much evil [*multa mala*]," in connection with the participation of women, until a definitive prohibition was imposed throughout France upon both the ritual and the banquet that followed by the kingdom's highest court of law, the Parlement de Paris, in 1538 (as reported by Lebeuf: see Harris, *Sacred Folly*, cit., 59–60 and n. 24; Wright, *Maze and the Warrior*, cit., 141).

[402] In his personal copy, Jung corrected this to *amictus*.

[403] (In Jung's personal copy this last sentence was deleted.) See Mead's translation of Lebeuf's "Lettre" (*Mercure de France*, May 1726), cit., 921–22, in Wright, *Maze and the Warrior*, cit., 139–40 (Latin phrases between single square brackets inserted by Wright): "When the ball (*pilota*) had been accepted from the newly received canon, the dean, or another in his stead (in former times with his head covered with his amice [[*almutia*, a cape, hood, or wrap, covering both head and shoulders; it generally had two ends hanging down front or back (96, n. 4)]], and the rest [of the canons] (in like manner), began to intone in antiphon the sequence (*prosa*), appropriate to the Easter festival, which begins: 'Praises to the

An older description of the ceremonies held on Easter Monday occurs in a twelfth-century Narbonensian manuscript, and it reads, "While the bells were rung for vespers, the whole chapter [*conventus*] assembled in the hall of the archbishop's house, where those gathered were attended

---

Paschal Victim.' [*Victimae paschali laudes*]. Then, supporting the ball with his left hand, he begins dancing [[*Tripudium*, generally a quick dance, but was also attributed to the Angels by Patristic writers (97, n. 1)]], in time with the rhythmical sounds of the chanted sequence, while the rest, holding hands, execute a choral dance (chorea) round the labyrinth [*circa Daedalum*]. Meanwhile the ball (*pilota*) was handed or thrown alternatively by the dean to the dancers, one by one or several at time, wreath-wise. When the chanting of the sequence and the dancing (*saltatio*) are over, the band (or choir, *chorus*), after the dance (*chorea*) used to hurry off to the repast (*merenda*). There, all [the canons] of the chapter, as well as the chaplains and officials, together with certain of the more distinguished citizens, used to sit on benches in the 'corona' or in the orchestra. And all, without exception, were served with [the repast]. And white and red wine was also served but in temperate and modest quantities, that is to say the cups were filled once or twice. During the meal the reader intoned a festal homily from the chair or pulpit" (Mead, "Ceremonial Game-Playing," cit. 96–98; cf. Harris, *Sacred Folly*, cit., 57–58). Mead then further cites Lebeuf, who observes that "the sequence 'Praises to the Paschal Victim' was chanted, accompanied by the organ, in order to make the singing more regular and more in time with the dance-movement"; the celebration was performed underneath the organ-loft, "at the place in the nave where, prior to 1690, was to be seen a kind of labyrinth, in the form of several interlaced circles, as is still the case in the cathedral of Sens. But the finest part of the proceedings was the 'circulation' of the ball, that is to say the passing (*renvoi*) of it from the leader of the company to the several players, and repassing of it back by them to the president, who was probably in the middle of the ring clad in all his distinctive vestements and ornaments" (ibid., 99). The "Praises to the Paschal Victim" was a venerable chant commonly attributed to the monk, chronicler, and poet Wipo of Burgundy (ca. 995–1048), prescribed for the Catholic Mass and some liturgical Protestant eucharistic services in Easter week, and one of the four medieval sequences included in the 1570 edition of the *Missale Romanum* (promulgated after the Council of Trent by Pope Pius V). The fact that Wipo's chant deals with Christ's resurrection after his Harrowing of Hell leads to these customs being regarded as "a mystery play representing the resurrection. The movement through the labyrinth represented Christ's journey into the underworld, his battle with Satan, and His escape from the Kingdom of the Dead. The dance, therefore, represented both the forces of good and evil. It revealed both conflict and resolution" (Mark Knowles, *The Wicked Waltz and Other Scandalous Dances: Outrage at Couple Dancing in the 19th and Early 20th Centuries* [Jefferson, NC: McFarland, 2009], 178; see also Reed Doob, "Labyrinth in the Middle Ages," cit., 125). The ball has been generally considered a representation of the sun, i.e., it has been interpreted as the orb of the rising sun on Easter morning, or the rising sun as Christ himself, the *sol invictus* or unconquerable sun (according to the definitions of Christ as *sol justitiae* and *sol resurrectionis*): see Karl Kerényi, *Labyrinth-Studien: Labyrinthos als Linienreflex einer mythologischen Idee* (Zurich: Rhein, 1950), 147; Reed Doob, *Idea of the Labyrinth*, cit., 127; for Rahner, the ball corresponded to "the sun triumphant" (Hugo Rahner, "Der spielende Mensch," in *Eranos-Jahrbuch XVI/1948* [Zurich: Rhein, 1949], 11–87 at 83); see also his "Das Christliche Mysterium von Sonne und Mond," *Eranos Jahrbuch X/1943*, 305–404, at 321ff. "Thus the flying pilota," Wright concludes, "symbolized [. . .] the instrument of salvation derived from the Greek myth, the rising sun from folkloric legend, and the harmony of the spheres from classical philosophy" (*Maze and the Warrior*, cit., 142).

to by servants [*ministri*] of the archbishop. When[404] the archbishop was absent, the prefect was to throw the ball."[405]

An even earlier description occurs in a ninth-century Neapolitan manuscript,[406] in which the relevant passage reads, "In memory of this event were celebrated every year certain ball games for the consolation and the recreation of the soul."[407] Already in the twelfth century the

---

[404] Orig. Ger.: *sobald* (once, as soon as); in Jung's personal copy replaced with *wenn* (when; but also "if/in case"), which includes a hypotetical connotation.

[405] Paraphrase from a twelfth-century manuscript from Narbonne's cathedral archives quoted by Mead, "Ceremonial Game-Playing," cit., 101, referring to a second letter of Lebeuf, dated January 2, 1727 ("Remarks on several curious Contributions to the *Mercures* of 1726, addressed to the Editors of this Journal," *Mercure de France*, March 1727, 494ff.), which in turn refers to a "five-hundred-year-old manuscript" from Vienne Cathedral described by a correspondent of Lebeuf. Mead explains, "Leubeuf had been informed by a correspondent that at Vienne in Dauphiné, that is Narbonne, the custom of 'throwing the ball' at Easter festival had obtained in ancient times." By the time of its recording, Mead goes on, "the ceremony was no longer celebrated in the sacred edifice itself, but in the hall of the archbishop's palace, where all clergy of the cathedral assembled on Easter Monday, while the bells were ringing for vespers. On such feast days the bells rang for a considerable time, during which the ceremonial repast was partaken in the prelate's palace, and after this meal the archbishop 'threw the ball'" (Mead, "Ceremonial Game-Playing," cit., 101). After quoting the document, Mead maintains that "[t]his custom must have been kept up for at least three centuries subsequently, for on the margin of the MS. there is the following Latin note in a handwriting that was judged to be two hundred years old: 'And it should be known that the prefect [*mistralis/ministralis*: a "mayor" chosen from amongst the canons] is to provide the ball, and is to throw it in the archbishop's absence'" (ibid.; see also *DAS*, 33. Cf. Harris's translation in *Sacred Folly*, cit., 55; see also du Cange, *Glossarium*, cit. s.v. "pelota," 6:253). Mead notes that the prefect was the most important official after the archbishop, and points out, differently from Lebeuf who connects the game with the tennis, that, because of the pelota's ample dimensions, the game required a certain skill. Moreover, because this "ball throwing" took place after the ceremonial repast, it must originally "have been of a sober and dignified nature" (ibid.).

[406] The reference is to the *Chronicon episcoporum Sanctae Neapolitanae Ecclesiae*, BAV, Vat. lat. 5007, an anonymous mutilated codex attributed to John, early tenth-century deacon of Santa Maria Maggiore in Naples, and Peter the sub-deacon. Modeled on the Roman *Liber pontificalis*, the text includes a list of all Neapolitan bishops from the legendary Saint Aspren (first century) through to Athanasius II (d. 898). See Luigi Andrea Berto's study *The Deeds of the Neapolitan Bishops: A Critical Edition and Translation of the 'Gesta Episcoporum Neapolitanorum'* (London: Routledge, 2023).

[407] Cf. Mead, "Ceremonial Game-Playing," cit., 103. The "event" alluded to in the quoted text is, as Mead notes, the successful exorcising of the devil (the sixth-century church of Santa Maria Maggiore was the first church in Naples devoted to the Madonna for her exorcism of the devil; besides, this leads to the inference that the game originated from the time of the construction of the church). Although the text does not provide a description of these games, Mead argues, following Beleth and Durand (see next note), that they "were played in the sacred edifice itself," and "had a religious or spiritual purpose"; moreover the playing of this Neapolitan *percula* proves that such ball games "were not confined to French dioceses" (ibid., 103–4).

educated clergy turned against these strange customs. Especially Jean Beleth,[408] a theologian of the Parisian faculty, wrote about this game and remarked that the bishop and his archbishop played a ball game with their minions. He added, "But although large churches like Reims keep up this custom of playing, it seems to me more praiseworthy not to do so."[409] Evidently because already at that time this game was considered somewhat pagan and worldly.

These examples suggest that this ball game was used directly as a religious symbol, and it was evidently believed to be dignified enough to be used by the clergy for the teaching and strengthening of souls.

If you have ever witnessed a Spanish bullfight, then you will know how, and how intensely, the Spanish take part in this game. The atmosphere is fanatical and religious, one might say, so it is hardly surprising that these bullfights represented a religious feast in classical Antiquity. For instance, such bullfights took place in the circus, and the god Mithras

---

[408] Jean Beleth (or Johannes Belethus) was a twelfth-century French liturgist and theologian, probably rector of the theological faculty of Paris. In the early 1160s he compiled the *Summa de ecclesiasticis officiis*, a key source on the church liturgy of his era, which was later printed in William (Guillaume) Durand's *Rationale divinorum officiorum*, arguably the most significant medieval treatise on liturgy and church architectural symbolism, which also served as an encyclopedic compendium and textbook for liturgists. Durand (ca. 1230–1296) was a legal scholar who graduated from the University of Bologna. During the last part of his life, when he served as bishop of Mende, near Montpellier (1285–96), he produced his influential eight-volume major work (later one of the first to be published using movable type, in 1459) with the aim of providing a *summa*, following divine indications, of every aspect of ecclesiastical liturgy and architectonic elements.

[409] The sentence cited by Jung comes from the following passage: "There are some churches in which every bishop and archbishop plays with their subordinates in the convents [*coenobiis*] stooping even to ball-play [*lusus pilae*] [. . .]. But although large churches like Reims keep up this custom of playing, it seems more praiseworthy not to do so" (Jean Beleth, *Summa de ecclesiasticis officiis*, 120 [2:223]; cf. Constant J. Mews, "Liturgists and Dance in the Twelfth Century: The Witness of John Beleth and Sicard of Cremona," *Church History* 78, no. 3 [2009]: 512–54). Equally, a century later, William Durand expressed a certain contrariety to such dances (Durand, *Rationale divinorum officiorum*, cit., 6.86.9 [1:445], in *PL* 202; both as cited by Mead, "Ceremonial Game-Playing," cit., 104 nn. 1 and 2; cf. Harris, *Sacred Folly*. cit., 55). There is a common consensus that the Christmas ball games mentioned by Beleth and Durand referred to an earlier and simpler version of the Auxerre Easter liturgical dance: i.e., a kind of "prototype, only later adapted for performance at Easter" (Harris, *Sacred Folly*, cit., 61; see also nn. 33 and 35). Both Beleth and Durand also connect these ceremonies with the aforementioned *libertas decembrica*, referring to a custom deriving from the Roman Saturnalia, for—according to Durand—"in former times it was customary among pagans that in this month the slaves, shepherds, and maids would enjoy a certain liberty and were allowed to tell their masters what to do" (cited in Rahner, "Der spielende Mensch," cit., 81. Rahner further suggests linking the origins of such ecclesiastical dances to "the legacy of Old-Germanic Easter customs steeped in ritual" [77]).

was the master,[410] the patron saint of bullfighters as it were. For this reason, bulls were often represented wearing an abdominal band, a kind of belt, so that the bullfighter, not unlike the Spanish toreador, could jump onto the bull's back to stab him to death by ramming a sword from the neck downward toward the heart. The ancient toreador did not place himself in front of the bull to thrust a sword through him, but mounted the bull from the rear and killed him from above. As you know, Mithras has been represented in art as a bull-slayer;[411] he can be seen riding atop

---

[410] Jung is referring here the Persian/Roman/Hellenistic deity Mithras. He was drawing primarily upon the groundbreaking works of the Belgian philologist Franz Cumont (see Cumont's *The Mysteries of Mithra*, trans. by Thomas J. McCormack [Chicago: Open Court, 1903 (1900)]), which summarized his broad, and at that time deeply influential, study of the subject). Mithraism was a mystery religion. Among the Persians it reached a form similar to that practiced in the Roman Empire, where from the end of the first century CE it spread widely, from North Africa to Hadrian's Wall in Britain, and from Gaul as far as Syria, reaching its apex between the second and fourth centuries, when it became popular especially among the Roman legions. It seems to have to disappeared shortly after its suppression by Theodosius in 391. Cumont's conviction of a direct lineage from Zoroastianism to Roman Mithraism has been contested. Andrew Fear, for instance, suggests that Roman Mithraism derived from a mixture of Graeco-Roman beliefs and a version of Zoroastrianism mingled with Catholic elements, in which Mithra was both bull-thief and bull-slayer (Fear, *Mithras* [London: Routledge, 2022]). For a discussion about the most extended account on Mithraism in Plutarch's work, see ibid., 56–60. For a passing reference on the affinities between the Mithraic mysteries and the Christian liturgy, see below, n. 572.

[411] Despite remarkable iconographical differences, there is "one element of the cult's iconography which was present in essentially the same form in every mithraeum [i.e, underground sacred cave used for the worship of Mithras] and which, moreover, was clearly of the utmost importance to the cult's ideology; namely the so-called tauroctony, or bull-slaying scene, in which the god Mithras, accompanied by a series of other figures, is depicted in the act of killing the bull" (David Ulansey, *The Origins of the Mithraic Mysteries: Cosmology and Salvation in the Ancient World* [Oxford: Oxford University Press, 1989], 6). In ancient Iranian mythology the ritual that in modern times has been named "tauroctony" was a creative act, because it was believed that from the bull's body the world was born. This cosmogenic model was taken up by the Mithraic mysteries, and Mithras's "original proximity to heaven and to the sun offered Greek thought, philosophy and astrology, the opportunity to understand Mithras even more definitely as a cosmic deity, which, united with Helios-Sol, rules the planetary system and to which the human soul ascends through several layers" (Hans Kloft, *Mysterienkulte der Antike: Götter, Menschen, Rituale* [Munich: Beck, 2003], 73, cited in Isolde Penz, *Wege zum Göttlichen: Die Sehnsucht nach dem Einssein mit dem Göttlichen in Mythos, Gnosos, Logos und im Evalgelium nach Johannes* [Berlin: LIT, 2006], 98–99). According to Roger Beck, the tauroctony "is a map (and calendar) for genesis and apogenesis, for the descent and ascent of the human soul" and represents "a *stage* or *scene* against which the two principal players perform their parts in the great drama which is playing out of those two opposed yet complementary principles of genesis and apogenesis," whereby such a symbolic conception of the sun and the moon, along with the projection onto them of the descent and ascent of the human soul, "is almost a commonplace, and it is attested within the Mysteries as well as in the wider learned culture" (*Beck on Mithraism: Collected Works with New Essays* [Aldershot: Ashgate,

the bull's back and piercing the animal's heart with a sword, his face turned away in pain, and not without a certain hysteria.

This game has survived up until our times. I would like to recommend a book by an American writer in this regard. He is utterly taken with the idea of bullfighting, one could even say religiously captivated. The book bears the title *Death in the Afternoon*—the time of day, that is, at which the bull dies.[412] The author brilliantly reproduces the whole atmosphere of the bullfight, and it is very obvious what bullfighting means for the Spanish. The toreador is a moral example of composure. For the entire meaning culminates at the moment of the greatest danger, namely, when the toreador must wait for the moment when he kills the bull. This utmost tension is what affords the bullfight its whole meaning.[413]

2004], 289). Beck refers to Porphyry (*De antro nympharum* 24), Julian, (*Oration* 5.172), and Plutarch (*De facie* and *De genio Socratis*), to substantiate the significance of the sun and moon as "the places of the soul's origin and final destiny," and refers to Plutarch's conviction that "it is from the Sun that the human intellect (*nous*) descends and to which it will return and on the Moon that it is both united with soul (*psyche*) for its incarnation on earth and uncoupled from soul following material death" (ibid.). Evidence of the Mithraic cult is abundant for the Roman Empire: along with some 420 sites, "there are about 1,000 inscriptions, and 700 depictions of the bull-killing (only about half of them complete); and in addition 400 monuments with other subjects" (Manfred Clauss, *The Roman Cult of Mithras: The God and His Mysteries*, trans. by Richard Gordon [Edinburgh: Edinburgh University Press, 2000], xxi). Nevertheless, the connection between the Mithraic cult and the tauroctony has been disputed in recent research: see Attilio Mastrocinque's *Des mystères de Mithra aux mystères de Jésus* (Stuttgart: Franz Steiner, 2009), 4, and *The Mysteries of Mithras: A Different Account of Their Genesis* (Tübingen: Mohr Siebeck, 2017), which also discusses the affinities between Mithraism and Christianity and argues that the cult was only practiced by Gnostic Christians; Roger Beck, "Mithraism since Franz Cumont," *ANRW* 2.17.4 (Berlin: de Gruyter, 1984), 2002–115 at 2026; and Robert Duthoy, *The Taurobolium: Its Evolution and Terminology* (Leiden: Brill, 1969). Jung speaks of the subject in the Zarathustra Seminar, recalling the ritual of the initiate who was covered in bull's blood, and adding that a place where the bull was slaughtered was close to where St. Peter's in Rome was built (Jung, *Nietzsche's Zarathustra*, cit., 1:226).

[412] Ernest Hemingway's lyrical volume of reportage, published by Scribner's in New York in September 1932 and enriched by photographs, hymns the heroism and poetry of the bullfight. In this panegyric, bullfighter and bull exchange their roles as victim and executioner. Though not well received at first, today the book is seen as a classic on this typically Spanish custom (which in July 2010 was prohibited by the Catalan parliament). For further details, see Miriam B. Mandel, ed., *A Companion to Hemingway's "Death in the Afternoon"* (New York: Camden House, 2004). On the Spanish bullfight more generally, see Alexander Fiske-Harrison, *Into the Arena: The World of the Spanish Bullfight* (London: Profile, 2011).

[413] In *DAS* Jung recounted some "synchronistic" occurrences related to bullfighting that took place at that time (35–36; 43–44), such as receiving a letter from a friend then in Mexico describing her impressions of bullfights: "The one point of supreme art in the whole thing," he quoted, "is the moment when the bull stops still, confused, and faces the matador, and the matador standing in front of him makes the gesture of scorn to show his

I would like to mention this to begin with, because this is also a game where the audience does not sit in a quadrangle, but in a circle, and where everyone is spellbound by the events occurring at the center of the circus.[414] Therefore, it seems extremely paradoxical to us that the benches [in the dream] are not turned toward the center, but instead face outward. Quite certainly, this means that no communion occurs. At the same time, we feel that the meaning of the theater does not exhaust itself. For this lack of communion could be represented in an utterly different way. There must be something else to it. Has anything occurred to you about what this strange arrangement of benches could mean, what it indicates? We must know this, because this symbol of space recurs in a peculiar manner. We must therefore already prepare for it now.

ANSWER: Couldn't the fact that the benches face the window[s] indicate that the person concerned is not facing inward, but outward instead?

Yes, this arrangement evidently means this turning outward, a general outward movement. Thus, whoever enters this room and takes a seat on these benches does not see the essence of the games, but instead gazes out of the window, shall we say, or at the wall. He does not see the essence, because he turns away from it. Since this refers not to the dreamer alone, but also to a hypothetical entirety and to a large number of people, then perhaps we may safely assume that this statement holds true for society in general, or indeed in particular for the company that the dreamer keeps. So let's assume that we are the company among which the dreamer moves, and this moreover to be a statement about this group; what would that mean?

---

complete mastery" (44). Jung commented, "And it seemed to me that that was the meaning of the symbol: one must have perfect conscious control, perfect style and consummate grace and daring, to live in the bosom of barbarism: if one weakens anywhere one is done for. That is why the bull-fight was the symbol of the divine. And the toreador is the hero because he is the only shining light in that dark mass of passion and rage, that lack of control and discipline" (ibid.).

[414] Cf. DAS: "The center of a social group is always a religious symbol. With the primitive, it is the totem; later it is a sacrificial symbol, like the Mithraic killing of the bull; and in higher forms of religion it is a sacrament. The center of social activity under very primitive conditions is dancing or magic ceremonial in a circle in the midst of the huts. Probably those ancient stone circles still found in Cornwall were such community places. And it was understood that when people came together, the ancestral souls were there too, watching them; not only their conscious was in communion, but their ancestors, the collective unconscious. The ceremonial was a symbolic game." In the Mithraic cult, Jung continued, "the bulls had a belt around the chest in different colors, and the toreador had to jump on the bull's back and stab him from above in the shoulder—not with a long sword. Mithras was supposed to be such a toreador, like a Jesus of the boxing-ring or the football match" (24–25).

ANSWER: We are all extraverted.

No, I wouldn't say that, [but] only in relation to this center, we all avert our gaze from the actual games and turn our backs, or at least we turn from where the essence is happening—that is, what establishes relationship and communion.

Let's let the matter rest for the moment and just note down this thought. The dreamer remarks that he takes his seat with his brother-in-law. At that moment it occurs to him that, unusually, he is alone with his brother-in-law and that the latter is not accompanied by his wife. How does that question arise all of a sudden? Well, for the same reason that we have already mentioned. The communion, which he already believes he has when he goes out, is with his wife. Whenever his brother-in-law goes out, he too takes his wife. Thus, if a communion exists, his wife must necessarily be present. She is not present, and he attributes this to the child's sickness. We have already spoken about the child. In his eyes, we have explained this as an attempt to somehow orient his mind; this attempt, furthermore, is pathological. For the time being, we have contented ourselves with his opinion. Matters now become somewhat more complicated, as you will gather. Because we find ourselves in a situation where no communion can occur. The whole layout of this gathering makes it impossible to establish communion. Doing so would require turning around all the benches. Consequently, he yearns as it were for the presence of the female, in this case his younger sister, with whom he has at least some sense of communion. She is not there, however, because the child is sick. In other words, that's the case because he has made an inadequate attempt to focus his mind. Do you think that communion would have come about more readily if his brother-in-law's wife had been there?

ANSWER: No.

No, of course not. But that's what affords him the semblance of communion, is it not? Marriage, for instance, affords us the semblance of communion with humanity. For this reason, the commonplace novel always believes that life ceases when one reaches that peak. The whole entanglement comes to an end when the couple marries, after which they live happily ever after. For the whole world consists of marriages. There is nothing but married couples, and that is the structure of humanity. Beyond that, there are only events at which married couples gather, such as concerts, theater performances, church services, and everything is in order. "And sure enough, all was well."

Regrettably, it's not like that in reality. Instead, all these marriages are merely reasons for separations. Because they afford a semblance of communion. But marriage is nothing of the kind, only a relationship between human beings, which is by no means a relationship with humanity, because it's no more than a relationship with one person and moreover no binding relationship either. Because there is no unconditional obligation to go to the theater or to a concert or to church, notwithstanding a long tradition.

Now, that is what the dreamer feels here. He explains the absence of the comforting feeling of alleged communion by the fact that, coincidentally, his brother-in-law's wife has been prevented from attending. The child obviously has a deeper meaning. If we translate this into approximate terms, this would mean that this inept attempt to gain orientation even functions as an obstacle to any kind of communion. He cannot establish a relationship to humanity in this way. And why not? Let us assume, for instance, that we were members of Christian Science or were followers of anthroposophy. Would that be a wonderful communion? Here the whole difference becomes evident. It's quite out of the question that we adopt any kind of idea and tell each other that it would be nice if we all worked together, and then sat down in pairs and claimed to be forming a community. That's not a human relationship, but rather an organization via an idea.[415] It's something conceived by the mind. It is doubtful, moreover, whether the heart is participating. Without that, no human relationship occurs. But at least the dreamer hears that the child is already feeling better, and that it has only one or two degrees of fever on the thermometer. Why do you think the child is suddenly feeling better at this stage?

ANSWER: Because he is with the shadow.

Indeed. At least he now goes with the shadow, that is, *he looks himself in the eyes*. He sees his dark side.[416] Consequently, his brother-in-law's

---

[415] Between the lines, Jung aimed at distinguishing an aggregation centered on the merely ideological and mental aspect from a unification based on a deeper and more mature human relationship. This resembles the distinction made by the famous German sociologist Ferdinand Tönnies in his work *Gemeinschaft und Gesellschaft* (1887). Tönnies, whose professorship at Kiel was abrogated in 1934, believed that *Gemeinschaft* (community) is founded on feelings of affiliation and the spontaneous sharing in the *res publica*, whereas *Gesellschaft* (society), the predominant form of association favored in the industrial era, was based only on rational organization and commercial exchange.

[416] Ger.: "Er sieht sich selbst in die Augen." Cf. above: 156 and n. 390; [I], 21–22. "To confront a person with his shadow," Jung notes elsewhere, "is to show him his own light. Once one has experienced a few times what it is like to stand judgingly between the opposites, one begins to understand what is meant by the self. Anyone who perceives his shadow

child feels better. Let's not forget that our analysis[417] has shown this to be his anima, the image of his soul. And their child must be related to this somehow. If he places himself on a better footing with his shadow, then he also places himself on a better footing with the other, dark side, the inner one; consequently, the child feels better. He has already begun to gain a certain orientation, which appears to be improving compared to when he was studying the occult. At this moment the scene changes very swiftly. He is no longer in a public building, but finds himself in his brother-in-law's gray house, which is monotonous because its outside is painted gray from top to bottom. Where is the dreamer psychologically?

ANSWER: With the shadow?

Where does that live?

ANSWER: In the unconscious.

In the unconscious. The shadow is his unconscious. He does not love the shadow. The shadow world, you see, is his unconcious; it is Hades. Thus he's in Hades, where everything is gray. For everything is nocturnal and bleak. And there he suddenly finds the child. It, too, is there. That is, the mysterious child of his anima is also in the dark, shadow-like inner world. He sees, moreover, how bad the child looks. It has been sick to death, so to speak. Now it feels much better. What follows next is this strange scene in which he wants to pronounce his wife's name to the child, who refuses to utter her name. The child thus has a manifest resistance to his beloved wife and categorically refuses to utter her name. In his attempt to break the child's habit, he yawns the name instead of enunciating it fully. Evidently, this is the point of the dream, and it has particular importance for the dreamer. You might have noticed that the dream begins with the problem of the sick child and ends with the same problem. At this juncture another element enters in. This cannot be related simply to the fact that this child was merely an attempt at gaining a worldview, for there must be something else; something, that is, residing within him that has a positive resistance to his wife. And what could that be? Do you have any idea?

ANSWER: That there is no inner community between the married couple.

Yes, he must realize, of course, that not only does outer community not exist, but also that community with his wife does not exist as he would

---

and his light simultaneously sees himself from two sides and thus gets in the middle" (Jung, "Good and Evil in Analytical Psychology" [1959], CW 10, § 872).

[417] *Analyse* in the protocol. In Jung's personal copy he replaced it with *Deutung* (interpretation).

wish and hope. Therefore, he proves this to himself by yawning when he utters his wife's name. Naturally, this can only mean, already when he speaks her name, what for him is a most unpleasant realization, that he is bored to death. What, then, would the child be?

ANSWER: A new life within him.

Probably, indeed. Here, ladies and gentlemen, I must admit that I do not know what the child is. We must accept certain things simply as they are. If we start guessing and interpreting, then we shall get into various vague generalities that actually serve us very little. We'll find ourselves entangled in verbiage. In such instances, I prefer a more primitive view of matters. You have already observed how primitive and simpleminded my thinking is in relation to dreams: for instance, my observation about his sister. I asserted quite plainly, "This is his anima." It's as if it were something like a person. Now, that sounds exactly as if a native chief said to me, "Last night, my friend, who lives on the other side of the mountain, appeared and spoke to me." He reports this event as if it had actually occurred. From time to time, one experiences the greatest difficulty in establishing whether these primitives have dreamed something or whether it is in fact true. Because their dreams are realities. They are strange realities. For instance, I remember the dream of a chief. He was an old man, who was already slightly too elderly to follow his herds, but who lived with them in a *participation mystique*—like a Swiss, so to speak. Somehow he possessed unconscious knowledge of his herd. He once told me that he had had a very beautiful, big dream, namely, that he had seen a black cow with a white patch on the right side of its chest down by the river, about an hour away from his kraal—obviously, he did not use the word "hour," but expressed himself differently—and there she stood with a calf, her calf. She must have given birth. He had sent down two men to the river to look for this cow. They told him that they had found the cow down by the river, and that she had borne a calf during the night. He had had no idea that this cow was pregnant. So he had dreamed it.

This case shows how a proper primitive herdsman lives with his herd and maintains contact with it so that he knows everything about it unconsciously. These people can count a herd of seven hundred animals although they cannot count with more than two hands, five on one hand and five on the other. Two hands are ten animals. The medicine man can count to seventy, and seventy is a hundred. A hundred is the greatest number of all. The medicine man is therefore respected by the whole tribe. Nevertheless, each herdsman counts his own herd! He knows when all his cows have returned, although he cannot count. When they have all returned, he knows: "I have not yet seen this one and that one." They call this counting. Naturally, this is possible only through an intense psychic coexistence with these animals. Otherwise, this would be impossible. These primitives have an absolutely concrete notion of dream figures. They assume without further ado that a dreamed figure is real—that is, a ghost or an image of that ghost, so to speak. They might go so far as to say that this one or that one "has shown me his image, and it appeared to me when I was asleep."

That is how dreamers think. They do not see any kind of abstract sentence, but instead they perceive a sentence to be saying this and that. The sentence itself is personified in the corresponding figure. These figures behave as if they were real beings. This is a fact. In dreams, the human figures move as if they were realities, so that we fear certain living figures. In the dream, we are where the primitive stands in his waking consciousness. In striving for a proper notion of dreams, we should entertain no scientific conception, but approach them through life. We cannot have an intellectual notion of dreams without missing the essence. We must create a living figure out of the dreams. We must behold dreams. As soon as we attempt to intellectualize dreams, their essence is lost. Consequently, my proposal is to abstract such figures only gently and with caution. For instance, we want to concede to our enormous development of consciousness that we do not claim straightaway that his sister appeared to him in the dream, but rather his anima. But this is the utmost that we can allow ourselves, and thereby we must say that this anima person in the dream is in fact a kind of living being, which might rise up from him, as it were, while he is asleep, that can behave quite autonomously and most peculiarly, thus rising out of him all of a sudden and entering a certain woman, causing her to be possessed and making him cling to this woman, as if he were possessed by her, or as if she were a werewolf or a witch. Naturally, this is most unscientific, as you will no doubt acknowledge. But this is a living view of events. This is how human beings have seen and understood such things from time immemorial. This corresponds to the real facts, while scientific abstraction simply does no justice to these facts.

* * *

LADIES AND GENTLEMEN,

The question concerning the evaluation of the persons in the dream is a fundamental problem, whose intricacy and complexity I cannot dwell on at length, unfortunately. In this respect, all I can tell you is that I am mindful not to violate nature in any way. I'm trying not to bring to the dream any viewpoint that is not itself given by the dreamer or his material. I'm trying, so to speak, to find, to think about, or to amplify material,[418] so that we can recognize its possible meaning. What might strike you as a very peculiar attitude is simply a phenomenological stance. I take these manifestations at face value. I neither imagine nor pretend to know anything beyond them, nor indeed dare I bring any external viewpoint to this material. Instead, I take the position that we must listen to this material in our profound ignorance. I am constantly aware that this dream is a message that should be heard, a message we must not approach from outside while bringing to bear the ridiculous curiosity of a so-called scientific viewpoint. For the dream is a message that doesn't want to be dissected, but instead wants to be *read*. The dream somehow wants to enter our soul, and it does. If we want to resist the dream, then naturally we had better rip this message into the tiniest shreds, or if we leave the meaning to one side and stick to the words, then we'll have to pick each word to pieces. In doing so, however, we'll surely not hear the message's meaning, for I know from experience that such misreadings of images fail the letter writer's purpose, and so we'd do much better to read a letter as it's meant, namely, as a message from one person to another; as a rule, moreover, we rightly assume that the person writing to us actually means what they convey to us.

This is how I understand the dream, and how I understand the figures appearing in it. I bestow on them the dignity of natural manifestations. Thus I confer on the appearance of this female figure, as just mentioned, the dignity of a natural being occurring in the dream. In the dream I'm not in the conscious sphere. Naturally, this figure does not appear in this form in consciousness. But *in the dream, I am as it were in another consciousness.* Therefore I'm in another world. For this [other] world is as

---

[418] This is the only (indirect) mention in the present seminar of the "amplification" which was crucial to Jung's practice of psychotherapy. "The *amplificatio*," he would explain for instance in *Psychology and Alchemy* [1944], "is always appropriate when dealing with some dark experience which is so vaguely adumbrated that it must be enlarged and expanded by being set in a psychological context in order to be understood at all. That is why, in analytical psychology, we resort to amplification in the interpretation of dreams, for a dream is too slender a hint to be understood until it is enriched by the stuff of association and analogy and thus amplified to the point of intelligibility" (CW 12, § 403).

my [other] consciousness sees it. In that world, too, evidently there are living figures: for instance, this female being. And there, too, female beings sometimes have children. Therefore, I note: in this dream, it's no great novelty that this alleged sister has a child. Why shouldn't she have one? She's married to my alter ego, after all. For instance, here she's the wife of the dreamer's brother-in-law. If she is married to this shadow, then it's possible that she also conceives from a shadow and has a child. I'm well aware that this is by no means an explanation, but it's only accepting as true the message in this letter: namely, the astonishing news that the anima has given birth to a child. And we had no knowledge of the fact that she has a girl. We believed she had given birth to a boy. In the dream, it's a girl. And this girl is sick.

We must accept this message as it is conveyed.

I remember a case that I treated many years ago. The person concerned was highly intelligent. He never wanted to accept his dreams as they were, and the dreams also came toward him in an affectionate manner. In the first few months of our acquaintance, he constantly dreamed of things that could be broken down into their basic components. Thus, if he dreamed of a table, it was a matter of verifying that it was his father's desk. That was pleasant. There was nothing uncanny. In the case of a particular book, he knew exactly which place and which library it was from. That was most satisfying. All the persons appearing in his dreams were also quite condensed figures.[419] Almost unmistakably, he discovered a little corner, somewhere, from which these figures could be drawn into reality. That, too, was most satisfying. And then something incredible occurred: that is, that a female figure, who came from nowhere he could think of, appeared in one of his dreams. I told him, "Now you must be faithful to your principle and inquire whether or not such a figure exists in reality." But he could not find anything, no matter how hard he tried. "This might be an accident," I said. Yes, indeed, that could be an accident. One cannot know that it's not an accident. We waited for a new dream to occur, and then this unknown female figure displayed such tenacity that she reappeared in a series of consecutive dreams. That drove him to despair, literally, and so he did his utmost to find the reality to which this figure belonged. But he couldn't work it out.

So I said, let's do something else. Let's assume, merely by way of a suggestion, that we're living in the Middle Ages. Let's assume furthermore that I am some obscure magician who casts horoscopes, and that you seek

---

[419] In his personal copy, Jung here replaced the expression "quite condensed" (*kondensierte*) with "concrete" (*konkrete*).

my counsel about these strange dreams. I would tell you, my dear man, that you have a succubus. Are you familiar with the term? The succubus is a medieval female figure, a woman of ill repute, who belongs to the group of silvans and female fauns and likes to live in forests where she seduces innocent youths and also visits them in their dreams at night. She is the counterpart to the so-called incubus, who played the same trick on women.[420] In the Middle Ages, as is well known, many witches and magicians were accused of having intercourse with succubi and incubi, and they made corresponding confessions.[421] The incubus and succubus originated in Antiquity and belong, as I mentioned, to the species of silvans. Originally, this figure was a forest sprite. In earlier times, she was a well

---

[420] The incubus (from Lat. *incubare*, "to lie above") names a malevolent entity, traditionally considered a son of Adam and Lilith, who molests sleepers, and often appears as a satyr. The succubus (from Lat. *succubare* [*sub-cubare*], "to lie below") is the female counterpart, which in the European medieval tradition comes to men (above all monks) in their sleep and leads them into sexual encounters. As Jung notes, such images are common in medieval demonology and survive in folklore (including both the Hebraic and Arabic traditions). The theme of diabolical and demonic copulation has long preoccupied theology. See in particular Augustine, *The City of God* 15.8 (PL 42: 876); *De Trinitate* 3. 23 (PL 41: 468); and Thomas Aquinas, *Summa Theologiae*, Ia, q. 51, a. 3, 43). Both "incubus" and "succubus" were terms widely applied by the Church Fathers to elementals, while Paracelsus understood them as parasitical creatures or *larvae*, connected with perverted human emotions, low appetites, or wretched thoughts lingering in the astral body. Aquinas and scholastic demonology tended to distance such concrete, sensible, personal entities from a much more abstract notion of the devil (Hans Peter Broedel, *The "Malleus Malleficarum" and the Construction of Witchcraft: Theology and Popular Belief* [Manchester: Manchester University Press, 2003], 44). By the late Middle Ages, Inquisition and clerical authorities had intensified their ban on magical practices and witchcraft. Pope Innocent VIII's bull *Summis desiderantes affectibus* (Desiring with supreme ardor), issued in 1484, condemned witchcraft as a form of apostasy and incited the Dominicans inquisitors Kramer and Sprenger to heal "the plague of heresy" in Germany by eradicating witchcraft and Satanism with the support of the—generally less enthused—local authorities; the papal bull would preface the *Malleus maleficarum* (see next note).

[421] According to the authoritative, (in)famous *Malleus maleficarum* (The hammer of witches; ca. 1486), such possession can be annulled through confession, making the sign of the cross, and exorcism. Bewitchment was deemed to happen in three ways: "First, as in the case of witches themselves, when women voluntarily prostitute themselves to Incubus devils. Secondly, when men have connection with Succubus devils; yet it does not appear that men thus devilishly fornicate with the same full degree of culpability; for men, being by nature intellectually stronger than women, are more apt to abhor such practices. Thirdly, it may happen that men or women are by witchcraft entangled with Incubi or Succubi against their will" (Montague Summers, trans. and ed., *The Malleus Maleficarum of Heinrich Kramer and James Sprenger* [Mineola, NY: Dover, 1971 (orig. London: Rodker, 1928)], Part 2, Q. 2, ch. 1, 164). Though anyone could fall prey to incubi and succubi, the authors of the *Malleus maleficarum* maintained that "witchcraft depended upon this intimate bond between woman and demon, close even to the point of identity" (Broedel, *The "Malleus Malleficarum,"* cit., 53).

defined, as it were tangible entity. Until recently, that is, approximately until the day before yesterday. Nowadays we allow these figures to go up in intellectual smoke, while our grandmothers and grandfathers were still haunted by them. In former times, one admitted, quite naively and reverently, that the whole world of the dream was a reality. Thus, I fail to understand why we don't do the same today.[422] We inhabit quite a different reality, after all, with quite different laws. Our consciousness occupies a completely different state. Thus, I consider it not only modest, but also appropriate for us to quite naively presume that these figures are capable of existing on their own. By no means are we required to relate this to our three-dimensional existence. It's an incommensurable quantity. While we happen to live in a three-dimensional world, there is nothing we can say about the size, weight, or spatial dimensions of such matters. Stupidly, however, they exist. It would be nice if we could measure the psychological with a yardstick—for instance, the length of a thought. Unfortunately, one can't do that. And yet one can't really deny that something like mental or conceptual realities exists. These are things that are. We cannot know how they exist in and for themselves. Therefore we must cautiously adopt a phenomenological stance, and say no more than that something happened in such and such a way. We know that once we start fingering these matters with our intellect, they will break. Somehow we bring them to light, and subsequently they are no longer the same things. We can observe them only in the twilight. So we had better leave them in the twilight, until we have really gained sufficient experience of the peculiar life of these things in our soul.

Thus I'd like to propose that we consider this simply as a piece of news conveyed to us: namely, that the soul has a child, for the soul is his female nature, and that consequently it can readily have a child, a little girl. We can even assume that this little girl has roughly the same nature as the anima figure. And if "la recherche de la paternité"[423] is not prohibited, then we can perhaps assume that his shadow, his alter ego, has probably fathered this child.

Now it's most unsatisfying to leave a matter in limbo in this way, to strike sail in the face of things that we feel we ought to hold in our hands. I shall quite readily admit that I share your dissatisfaction in this respect.

---

[422] To this sentence Jung added *können* (be able to) in his personal copy ("und ich sehe nicht ein, warum wir das nicht auch tun können"), such that the translated sentence would read, "Thus, I fail to understand why we could not do the same today."

[423] Fr.: "paternity suit."

Practical experience of grappling with such matters has taught me patience so thoroughly that I can leave these things be if I don't understand them, and simply wait until they resolve themselves. I have no knowledge of a strange adventure that is being played out in the background. If it's taking place, then this comedy will continue. We shall have further opportunity to observe events, and to gather what transformations occur, as well as what deeds and misdeeds are committed. Only the entire series of events will then perhaps illuminate the meaning of this strange piece of news that the soul has given birth to a child that is ill. We must leave that be for the moment. But we must bear in mind that the last sentence of the dream rubs the dreamer's nose in the fact that he is obviously utterly bored with his wife. He must somehow take this to heart. That seems to be the purpose or the lysis of the drama in this dream. We can assume, quite safely, that his tendency toward correctness has made it impossible for him to realize that he has this feeling, and that it's of the utmost importance that he make this clear to himself and acknowledge it. Now you might consider this to be a most peculiar fact given our assumption that the Old Man is writing him a letter. The Old Man seems to be walking him down a funny path. Can you recognize anything significant in this curious message? What sort of value should attach to the fact that he must realize that he is bored?

ANSWER: It constitutes progress from the subjective to a more objective evaluation of himself.

Where does his subjective self-evaluation lie?

ANSWER: In the one-sided conscious evaluation.

And what would constitute a more objective evaluation?

ANSWER: The end of the dream more or less enables him to laugh about or classify his weakness for the first time.

One could say that is definitely a correct thought. That's an objective statement. It is indisputable that he has grown bored, and he must admit this to himself in spite of his protest. We recognize a deeper meaning in this fact. It makes no difference whatsoever whether this man admits that he is bored with his better half or not. Surely, that's of no interest to us, or what would you say?

ANSWER: But this admission connects him with generality.

Absolutely! This recognition, you see, connects him with humanity. Your diaphragm stirred immediately when you heard this liberating truth. So he is connected with the community. You have demonstrated it in the loveliest way. He is accepted. Every man understands him. Besides he feels somewhat embarrassed. It's most unpleasant to have to admit as much. From this he learns that, notwithstanding all his correctness, he is not as wonderful as he believed. It teaches him, moreover, that his whole correctness is not worth a straw, and that acknowledging this simple fact in itself makes him human. Where does the Old Man seem to want to lead him?

ANSWER: Into the community.

Good heavens, that's far too sweet.

ANSWER: From correctness into nature.

Yes, as if the Old Man had said: Just admit it, just this once. You are simply bored, and the first thing you must do is admit this. Be simple and recognize what lies nearest. Then the child of your soul will no longer be ill. It will already feel better once you have noticed this. Then it will have only one or two degrees of fever, and the reconnection with your soul will even assist the child's convalescence. This is a mysterious afterthought. No doubt it's very hard for us to understand how this commitment to general humanity should bring about the recovery of his soul's child. Can you see that? This is very hard to understand.

ANSWER: The condition of isolation has made him anxious.

First, through this commitment to humanity his isolation in society ends. He is accepted. Obviously, something pathological can fall away from his nature: for instance, certain feelings of inferiority, or his sense of self-consciousness.[424] Self-consciousness can fall away from us if we manage to let something foolish to happen, such as spilling a glass of wine on a tablecloth. Then one can hold one's ground.[425] Something foolish must happen first for one to lose this self-consciousness—for which there is no German word. Incidentally, it's typical that precisely the English have a fitting expression for it. But how do you imagine the connection between his recognized humanity and the sickness and recovery of this child? Have you any ideas?

ANSWER: The child represents his worldview, which is already led thereby onto a healthier path.

You realize, don't you, that we can also see this child in connection with his effort at reconceiving his outlook? If the Old Man is actually inclined toward bringing him to recognize his all-too-humanness, then evidently the tendency also exists somehow to reform his worldview. For if he

---

[424] "self-consciousness": as previously, both here and below, in English in the protocol.
[425] In Jung's personal copy, this last sentence has been deleted.

himself becomes more natural, then he will, of course, be much more likely to attain a better notion and view of the world and everything in it. It almost seems as if the Old Man renders homage to a Rousseau-like philosophy of nature by observing, "Go back to nature and you will get well."[426]

Thus, these are approximately the expectations that we can deduce from this dream. If we don't arrive at a conclusive interpretation of a dream, but place a question mark after it, that would constitute an appropriate response, since then we are taking the dream as the beginning of a correspondence with the Old Man. We place a question mark after his messages, and once again direct his question toward the unconscious: for instance, "What do you mean that I must understand this, and that then this child would have recovered?" How do you think the unconscious will respond to that question? It will produce another dream, because it has been asked a question. For this reason, it often happens that if you are pondering a problem to which you find no answer—and you actually turn it into a matter of conscience—in fact you'll have a dream. The unconcious responds to that. An answer comes in the dream; it's never the final answer, however, but merely the beginning of one, which again leads to another question, which in turn receives an answer. If it ever came to an end, and sometimes it does, then life, too, has reached its end. The last dream contains the last answer. And thereafter one is dead.

Given all this knowledge, what do you think might occur in the next dream? How would you expect matters to continue? What would you most like to hear? Of course, once again you'll be thinking that this is one more strange question, right? But in my analytical seminars I tend to ask my audience these kinds of questions. In discussing a series of dreams and visions, I ask, "What will happen in the next vision?" or "What fantasy will come next?" And it's astonishing how often I hit the bullseye. Very often people to whom one wouldn't give much credit have excellent intuitions. They can follow the strange melody of the unconscious and all of sudden can imagine the next structure to emerge from the night.

ANSWER: I'd like to have an accurate diagnosis of the child's sickness.

---

[426] The return to nature is thematic in the works of the Swiss-French philosopher, musician, and writer Jean-Jeacques Rousseau (1712–1778). Radical critic of the customs of his time, and nostalgic for a model of civil cohabitation inspired by genuine feeling, Rousseau adopted a new perspective on the myth of the so-called *bon sauvage* (noble savage), which had developed in French writing since New World exploration, idealizing primitives and their "state of nature." See Jung's reflections in "Yoga and the West" [1936], CW 11, § 868 (recalled also above [I], 45–46 and n. 146).

You are interested chiefly in the child's sickness. Indeed that's a very good idea. Does anyone have a better question?

ANSWER: I'd be satisfied if the child got well again.

You'd like to hear that the child makes a recovery? Well, that's asking too much.

ANSWER: I'd like to meet the wife.

But that's an illegitimate wish.

ANSWER: I would like some kind of community to occur.

What kind?

ANSWER: Either general or personal, or indeed any kind.

Could someone perhaps take notes or remember these points? Two people are interested in the child's health, or rather its sickness or indeed fate; someone mentioned community, and a further point was made, too.

ANSWER: I'd like to see whether he continues to follow the shadow so willingly or opposes the shadow instead.

I can already divulge that now, since the question has been answered. He has decided to go with the shadow. I would think it unlikely for the shadow to reappear right away; what would be more likely?

ANSWER: Whether the people sitting on the benches turn around, and what is happening in the room.

Indeed, that also interests me a great deal. In fact, it interests me most. But I know from many years of experience that if I have a particular wish, then it won't come true. Things never turn out the way you expect. I entirely agree, and would also very much like to see these benches turned around so that the proceedings can be observed. Although that's our innermost wish, it will not come true.

ANSWER: I'd like to see the shadow leave the gray house.

You'd like to see the shadow go out of the gray house. Well, I'm not sure that would be desirable. It seems that most of us are interested in the problem of the child, and someone has expressed interest in the wife: specifically, that she appear in the Old Man's next message. Let me report the contents of the next letter:

My wife asks me to visit a seamstress with her; the seamstress is suffering from tuberculosis and must work in an unhealthy, squalid hole. I visit the girl and tell her to work outdoors, in front of the house. She replies that she has not been granted permission to do so.[427] *I suggest that she come and work in our garden. She replies that she would not have a sewing machine there. I offer her ours.*[428]

So, here we see the wife appear.[429] She makes an active appearance in the dream. Moreover, we have a diagnosis of her [the child's] sickness. It is tuberculosis, tuberculosis of the lung. The child is evidently not cured yet, but the question remains how it could be cured. But it's not a child that comes up; rather, a strange event, which is quite characteristic of dreams as such, has occurred instead: the figure of the anima [in the seamstress] has coalesced with the child. They have become one. The mother has absorbed the child's sickness, but the child has also rejuvenated her [the anima] as it were. It's no longer a mother, but a very young girl, who is suffering from tuberculosis. Evidently, that's the illness in question. So the situation is quite askew. I fear that we won't have any more time to discuss this dream in its entirety. Nevertheless, let me give you the dreamer's material. These are his comments regarding the dream:

> *My recollection of this dream is very unclear. I seem to have forgotten important parts of it. The whole dream left behind an erotic impression, although the wording of its rendition by no means conveys this. I seem to recall thinking immediately that an interesting acquaintance could arise from this visit. However, this idea fails to recur in what follows. My wife thereby plays a completely passive role. I have the feeling that she is present, but I take action alone, without consulting her. I can't remember what the seamstress looked like. She was wearing dark clothes. Tubercular persons are said to be highly erotic. She was desperately impoverished, and I felt that there was something I could do for her.*[430]

He now starts thinking himself. He asks himself something about this dream, for he says,

> *Is the girl wearing dark clothes supposed to represent my own erotic feelings, which are sick, because they are at work, furtively, in a dark*

---

[427] This reply is missing in the corresponding exposition of the same dream in *DAS*, 87.

[428] Cf. *DAS*: "I tell her that she can have my wife's machine" (87).

[429] The pronoun references in this paragraph appear to be unreliable with respect to identities of the persons involved: the wife, seamstress, anima, and the ill child. That is, the transcript appears to confuse the wife with the seamstress, and then the mother with the dreamer's anima. The bracketed references resolve the matter in the light of Jung's remaining remarks (in the next seminar session). [T.N.]

[430] In *DAS* an erotic component in the expectation of the dreamer toward the seamstress emerged more strongly (87).

hole? I summon them to move freely outdoors. But that is not al-
lowed. Thus, feelings must set to work in our own garden, with our
sewing machine. When I offered the sewing machine, I remember
feeling that it actually belonged to my wife, and that she should have
a say in the matter first of all.

From the dreamer's speculations, you will glean his understanding of
the dream. He tries to shift the entire matter onto an erotic adventure,
naturally to rationalize it as far as possible. For one tries to draw these
messages into daylight as far as possible. That's human. The totality of
consciousness must always be preserved as far as possible. It's characteristic of this dreamer that he soon began speculating about his dreams. As
you know, that's thought to be illegitimate. Dreamers shouldn't speculate
about their dreams, but gather the pertinent contexts. I always let people
do what they enjoy doing most. Then I can be sure that they'll go along
quite naturally. For it's entirely natural that I begin speculating about
something that eludes my grasp. Someone else might not do that. Instead,
he'd consider the matter from all sides, but how could I then inquire about
his thoughts? I would observe the dreamer's temperament, and then say,
well, it could be this or that.

Our dreamer's temperament leads him immediately to engage in speculation. If one tried to forbid this, a technical error would have occurred.
One would have interrupted his activity. Immediately, he began consulting books and asked whether it was correct for him to read books during
his analytical work. I told him, "Read as much as possible, as much as
you can lay your hands on. That will help us. It invigorates the mental
process. You'll be fully involved if you contribute to the thinking and speculation. You'll afford yourself the best possible opportunity of understanding by yourself what you must learn to understand."

I utterly object to withholding knowledge from such people. By all
means they should be given the instruments. Because, after all, they could
fetch the instruments themselves. One cannot deny anyone the right to
acquire that knowledge. Nonetheless, his relationship with his dreams did
change thereafter. I had trouble making him stay in context. Yet these notions make an extremely valuable contribution to the contents of the
dream. One can observe how a certain tendency is constantly at work in
his speculations, leading him away from the actual meaning of the dreams
into their own meaning, as it were replacing the objective meaning of the
unconscious with their own meaning. That's the characteristic hubris to
which we are all somewhat subject: namely, of replacing objective psychic consciousness with our own self-consciousness.

Lecture 3

Ladies and gentlemen,

Yesterday, you recall, we were considering the seamstress when we adjourned our discussion of the dream. It sounds exceptionally simple and suggests, as it were, that we try to see the dream in terms of a wish being fulfilled. Moreover, the seamstress, and the erotic background evoked by her, appear after the first dream has faded. Thus here a wish comes true that was already latent in the first dream, as indicated by the boredom.

Obviously, one *can* bring to bear such a point of view. But the question is whether this corresponds to the dream or not. If it is a message, then we must accept it as it is, and observe no more than what it states. So let us consider it in these terms.

It begins with the following remark: "My wife asks me to visit a seamstress with her; the seamstress is suffering from tuberculosis and must work in an unhealthy, squalid hole."

If this dream were nothing other than the fulfillment of an erotic wish, then this request would be an evident untruth. It's precisely his wife who would not ask him to come along for that purpose. If the dream says that his *wife* asks him to join her, then we must assume that this is indeed what is meant. It's possible, after all, that he should actually attend to this seamstress. Under no circumstances should we impute that this is not what the dream means, and that it is cheating. We must assume, instead, that something quite natural occurs, and that precisely *that* is happening which the dream says is happening. If a doctor examines a person's urine, and notices that this urine contains protein, then we must accept that the urine contains protein, and not sugar.

We must confine ourselves to the fact that the dream means what it says. In this case, the dream says, "My wife asks me to visit a seamstress with her." We can only understand this correctly if we know who his wife is, and what it is that she asks him to do. We can be sure that she doesn't ask him to have an erotic adventure! Although the idea of such an adventure would involve an understandable relationship with the seamstress, we must confine ourselves to the startling fact that his wife issues this request; and accept this as the initial condition.

What is the meaning of his wife in the dream? There are two possibilities: either she is an imago or indeed she is his real wife. The function of imagos is to embody certain attributes.[431] The rule is that if we dream of unknown persons, then we can quite safely assume that these are imagos. These figures do not correspond to actual living persons. If, however, we dream of people who are closely related to us, then we must acknowledge that *the real persons* are in effect involved. Given that his wife asks him to join her, we must confine ourselves to the fact that such a wish emanates from her—as an objective statement. In this case, it would be an unnecessary overstatement to assume that it's merely a question of an imago. It is good, rather, that his wife wishes that he busy himself with the seamstress.

What does the seamstress mean in the dream? There is no human or police evidence of this figure. She is a complete and utter invention of the unconscious, based on no corresponding facts in reality. Consequently, this figure is created artificially by the unconscious, an imago, a creation, which has come about for a very specific purpose. From which material has this figure been made? The integrating components are her illness, her unfortunate situation, and impoverishment, a conglomerate that calls forth certain reactions. This dreamer has a particular inclination toward protecting and pitying others; he exhibits a certain willingness to assist, support, and care for others—thus, his feeling side is called upon. In fact, this figure has been created by the unconscious to evoke his feeling, indeed it even represents his feeling. In his case, as a thinker and intellectual person, feeling occupies the background, and it is also inferior. "Inferior" is not meant here as a moral judgment. What it means here, rather, is that feeling is less differentiated and less subject to the arbitrary control exercised by consciousness.

In this respect, I have often been asked how it is that intellectuals are not "cold" matter-of-fact persons, but have exceptionally strong affects.

---

[431] *Imago* (Lat.): "image"; the term is normally applied by Jung to the subjective perception projected onto objective reality as a characteristic expression of unconscious contents. "The object-imago itself is a psychological entity that is distinct from the actual perception of the object; it is an image existing independently of, and yet based on, all perception, and the relative autonomy of this image remains unconscious so long as it coincides with the actual behavior of the object [. . .]. This naturally endows the object with a compelling reality in relation to the subject and gives it exaggerated value" ("General Aspects of Dream Psychology" [1916/1928], CW 8, § 521).

Precisely that, however, implies the inferiority of feeling, and it's a sign of its particular primitiveness. Intellectuals are subject to all sorts of moods. Their will is unable to prevail against this. There is no discipline or submission to higher considerations, but instead sheer occurrence. Feeling behaves impulsively and tends to engross one. Consciousness cannot resist this. Sometimes the function is present, sometimes not; nor does it come when summoned; nor does it go away when it should. It is a spontaneous and unrestrained phenomenon, which requires the most unreasonable effort to be governed. It manifests itself like a pathological symptom, such as coercion, fear, and so forth. It appears as an unruly power, which can make an ill-fated intervention in our life, launching it to suit its own designs.

The sick seamstress represents the inferiority of his feeling. It is now for the dreamer to consider what he should do with this feeling.

He begins: she should work in his garden, outdoors, in the fresh air, not in the dark. The unrestrained function is drawn out into the daylight. He should use his feeling in an appropriate place. The seamstress replies that she hasn't been granted permission to work outdoors. In effect, she wasn't allowed to exist. One forbids oneself from having uncontrolled affects, which, however, doesn't prevent them from being present. If we can avoid having them, then we suppress them; we suppress them, moreover, because they are socially harmful. For this reason, we witness the attempt to control inferior feeling already at the earliest stages of human history. From time immemorial human beings have lived in profound fear of this inferiority. If you are interested in the rules adopted by the primitives in human intercourse, then you will discover that precisely the primitives are often exceptionally courteous, because they are terribly frightened of possible affects. That's perfectly understandable given how unrestrained these affects are when they break out and with what catastrophic consequences. I recall a bushman who showered his five-year-old boy with customary parental affection. On his way home, he was greeted by his boy running towards him. Instead of being pleased, he vented his irritation at the boy, and strangled him. Naturally, he despaired of what he had done afterwards. If a mood overcomes the primitive, then matters may become dangerous.

I experienced the following case myself. One night we were staying with an African tribe that we knew nothing about. The chief, a young man, asked us whether they should dance a "N'goma"[432] for us in the evening.

---

[432] Bantu term (also *engoma*, *ng'oma*, or *ingoma*) for "dance," and connected meanings such as "music," "drum" (in most of Bantu-speaking Africa its root *goma* refers to drum)

We accepted the offer, and soon sixty to seventy people gathered—men, women, and children, armed with spears, bows, and arrows. Although it was scorchingly hot, an enormous fire was lit, and we almost evaporated. Then they danced until midnight, with ever-growing excitement. It began to feel uncanny. I recalled a story in which the dancers had become so agitated that they eventually hurled their spears at each other. I saw this event swiftly approaching and told myself that something would have to be done. I took out some matches and cigarettes, and said, "You may have these if this is the last dance and you go back home afterwards." That was easier said than done, however, because they continued dancing and became increasingly excited. So I took hold of my rhinoceros whip and swung it over their heads, obviously without striking them because that would have been terribly forceful. They scattered, clamoring wildly, and finally disappeared into the night—without resenting me, by the way.[433]

---

and related instruments. Along with dance, drumming plays a significant role in healing rituals for it is considered to be "the voice or influence of the ancestral shades or other spirits that visit the sufferer and offer the treatment" (John M. Janzen, *Ngoma: Discourses of Healing in Central and Southern Africa* [Berkeley, CA: University of California Press, 1992], 1). Janzen's book is presented as "the first comprehensive study of the discourse on misfortune and healing in Central and Southern Africa in connection with the institution of Ngoma" (ibid., xii). Ngoma dances as widespread ritual institutions are attested not only in Central Africa, but by and large across the whole continent, although they are not always associated with rites of healing (ibid., 4).

[433] For further details of this episode, which occurred in Sudan during Jung's North African expedition in 1925–26, see Hannah, *Jung: His Life and Work*, cit., 176–78, and *MDR*, 270–72. Initially Jung, along with his companion Peter Baynes, gladly accepted the tribe's proposal to have a big dance, though mindful of his conviction that "the natives easily fall into a virtual state of possession" (*MDR*, 271); but when the situation seemed to be getting out of hand, Jung exaggerated his performance by cracking the whip and urging dancers in a semi-jocular manner—in Swiss-German ("for lack of any better language")—to stop, which secured the desired effect. Burleson notes that such dances were "highly regulated or forbidden by colonial governements in Swahili-speaking East Africa," and that by joining (and even encouraging) the dance Jung had "naively committed a political blunder" with respect to the local authority, considering his "semi-official status with his *askari* escort from the *ma'mur* of Nimule" (*Jung in Africa*, cit., 195). This episode (which, incidentally, inspired Peter Gabriel's song "The Rhythm of the Heat" [1982], which was provisionally entitled "Jung in Africa") has been variously commented upon also in light of Jung's supposedly racist, Eurocentric, colonialistic tenets, and even his alleged fear of encountering his own shadow side (i.e., that the primitiveness could "invade and overwhelm the consciousness of the European": Jo Collins, "The Ethnic Shadow: Jung, Dreams, and the Colonial Other," *The Birmingham Journal of Literature and Language* 1, no. 2 [2008]: 22–30 at 25). See Roger Brooke, "A Critical Discussion of Jung's Experience in Africa: The place of Psychological Life," in *Jung and the "Other": Encounters in Depth Psychology* (London: Routledge, 2018), ch. 4. On Jung's fear of "going black under the skin" ("a spiritual peril which threatens the uprooted European in Africa to an extent not fully appreciated": *MDR*, 245 and passim), see Burleson, *Jung in Africa*, cit., 197–98, 262–64. See also

Another white man, a district commissioner, who traveled to the village a few days later, was not as fortunate. He was impaled, literally,[434] because he failed to strike the right note with them. *I found them to be the most charming people I had ever met. One must simply know how to deal with them properly.*

This is one of these unrestrained functions. It instills greater fear in primitives than do wild animals. Those one can elude by leaping to one side or with the help of magic, but one cannot evade affects. They make everyone scatter within seconds. Therefore, primitives are very strict. At a palaver, no one is allowed to raise their voice, for if two people raise their voices, an altercation can occur and they can become emotionally possessed right off; it can happen, then, that a whole tribe destroys itself. Whoever raises his voice is given the whip—hence the pronounced courtesy among primitives.

We imagine that we are superior to such states. But if we observe ourselves closely, we realize that we too are prone to such incidents. As we know, it is not uncommon among us for pleasant circumstances to become transformed into poisonous dragon lairs by unrestrained affects.

Our dreamer also experiences such difficulties, namely, a rather sickly-looking feeling that is condemned to working in a squalid "hole," and which has no chance of developing. He must bring it into the daylight so that it may recover. The seamstress must work in his garden.

What does the garden represent? What is the garden in contrast to his sphere of activity? It represents a congenial place which affords pleasure and a mellow feeling. The flowers in the garden represent feelings. Thus, the flowers in the garden denote the emotional side of his situation. In the garden, he may indulge in his feelings and fantasies. Here, there are various kinds of flowers and much other beauty—that is the feeling part. And that is what he has at home. He is married, after all. Almost everyone

---

Jung's "The Complications of American Psychology" [1930], *CW* 10, and the classic Laurens van der Post, *Jung and the Story of Our Time* (New York: Pantheon Books, 1975). For an evaluation of Jung's controversial stance on racial matters (including his notion of the cultural, ethnic, racial unconscious) and a (post-)Jungian perspective on contemporary racial issues, see respectively Michael Vannoy Adams, *The Multicultural Imagination: "Race," Color, and the Unconscious* (London: Routledge, 1996), and Fanny Brewster, *The Racial Complex: A Jungian Perspective on Culture and Race* (London: Routledge, 2019). A thorough study of the social, anthropological, and intellectual context at the turn of the twentieth century and after is Clelia Brickman's *Aboriginal Populations in the Mind: Race and Primitivity in Psychoanalysis* (New York: Columbia University Press, 2003). See, finally, Giovanni Sorge, "Between Stereotypes and Hermeneutic Quest: C. G. Jung's Approach to 'Primitive Psychology'," *Journal of Jungian Scholarly Studies* 18, no. 1 (2023), 116–43.

[434] In Jung's personal copy these words are replaced by "imprisoned" (*gesperrt*).

yearns for their own house and garden. Mr. and Mrs. Such-and-Such, who indulge in this garden and have a nice relationship—that is the element of the world and of the state, and it is good and proper this way. For this reason, the dreamer proposes to his feeling that the seamstress should come to this garden and recuperate there. She tells him, however, that she cannot work in the garden because there is no sewing machine, whereupon he offers her his wife's machine.

This is yet another strange symbol. I have tried to establish why this anima is a seamstress, and what the symbol of the sewing machine means: that is, what the tendency and meaning of this dream are. I have become very apprehensive in this respect, and avoid interpreting anything into the dream. What matters is that we know what the dream *in itself* says. If the dream says it is a sewing machine, then we must under no circumstances assert that it is a motor car or something of that sort, but a sewing machine.

Sewing machines are used to sew together pieces of cloth to make clothes. Thus, the seamstress makes clothes from stitching together pieces of cloth. This truth is most banal, and it tells us precious little. The dreamer places great emphasis on the machine. It represents a method of sorts, a concentrated intention or a human method of production. It also involves a complicated intertwining of mechanisms through which success is achieved. The machine is needed to perform work, and this is also the thought: the seamstress herself says that she wants to work. The dreamer suggests that she go out into the garden, to which she responds that she cannot because there is no machine there. Thus, it's about work. Would you have any idea what this need for the sewing machine could mean?[435]
ANSWER: The possibility of exercising the feeling function is missing.

Evidently, the dream inclines toward affording the female figure a meaningful activity. A dress—what does that mean? Women frequently have such dreams. I myself often dream about being a lady's tailor.

The seamstress assembles clothes—that is, the inferior function has a synthetic ability. It is even her destiny to have a synthetic effect: sewing together pieces of cloth and clothing a person; in other words, giving shape to his behavior and presentation in the world. Clothes make the man.[436]

---

[435] In *DAS* Jung regarded the sewing machine as the facilitator of integration ("The feelings of a respectable man cannot work in the open, hence 'in his own garden' means pressing his feelings back into his marriage"; 87) and as "the method of his wife [. . .] to join that which has been separated [. . .] [namely,] the conscious and the unconscious" in the man (92).

[436] An allusion to Mark Twain's variation on a proverbial chestnut of great age, traceable back from Carlyle's Professor Teufelsdröckh in *Sartor Resartus*, ch. 5, to Shakespeare's

Human beings, and women in particular, express their character through clothes. She chooses what suits her from factory-made clothes and a large number of prefabricated hats. She makes a statement about who she is by wearing a particular dress and donning a certain hat, just as she does with her choice of shoes and stockings. Every piece of our clothing is symbolic. The hat speaks a particularly distinct language, as it covers our head and serves as the roof over the whole house, the uppermost part, where our central office, our headquarters, is located. For this reason, hats are particularly characteristic of persons of rank. They indicate the general idea of a person. Nowhere in the world is this emphasized so strongly as in Germany. In England, for instance, a newly elected minister is also judged on the basis of his professional competence; in the case of a negative assessment, one adds, if applicable, "But he's a gentleman." That was said of the Labour prime minister. Titles play a particularly significant role in Germany. As a Swiss, however, I shall not speak about German affairs; our obituaries, for instance, speak of "former lieutenant" if the deceased had long since ceased to hear the rattling of sabres.

Vesture, for instance, a wedding dress or a mourning gown, is also an expression of personality, of a person's attitude, idea, and willingness. In our present case, confidence is placed in the inferior function to serve as a vesture; thus, it's impossible to have the right attire, or indeed the correct attitude, without the assistance of the unconscious. The vesture does not lie ready at hand—it must, instead, be established time and again with the help of the unconscious. For instance, we'd like to be as pleasant as possible to someone—on some occasions we succeed, on others we do not. We imagine that we have the vesture, but our anima plays a trick on us behind our backs. We discover that it's not a function of our conscious intention that produces our vesture, but instead that this requires the collaboration of the unconscious; otherwise, we're scuppered. The dreamer thus receives no support from the unconscious with the work in his garden—the anima would be able to provide him with the correct attitude, but she is unable to do so, since she is suffering from tuberculosis, moreover she would have no sewing machine to work with in the garden.

The idea that the sewing machine belongs to his wife implies that his feeling also belongs to her, and that he is not entitled to hand it over to another figure. In his marriage, his feeling has become an automatic

---

Polonius in *Hamlet*. Erasmus's *Adagia* has "vestis virum facit" (3.1.60), citing Quintilian. [T.N.]

function—whenever machines occur in dreams and fantasies, it is a matter of functions that have become automatic.[437]

From this allusion, we can learn that the synthesis of the attitude cannot be produced by consciousness, but only by the unconscious. We are unable to influence our emotional attitude by our will, which we can use to command only our outer actions.

The primitive cannot do even that much. "It" must enact something through him, otherwise he will do nothing whatsoever. If bushmen are hungry, they tighten their belts; before they decide to go hunting to bring down game, they can almost starve to death. They are much too lazy to seek food. Only when their hunger becomes intolerable do they go hunting. They bring themselves to search for game, and can track it for a long time before they kill it. This makes no difference to them. They hunt down the game, devour ten pounds of meat on the spot until they almost burst, stretch out in the shade to digest, and it occurs to no one to hunt game to build up a stock of meat. One waits for the next bout of hunger, and matters start all over again. We believe, of course, that we have moved infinitely beyond this level of civilization. But not in all respects. In some respects, we haven't yet arrived. But we don't know that.

The primitive must always bewitch himself first to be able to do anything, and that requires work. Therefore, one dances for days on end before setting off to hunt. Buffalo hunting is far from being an art; what it amounts to, however, is proper preparation. Buffalo hunting in itself is a minor matter. Preparation is everything. Thus, if the primitives wish to go hunting, then they must perform a *rite d'entrée*, that is, an initiation rite. This can be a dance, for instance. They dance the buffalo dance to prepare for hunting. In this dance, the same persons play the buffaloes and the hunters. Everyone wears a buffalo's head, with horns, and holds an arrow to show that he's a hunter. Then, the buffalo is shown grazing at a distance. The hunters surround him, at a distance, and make some kind of noise to startle the animal. Then they all raise their heads and begin to dance. Quite unlike us, they dance somehow in noble exhilaration,

---

[437] From *DAS* we learn that the analysand linked the "method" of the sewing machine with a mechanical approach to the sexual sphere: "That is the way he looks at sex [. . .]. For most men the idea of sexuality is purely mechanical and unpsychological." The fact that his wife owned the sewing machine indicated that "the sex mechanism belongs to his wife" (94). Jung's full account implies that the wife's machine, if loaned to the inner figure of the young seamstress, would move the man's mechanical outlook into the hands of neglected feeling.

indeed true to the Latin saying: "Nunc est bibendum, nunc pede libero pulsanda tellus."[438]

Their dancing really involves working upon the earth by treading the soil hard with the right heel *([Jung] beats rhythmically on the table to demonstrate)*, using one's full force. This is the only work that these people undertake. This is where our notion of work comes from. This is the rhythm of work, as suggested by the accompanying drums and song. And once this is in full swing, and when one thinks that now they're all utterly exhausted from these preparations, from fasting—and they do some serious fasting and dancing, believe you me—only then does the hunting commence. But that proceeds of its own accord, because, after all, one has attained an elevated state.

For a long time, these things, specifically the psychology, were not understood. But the matter is extraordinarily profound. These people first enter into an identification with the sun by exposing themselves to it for hours, and by turning with the sun. Thus, they become sons of the sun, that is, sun-like. In this sun-likeness, they are purified and can dance as gods, as the sun, and as such they are allowed to kill buffaloes. They are the sun-buffaloes, they are sacred animals, who are allowed to kill other animals by observing the formulae of politeness. That is why the chief must greet the buffalo as a brother at the very moment when the hunters have sneaked up on the buffalo. Then the killing constitutes no sacrilege. When they have sneaked up on them, the chief lights a pipe and takes the first puff, blows the smoke at the buffaloes, counts to five, and offers the buffaloes the pipe. Only then do the hunters open fire. This is similar to the gesture of the French soldiers at the Battle of Malplaquet, where they said to their English counterparts, "Messieurs les Anglais, tirez les premiers!"[439]

---

[438] "Now is the time for drinking, now let the foot pound the earth freely": Horace, *Carmina (Odes)* 1.37. Jung cited this verse in *Transformations and Symbols of the Libido* with reference to a Pueblo foot-stamping dance which he may have witnessed: "'calcare terram'—a persistent, vigorous pounding of the earth with the heels (nunc pede libero pulsanda tellus)" (*CW* 5 [1912/1952], § 480). Horace recast the famous line "νῦν χρῆ μεθύσθην" (Nyn chre methysthen; Now one must get drunk) by Alceus (Fragment 332.5) in praise of Augustus, Western guarantor of peace against a corrupt East at the battle of Actium (31 BCE), which led to the taking of Alexandria and the deaths of Antony and Cleopatra [with thanks to Ernst Falzeder].

[439] The French challenge ("You shoot first, English gentlemen!") was uttered not at Malplaquet, but at Fontenoy, in 1745, by the army of the Marechal de Saxe to the Anglo-Dutch force led by the Duke of Cumberland. Having nothing to do with politeness, in fact this gesture is a ploy, since those who first opened fire faced greater risk [with thanks to Ernst Falzeder].

These primitive *rites d'entrée* illustrate how those matters that we cannot enact with volition are performed. Ominously, we always imagine that if we exercise a bit of will, then that takes care of things. Nothing could be further from the truth. Everything goes wrong. That's a fact, and we may safely admit it. This is the run of things in our human relations. There is a lot that we cannot achieve, although for two thousand years we have been imagining that something could be accomplished if, indeed, if the other fellow . . . of course, if only the other fellow. . . . Stupidly, there is always some other person, and consequently this absolutely correct ideal has not yet enjoyed its breakthrough. It occurs to no one, of course, that we ourselves are that other person.

Exactly the same holds true for disarmament conferences.[440] Mankind has been wanting to disarm for two thousand years. We get annoyed about the French and the English. They get annoyed about the Germans, and God knows whom else. One blames the other, because no one does what he demands from the other. These are things we fail to achieve. There is not one single person who would not like to disarm, and who fails to agree that this nonsense, this widespread butchery, must cease. Or at least an overwhelming majority holds that view. Well, what happens? Nothing. No one can do anything, not even the simplest of things. Not even currencies can be stabilized, and there we are speaking of good intentions, of goodwill, and everything that could be achieved with the will. From this you

---

[440] Jung was most likely thinking of the World Disarmament Conference held at the League of Nations in Geneva from February 1932, sponsored by the USA and Great Britain and attended by delegates from over sixty coutries. This attempt, following two other international efforts aimed at arms reduction at the Hague Conferences of 1899 and 1907, was destined to fail in 1934. Chafing under restrictions imposed by the Versailles Treaty, Germany refused to place itself on a par with the minimum level allowed (as firstly proposed by the democratic-liberal parliamentary majorities); in effect it was granted a de facto legitimation for rearmament. Jung's disheartened appraisal may also stem from the recent suspension of the Geneva Conference in June 1933, in favor of the International Economic Conference in London. Also, on June 7 the "Four-Power Pact" between Italy, Germany, France, and Britain had been initialled, due for signing on July 15 at the Palazzo Venezia in Rome. Proposed along vague lines by Mussolini to foster mediation among the major powers, and raising hopes in the international press, it was refused ratification by the rivals France and Germany. On October 14 Germany would withdraw from both the Disarmament Conference and the League of Nations (see also Jung, "Psychology and National Problems" [1936], CW 18, § 1333). Almost exactly a year later, Jung would affirm, regarding rearmament, "The political analogy to this is the secret will to destruction all over the world, not only in Germany [. . .]. Our actual collective unconscious seeks the destruction of millions [. . .]. And why can nobody put a stop to it, damn it? We can only explain it by the fact that there is a superior will which forces all heads" (*Nietzsche's Zarathustra*, cit., 1:93). On Jung's later view concerning the question of world peace, see "Techniques of Attitude Change Conducive to World Peace: *Memorandum to Unesco*" [1948], CW 18).

may gather that sheer determination alone does not make such things happen.

Let us move on to a different subject. Our dreamer also believed he could succeed through sheer determination. It is precisely in such a case as the present one that this will not work. Something else is required instead. Yes, indeed, this little seamstress, this sickly creature, the thing you regard as pathological, which suffers within you, that is what makes this go, performing this rhythmic work with its feet. It is a sewing machine—that is, a machine operated by regular movements of the foot. If you observe people who are seeking to push through a point, for instance, at an assembly, only to discover that they are unable to, then here is what they will do. *([Jung makes an] explanatory movement.)* They will either break into rhythmic drumming or tap their toes. *([Jung] drums his fingers on the table to demonstrate.)* These are rhythmic movements, the beginnings of dance. They would like to start dancing so that movement finally occurs. Now if movement fails to occur among primitives, they will start dancing. Then matters move. The *rite d'entrée* must always be performed to achieve movement. Our seamstress is this *rite d'entrée*. She treads long and regularly, and the machine makes a regular noise, and this regularity, this rhythmizing of work, is the primitive dance, of which we no longer have any knowledge, and have long forgotten. Primitives experience our machines as dance machines. I have witnessed this myself in southern Sudan. I came across a motor vehicle that was supposed to collect cotton from the natives. We also used it to transport our baggage for some distance. The vehicle drove past a tribe. When the vehicle stood still for a while and made a noise that we all know, and when the primitives heard it, first the women drew closer with the children and cried out "N'goma." That was wonderful. *([Jung] makes rhythmic movements.)* They performed a kind of jazz dance, as it were, because they had evidently and instinctively sensed the kinship with these rhythms.[441]

* * *

---

[441] This episode is not mentioned in *MDR*. See above, n. 432, concerning Ngoma; cf. Jung's considerations on African dance and jazz in "The Complications of American Psychology" [1930]; for example, "You can dance the Central African *n'goma* with all its jumping and rocking, its swinging shoulders and hips, to American jazz. American music is most obviously pervaded by the African rhythm and the African melody" (*CW* 10, § 964). In *Symbols of Transformation* [1912/1952], Jung conceded that despite his initial conviction, following Freud, that the various modes of rhythmic performance and patterning were somehow "a substitute for some form of sexuality," he later realized that more was at issue. "The rhythmic tendency does not come from the nutritional phase at all, as if it had migrated from there to the sexual, but [. . .] is a peculiarity of emotional processes in general" (*CW* 5, § 219).

LADIES AND GENTLEMEN,

My remarks on primitive dance and its meaning for the work performed by primitives provide further insight into this dream. The inferior function forms part of what the will cannot achieve. Therefore here we encounter an obstacle. We cannot be as we wish to be. In this respect, we can only make an effort without being certain that we can also achieve our goals; in actual fact there are more defeats awaiting us than victories in this regard. We can be wrong, after all. We are incredibly ready to entertain illusions. We can simply imagine that we have achieved something. I have known countless people who have assured me that they have achieved it. Naturally, that's a deception. One can never achieve it. One has only accomplished it in relative terms. But all this has taken great pains, experiencing so much and working so hard.

Under no circumstances can one really leap from one attitude to another, and then simply be converted. We have gradually come to entertain this idea, however, on account of the history informing our education. This education somehow took shape at the beginning of our calendar, shall we say, during the day and age of Tertullian.[442] As is generally known, he was quite surprised that persons who had been baptized according to the proper rite could still sin. He concluded that the rite of baptism had evidently not been performed *lege artis*.[443] Something, he assumed, had been lacking. Consequently, they had to be baptized all over again, but this time in the correct manner. Thereafter they neither could nor would sin any longer.

---

[442] Quintus Septimius Florens Tertullianus (ca. 155–after 220), was an African Latin apologist known for his polemics against both pagan and Christian adversaries of his theses. He was the first Church Father to write extensively in Latin and is considered the founder of Western theology. In *De baptismo*, written between 190 and 200, he called baptism "our sacrament of water": contrary to a certain Manichean dualism, according to which matter is essentially negative, he insisted on the necessity of both water and faith for such a sacrament. The symbolic theme of water's cleansing power over sin was central to baptismal theology from the days of the early Church. In the Latin Church baptism was by immersion, still customary in the Greek Church. Jung here refers to the proto-Christian era, and its severe dominant conception of sin and guilt, prior to the common practice of infant baptism beginning at the earliest in the second or third century. In the text cited above, Tertullian opposes infant baptism because children cannot grasp its meaning. Tertullian also wrote the first description of penitential practice in the early Church (*De paenitentia*), in which he affirmed that God grants remission of sins (while demanding confession and repentance) only twice: in preparation for baptism, and for sins committed after baptism. This view was common in early Christianity, and had already been explicated in *The Shepherd of Hermas* (ca. 140). The practice of the second repentance began only with the sixth century, although according to *Hermas* this is possible only once in a lifetime (*Pastor Hermae* 29.18). Tertullian held the same view (*De paenitentia* 7.10).

[443] Lat.: from *lex* and *ars*, signifying "according to the law of the art"; commonly "authentic" or "to the letter."

He ordered such rebaptism in a number of cases, and then discovered that even individuals who had been baptized twice had also sinned. From this he concluded that human beings were beyond help. They were destined to fry in hell, and all further help was in vain.

You see how this idea suggests that one held the view that some momentary enthusiasm was required, a leap so many meters high and so many meters wide, to attain a different state and be a quite different person. It became embarrassing only when it emerged that this was an illusion. We encounter this both in general and in particular: namely, that we cannot accomplish through sheer will what we wish and principally strive to make of ourselves. To accomplish this, we instead require quite a specific kind of work, and that a certain element should come into play that helps us do so. Neither goodwill nor reason will suffice; a third quality is needed, instead, something irrational, and I have no idea where this should come from. But it must enter into things. Otherwise it cannot develop.[444]

This dream tells us that the seamstress, who who was supposed to make our clothing, is thus, as it were, this despised inferior function that has the power of synthesis. The idea of the dream is that this function should somehow rise to the surface. He [the dreamer] should somehow allow his inferior feeling to emerge into freedom, to engage in open discussion, let us say, to put his cards on the table or at least discuss matters, so that it has the chance to move around outdoors, or else suffer depletion. The young tubercular seamstress, who suffers from a pulmonary illness, typifies early withering and emaciation, and so forth. That's what would happen to his feeling if it failed somehow to rise to

---

[444] Almost a year later, reflecting on the same subject—the belief that religious conversion could or should create an entirely different person—Jung mentioned the insistence of religious groups such as the Christian Scientists, Methodists, or the Oxford Movement upon their possession of "the living truth which changes people completely," and this very change in itself providing the proof that "there is some secret magic in these forms or convictions" (Nietzsche's Zarathustra, cit., June 6, 1934, 80). Interestingly, he continued by paralleling such proto-Christian forms of belief with Nietzsche's notion of the Übermensch: "the idea that the Superman would be the lightning which would upset the dormant condition of his world, so that man could change. It is not exactly the Christian transformation through the grace of God or baptism. It is due to man, because when God is dead he appears next in the one who kills him; then the divine creative faculty needs must dwell in man. And then man has the faculty of transforming into the Superman, by which is not meant a man of greater virtue but a man who is simply beyond this man of today, a different creature obviously, a man who can deal with the darkness in human nature" (ibid.). Cf. "Psychology and Religion" [(1937) 1938/1940], CW 11, § 142ff.

the surface. If it surfaces, however, it does so to become effective in its own synthetic way.

Naturally, it's extremely hard to understand what constitutes this power of synthesis. Here once more we come up against a question mark where I must, without further ado, admit that I cannot set forth matters in scientific terms. I simply accept the fact that the dream already makes sense whether or not we produce a complete explanation or interpretation. The dream is here. This very thing, which the dream has produced, is working away somewhere; what we can do is to place no difficulties in the path of this function, but instead promote it as far as possible.

Accordingly, I told the dreamer, "If you ever have the opportunity to discuss your feeling, which you otherwise somehow conceal, with your wife, then you may do so safely. Raise the subject and discuss it. You might perhaps await an opportune moment. But the matter must be aired and brought to the surface. Try to attend to it as well as possible, because this will enable you to gain a decent attitude. As you can see, the seamstress is at work assembling the clothing."

That is simply a symbol, and I allow this symbol to be, on the assumption that if we have failed to sufficiently understand anything about it, the next dream will tell us something that more closely explains that symbol.

Therefore this would pose a question mark for me. For instance, if I did not know what followed, I'd nevertheless say—as I told the dreamer at the time—"Something is dark here." This *rite d'entrée*, as represented by the rhythm-making machine, is something dark. Here there is either work or the need for work that finds no other expression than in this symbolic form.

For me, this is a symbol rather than an allegory or a sign. It's truly a symbol: that is, the best possible expression for a matter that we are unable to grasp entirely as yet. For instance, if you take the signal fan of the railwayman to be a symbol, then that's obviously nonsense. It is a semeion, a sign. It is not a symbol.[445] Please don't forget that the original use of the word *symbolon* is to mean "credo," that is, faith, which is also referred to as a *symbolon* in the early Church. Faith is the symbol

---

[445] See Jung's classic distinction between sign and symbol in the "Definitions" in *Psychological Types* [1921], CW 6. In Greek, the *symbolon* (σύμβολον) originally referred to an object reconstituted from two halves. For Latin speakers, moreover, the *symbolum* referred to the ring that sealed documents with wax and, by extension, the image engraved on such a ring or seal.

par excellence.[446] Faith is the expression for what we cannot recognize and what nevertheless we strive to give some kind of shape to. That is the symbol.

Thus this sewing machine, that is to say, this peculiar rhythmic activity, for which we find no other expression than the means available to us, and for which we haven't yet found any expression, is a symbol. Indeed, if I say it's the primitive dance, then that is a symbol, of course. I simply return to a time before the machine. I flee from our machine age to that age when rhythmic activity had not yet concretized itself in the form of engines, when the human being was himself still an engine, and when he sought to accomplish what he longed for and aspired to through this rhythmic activity.

From the appearance of this rhythm-symbol, we can deduce that the dreamer here encounters a great difficulty in his nature: namely, the question about his attitude, which he cannot resolve without having recourse to this ancient rhythm or rite. In the first instance, I'd say that it's something like a ritual activity that must help him attain a proper attitude toward the world, human beings, and things. For when we have the right attitude *somewhere*, then that holds true for the universal.

Are there any further questions about this dream that you wish to raise? If not, then we can proceed to the next dream.

The next dream says the following:

From a bird's-eye perspective, I see a road-making machine, a kind of road roller, which creates an arabesque-like figure from continuously intertwining paths, a kind of labyrinthine garden.[447] *It occurs to me that this somehow resembles analysis.*[448] *I am standing at a crossroads, undecided about which road I should*

---

[446] Jung refers here to the Christian usage that yokes the apostolic symbolon (*symbolum apostolorum*) to the profession of faith or creed, which was ratified at the first Council of Nicaea (in 325) and nowadays prevails in Christian liturgy. He also alludes to the proto-Christian meaning of *symbolum* as the bond of faith in which members of the community recognized the Sacred Covenant. To legitimate the cult of icons, the second Council of Nicaea (787) established the possibility of paying homage to the person of the emperor (and consequently of Christ) by honoring the image that replaced his physical presence (or epiphany). In that way, through both image and prayer the unity of the symbol was reconstituted.

[447] In *DAS*: "forming a particular pattern like a labyrinth" (97).

[448] In *DAS*: "And in the dream he thinks: 'That is my analysis;' and then he is in the picture which he has looked at from above" (ibid.).

follow. The place is situated in a forest, but I cannot recall any other details.[449]

He offers the following comments on this dream:

If one observes things from above, one sees their connection, meaning, and purpose much better. What seems incomprehensible and without meaning close-up becomes, when seen from the correct distance, a symmetrical, pleasant drawing.

About the steamroller, he says,

I was recently reading about an American road-making machine, with which the best tar macadam roads can be built in a relatively brief period of time. In actual fact, there is no purpose in building roads that lead nowhere, and which do not connect two separate points in the most direct way possible. But I did not notice this in the dream. I only saw the resulting drawing. Such drawings often appear in magazines as puzzles. If one can exercise the necessary patience, one can perhaps find a hidden path to the goal. I also think that analysis requires much patience to reach the goal.

With regard to the crossroads, he refers to a conversation between us, in which I told him that no one was forcing him to somehow follow a difficult path. The sewing machine gave him the creeps. I explained that there was no need for him to subject himself to any kind of difficulties. No one was forcing him to have himself treated. He should by all means take his time to consider whether he should perhaps stop again and take sides with his resistance rather than with his inclination. I'm completely open-minded in this respect. If a patient has any kind of resistance, my view is that he should maintain that resistance as long as possible. It's his resistance, after all. Why shouldn't he keep it? Naturally, I don't want to have it. I have no use for it. It's his property. He has resistance. He should enjoy it. If one possesses something, then that is something beautiful. Therefore, I leave his resistance in his possession, if it goes so far as that he'd rather do something other than consult me. I told him, "You'd be stupid if you consulted me, and did not like doing so. If you do so nevertheless, then that's another matter."

---

[449] In *DAS*: "At first he did not pay much attention to the arabesque the machine was making" (ibid.). Cf. the comment on this dream in Michael Vannoy Adams, *The Mythological Unconscious*, 2nd rev. edn (Putnam, CT: Spring Publications, 2010), 264–67.

I have very little to do with resistance, because I also hold resistance in such high regard. It's a strange beast, which the creator has brought forth. Some people live in constant resistance. Evidently, that's the way it has to be.

One must also accept one's own resistance, since that is often an important guide and an important instinct. It would be quite wrong to oppose resistance. One shouldn't resist one's resistance in any shape or form, but accept it and see where it leads one.

I spoke to him in those terms. The crossroads refers to our conversation. That is, the place to which the crossroads leads in the forest, as he phrased it. At this point, he has a strange association. He immediately thinks of the forest at the beginning of the *Divine Comedy*.[450] This indicates that our guesswork is not that far off the mark. The dream now says that he sees a road-making machine as if from above, a bird's-eye view. Thus the Old Man is saying, "Remove yourself from things, place a great distance between yourself and these things which have been discussed, between the sewing machine and all these stories. Consider matters from a distance, and be mindful of Lao Tse's words, 'The man who sees from afar, sees clearly.'"[451]

---

[450] "Nel mezzo del cammin di nostra vita / mi ritrovai in una selva oscura, / ché la dritta via era smarrita" (Midway upon the journey of our life / I found myself within a forest dark, / For the straightforward pathway had been lost [H. W. Longfellow's translation]). Cf. *DAS*: "the patient thinks also of another old story, dating about 1450, of the monk who lost his way in the Black Forest and a wolf becomes his guide to the Lower World" (99).

[451] In Jung's (old) German: "Klar siehet, wer von ferne siehet." The adage continues: "und nebelhaft, wer Anteil nimmt" (and mistily, he who takes part) (Lao-tse [Laozi], "Der erste Spruch," in Alexander Ular, trans. and ed., *Die Bahn und der rechte Weg des Lao-Tse* [Leipzig: Insel-Verlag, 1921 (1903)], 7). Cf. *DAS*, 100ff., where Jung emphasizes the schematic symmetry of the symbolism inherent in nature's generative power, adding, "There is also a peculiar rhythm in the design, in and out, approaching the center and leaving it again, such a pattern as people would make in dancing, if you marked it out on the floor." Here he seems to evoke Goethe's dynamics of systole and diastole, along with the danced mandala or *mandala-nritya*, which he had described in his "Commentary on *The Secret of the Golden Flower*" ([1929/1957], CW 13, § 32), thereby linking back to the labyrinth theme. "Remember the pattern," he continues, "of throwing the ball back and forth, associated with a previous dream? We get some valuable contributions to the idea of patterns, the unconscious source of them, and what the machine is meant for." Then he reflects on the spiral's movement that expresses "one of the fundamental laws of natural development." The labyrinth motif appeared indispensable for reaching the "true law of nature." He recalls the epitaph carved on the tombstone of his fellow-Baseler the mathematician Jacob Bernoulli (1654–1705), best known for having investigated, after Albrecht Dürer and then Descartes, the geometric progression of the logarithmic spiral: "Eadem mutata resurgo" (Though changed, I arise the same). Jung's comment alludes to Heraclitus's famous river (fragment DK22B91) which though the self-same always changes: "Psychologically you

In the dream, at least he seems to have succeeded in somehow looking down on this story of the machine. Again, a machine is making a path and plotting quite a specific one. By making this path, it produces a strange pattern in the landscape, something that resembles, as he phrases it, a labyrinth garden. Naturally, since the word "labyrinth" turns up, our analytic sessions come to mind for him, which seem to have confused him a great deal. As if finding himself in a labyrinth garden, he is trapped in various trains of thought, which are necessary, unfortunately, if one wishes to consider things from all sides. It is quite inevitable that one repeatedly comes back, to consider the matter time and again from another angle. If this path could be described, it would look something like the famous autobiographical life plan of Hieronymous Jobs the candidate, which was included in the original edition of the *Jobsiad*,[452] like the map of an insect's flight.[453] If you have ever seen such a map, then you can imagine what the life of Jobs the candidate looked like and also what analysis seems like to the dreamer. It's the kind of picture in which he can no longer find his bearings. He's lost in a labyrinth. Naturally, that has also caused him to develop a certain resistance, so that matters had evidently reached the stage at which he thought, "What Doctor Jung is saying is all quite mad."

That's the prevailing assumption. If someone experiences difficulty, he thinks the other person is crazy. If a psychiatrist fails to grasp what the insane person is saying, then he considers that person confused, instead of thinking the opposite. And when psychological matters are concerned, one assumes one always knows better oneself.

So he thought I was confused. That's what happened to him. He actually considered whether or not he ought to go any further. Therefore, the Old Man led him up onto a high mountain and said, "Now take the long view of things from up here for a while." And he saw that this machine, which he hadn't grasped in these terms, did something. It made a road. And this

---

develop in a spiral, you always come over the same point where you have been before, but it is never exactly the same, it is either above or below" (*DAS*, 100).

[452] Reference to illustrations in *Bildern zur Jobsiade* (1872) by poet and pioneer satirical illustrator Wilhelm Busch, inspired by German physician Carl Arnold Kortum's *Leben, Meynungen und Thaten von Hieronymus Jobs dem Kandidaten* (1784) about the mishaps of a clumsy prophet, expanded and reissued as *Die Jobsiade: Ein komisches Heldengedicht in drei Theilen* (1799).

[453] In *DAS*, after recalling the sexual component of the problem—confirmed, in his view, by the reappearance of the motif of the machine—Jung proposed that an aerial view could indicate a way to "see himself and his problem more impersonally as Mr. and Mrs. Ant who are having some dispute about ant sexuality and Mr. Ant's interest in another ant, and then he can look at it easily" (99).

made a lot of sense to him. For that was precisely what he couldn't see. He had seen neither that the machine makes a road nor where that road leads, because he had assumed that all path-making consisted of drawing a line between two given points; that is, of building a road from one place to another. That was the meaning of the way, surely. What he is being told explicitly is that there exists a path that does not simply link two points, but instead delineates a pattern that he's never encountered before.

It has never occurred to us that we could build a road that actually leads nowhere. Parallels exist here, however: as you know, such matters were toyed with in the Middle Ages, when labyrinth gardens were constructed to allow one to indulge in taking a stroll and surprising a lady on a solitary bench. Since it is most appealing to have no knowledge at all of where one actually is, that was done out of sheer playfulness.[454]

---

[454] In general, for the Christian faith, the labyrinth (a term possibly deriving from *labrys* [λάβρυς (Lat. *bipennis*), a Lydian word for the double-bit axe, according to Plutarch, *Quaestiones graecae* 2.302a], or from *labra* [cave with burrows and corridors]: see Umberto Eco's preface to Paolo Santarcangeli, *Il libro dei labirinti: Storia di un mito e di un simbolo*, 2nd edn [Milan: Frassinelli, 1984], viii], represents an inward effort (*labor intus*) and a reconfiguration of mappings of the world (*mappae mundi*). In other words, it was a cosmological model, representing a human descent into the underworld and a metaphysical ascension toward salvation—a symbol of the corporeal world located in the celestial city as described in the commentary on the Book of Revelation by Pseudo-Dionysius the Areopagite (fifth–sixth century CE) and embodied in the architecture of the Gothic basilica (see Tessa Morrison, "The Dance of Angels, the Mysteries of Pseudo-Dionysius and the Architecture of Gothic Cathedrals," in Anna-Teresa Tymieniecka, ed., *Metamorphosis: Creative Imagination in Fine Arts between Life-Projects and Human Aesthetic Aspirations* [Analecta Husserliana 81] [Dordrecht: Kluwer, 2004], 299–319 at 312). With its new conception of the centrality of the human being, its growth of a scientific worldview, and its increasing interest in classical literature, aesthetics, and arts, the Renaissance marked a shifting of the metaphysically freighted meaning of the labyrinth to a more earthly, sensual perspective. Consequently the labyrinth "lost some of the Christological associations it had acquired during the Middle Ages. It became, as it once was long ago with the patristic fathers, a simple metaphor for the route along which the solitary soul must pass" (Wright, *Maze and the Warrior*, cit., 213). So geometrical gardens, many in labyrinthine patterns, flourished in Italy, France, and England (ibid., 216). The first labyrinth garden dates to at least 1311 in France, apparently already serving as a *divertissement*. With Humanism, the trope of the labyrinth as an allegory of love's entanglements and adventures became diffused: in Petrarch the *laberinto* comes to stand for self-abandonment to the pains of love ("Un lungo error in cieco laberinto": *Canzoniere*, Sonnet 224), or *inextricabilis error* (a phrase originally from Virgil's *Aeneid* [6.27; cf. *inremeabilis error*, ibid., 5.91]); that is, "inextricable wandering" (cf. It. *errore* [error] and *errare* [to wander; to err]). Similar examples occur in Boccaccio's *Corbaccio*, the versions of *Orlando* by Ariosto and Tasso, and the *Hypnerotomachia Poliphili* of Francesco Colonna (see below, 251ff.). In Fontanellato near Parma, today's world's largest labyrinthine garden traces a three-kilometer (1.9-mile) path among twenty-five species of bamboo and houses at its center a precious book and art collection; it was conceived by bibliophile and publisher Franco Maria Ricci in honor of Jorge Luis

Nowadays, nobody would use a steamroller to make a road that actually leads nowhere. Moreover, it struck him as utterly stupid that this machine would engage in something quite as pointless as building a labyrinth garden. That ill suits our consciousness. The Lord works in mysterious ways, and they don't match up with human reason. In such a case, we are so awestruck that we don't think that the Lord is confused, but cite rational arguments to excuse him on other grounds.

I advised the dreamer to draw the picture of the maze that he had seen in the dream, so that we could see what he had seen there. His description was slightly unclear, but he nevertheless furnished a drawing. Let me show it to you [see Fig. 4].

The picture of the maze failed to convince him that we were somehow on a reasonable track. In the dream, he therefore comes to a standstill at the crossroads, where he must decide whether to return to the sensible upper world of bright daylight, to reasonable people who live in small houses, small gardens, and maintain the correct relationships—or whether to submit to an adventure on the off-chance. He had evidently feared this adventure, and justifiably so. I made no attempt to coax him into undertaking this adventure. In the dream, however, he observes that the crossroads already lies in the forest, in the very forest of the *Divine Comedy*,[455] and at the very point where Dante and Virgil's way commenced—and where was that?

PROFESSOR ZIMMER: Where the right way had been lost.

Indeed, that is where Dante lost his way. And what did he encounter?

Borges, who told Ricci that the world's greatest labyrinth still remains the desert. See the superbly illustrated Franco Maria Ricci, ed., *Labyrinths: The Art of the Maze*, with foreword by Umberto Eco, texts by Giovanni Mariotti and Luis Biondetti (Milan: Rizzoli, 2013). Another labyrinth dedicated to Borges is at the Cini Foundation on the island of San Giorgio Maggiore in Venice; recall too Borges's many labyrinth stories, including "The Library of Babel" ("La Biblioteca de Babel," 1941), collected in *The Garden of Forking Paths* (London: Penguin, 2018) and "The Aleph" ("El Aleph," 1945), collected in *The Aleph and Other Stories* (London: Penguin, 2004). On this subject and its connections with the notion of time in the work of the Argentinian polymath, see Pio Colonnello, "Il labirinto, il tempo, la danza," *Rivista di filosofia neo-scolastica* 111, no. 1 (2019): 13–30.

[455] Dante Alighieri's world-famous epic poem *Commedia* (supplemented by the adjective *Divina* by Giovanni Boccaccio) was composed between ca. 1308 and shortly before the death of the Florentine "sommo poeta" in 1321. It depicts the soul's journey towards God, its ascent progressing through three *cantiche*, from *Inferno* through *Purgatorio*, to *Paradiso*. It is deeply rooted in the tenets of Thomas Aquinas's *Summa Theologica* and furthermore presents a colorful representation of Dante's contemporary world with its various historical figures, whom the poet candidly portrays, located within his magisterial literary cosmology, according their vices and virtues.

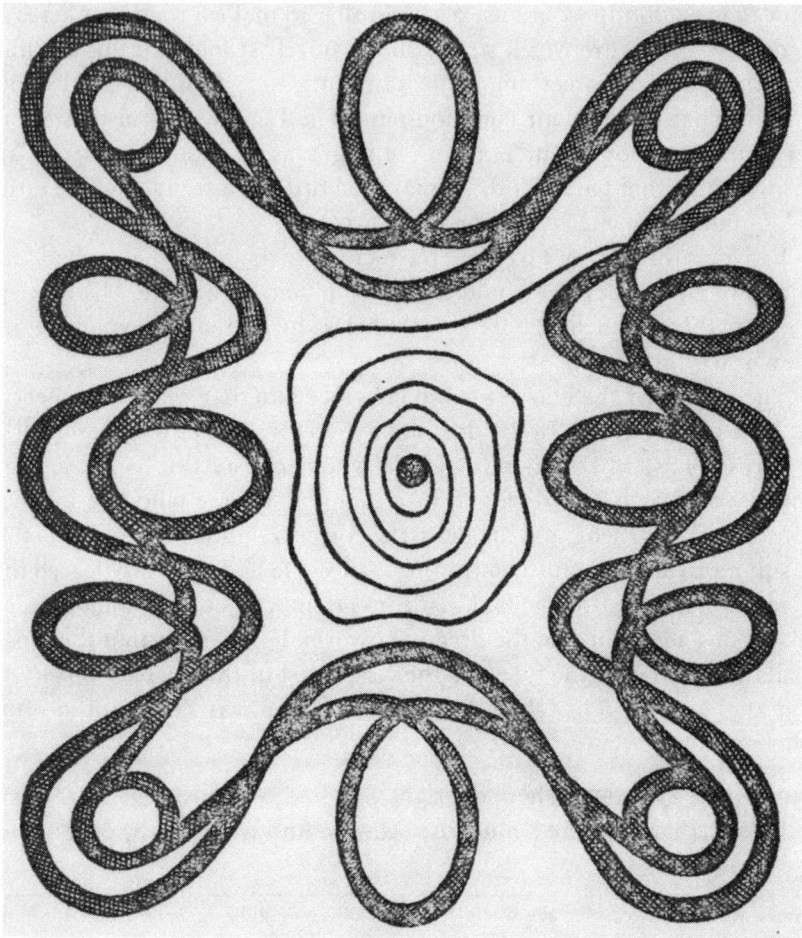

Figure 4.

PROFESSOR ZIMMER: The animals.[456]
  Yes. And then?

PROFESSOR ZIMMER: The guide.[457]
  And what came next?

---

[456] The three "beasts" (*fiere*) are a leopard, lion, and she-wolf: symbols of fraud, violence, and lust—the primary sins, that is, that burden the denizens of Hell (*Divina Commedia: Inferno* 1.31–60).

[457] The figure of the "supreme poet" (*sommo poeta*) Virgil (*Inferno* 1.61ff.). Ger.: *der Führer* in the protocol.

PROFESSOR ZIMMER: The descent into the underworld.[458]

Indeed, the underworld. Thus, he has already reached the point where the way ends, and where one begins to lose one's way, where the wanderer's panic sets in, and where one must come to realization.[459] Now Dante can no longer go on by means of his own strength. Now comes the great adventure, and then the helpful animals appear. At this moment the first instinct awakens. As long as we cannot lose our way, we do not have the faintest idea about instinct, but rather everything goes by the book. Only when we lose our way can instinct become active. After the animals comes the guide, who does not lead him back to the clear, sensible upper world, as we believe who always seek a guide to lead us to clear and rational circumstances; instead, this guide leads him into another world, into a world that is completely dark to us.

We assume that the *Divine Comedy*, which is still playing and will play forever, is [exclusively] a poem written by Dante. But no, ladies and gentlemen, Dante wrote his divine comedy about an experience.[460] He wasn't the only person to have this experience, however, for he shared it with many others, who were unable to commit it to poetry.

Already I can let you in on the fact that this path will go on resolutely: helpful animals will occur in the next dream, just as the guide will appear later.

Ladies and gentlemen! These are eternal truths, which always repeat themselves in the life of the individual: that is, when we reach a point on the way where a way no longer exists, when we no longer know where we should go and how we can continue. Then the way of the *Divine Comedy* commences, precisely when the human comedy has reached an end. We have forgotten this. We've come to believe that we can set the sciences over and above that, or just go on producing science. But science ends there and experience begins. *This is where life begins.* And this can only inspire us with awe. One can do little more than observe this and accompany such a

---

[458] In his personal copy Jung underlined Zimmer's three answers and added the following remark: "Das war 1933!!" (That was 1933!!).

[459] Cf. *DAS*: "The dream says, 'Here you are in the mythological situation of the hero, like Hercules. You are in the wood of the *Divine Comedy*'" (101).

[460] Jung believed that masterpieces such as Dante's *Divina Commedia*, Goethe's *Faust II*, or the *Hypnerotomachia Poliphili* (which he will discuss later in the Seminar), in all their visionary magnificence, derived from their authors' special capacity to dig down to and channel profound, transhistorical layers of the collective unconscious (see "Psychology and Literature" [1930/1950], CW 15, § 151ff.). That is, they arise not from a merely literary experience, but from a total human involvement, attaining gnoseological, imaginative, and spiritual depth beyond the "spirit of the times." In this regard, see Tommaso Priviero's recent comparative analysis *Jung, Dante, and the Making of the Red Book: Of Fire and Form* (London: Routledge, 2023).

person. We can place our entire science at the service of this movement. We can assist it with our reason, insight, and knowledge, but no science *about* it exists. For this experience is greater than all science, than all human ability. If we wish to make correct use of any kind of psychology, if we want to fulfill any purpose with such a study of the human soul, then it must serve to bring about this experience, from which true transformation arises.

However, we shall never achieve this with so-called scientific knowingness.

At this point, this man naturally began to feel slightly uneasy. For suddenly the notion struck him that amidst a culture apparently as safe as ours—please think of Switzerland—there might be something that could not be grasped with rationalism, that problems could be presented that we hadn't already read about somewhere, perhaps in an encyclopedia, and for which not a single expert existed whom one could consult for advice. This proves that those things which were previously thought to be purely digestive systems or the like are no mere viscera, but keenly inward matters, and that *suddenly something speaks to us* from within, where we had assumed nothing other than the existence of physiology and anatomy.

This place, which lies in the forest, is where the dreamer feels that only two possibilities exist. Either he makes a resolute return to the rational world and to rational measures: that is, to a world where if there is no sunshine, then at least a gas lamp or an electric light will shine, and where one can observe that everything proceeds along rational lines. Or he embarks on an utterly hazardous enterprise, where no streetlight serves as an absolute guarantee that the place is not haunted, or where no railway, army, or police prove that we live in an absolutely clear and rational world: the route where unimagined adventures are possible.

If he doesn't choose this second way, then I don't quite know what awaits him. I'll shake his hand and tell him, "Well then, farewell, it's been a pleasure to meet you, quite charming. It's very sensible that you've found your way back into the clearly delineated upper world. Have a jolly good time."

An enormous yawn, towering as high as a house, will overcome him as soon as he sets off home. What does one do now? Nothing, in actual fact. Nothing will improve. Everything will go on just as it always has.

In the other case, you would say, Well, I wonder what will happen now? It's utter madness of me to assume that something new could now occur. For nothing of the kind has been hinted. One must constantly remind one's instinct and reason, "I live in a world where some kind of adventure remains possible, where something unknown can still happen. I actually live in a fairytale forest, where animals can talk, where trees can talk, where the strangest things are possible, and where I too shall do things a normal person can never do."

If he doesn't yield to this possibility, then nothing in his life will change. He will never see any world other than a three-dimensional one, where the sun shines during the day and streetlights at night.

Lecture 4

Ladies and gentlemen,
I have a question on my desk, but the sheet bears no signature. However, perhaps I could answer the question, since it concerns the problem we have been dealing with: namely, that of the seamstress. The questioner asks whether the figure of suffering, as represented by the seamstress, might be found less in the dreamer's soul than in his wife's. One must give an unconditionally affirmative answer to this question. Most probably, this figure can also be found in his wife's soul. But what he dreams about is this poor woman. For if we interpreted this as being within his wife, this projection, which exists anyway, would be intensified even further. Consequently, the dreamer would develop only a stronger resistance toward his wife, because then he'd be bound to her even more. Such a relationship is not healed by further encouraging these projections, but rather by making them utterly conscious as projections. Thus, if a person has a resistance to his spouse, or friends and acquaintances, or indeed his enemies, and we agree with him—yes, indeed, it's the others, they are like that, they're beasts, black sheep—then in fact we are depriving him. Then we take away a precious piece of his own soul. For instance, if he projects a black animal into the other, then we deprive him of *his* black animal. Naturally, he is very pleased about this. We're always satisfied if we can discover that it's the other. Therefore, when we watch a stage play featuring villains, we can proudly say, "It's the others, this actually has nothing to do with me." That's why we find reading reports on crimes and trials so satisfying. It is very nice to know that it's other people.

But one doesn't know what kind of loss one suffers in this case. Possibly one has lost one's evil, and that consists of quite a lot of libido, which can be converted into dollars and cents. If we lose that, then we lose our shadow, and we discover that we are a man without a shadow.[461] We have lost our three-dimensionality and are merely painted onto the wall. Thus, many people walk around as sheer frescoes, utterly bodiless. That's not much fun. These matters really happen. I knew the case of a young woman aged twenty-eight, whose symptom was ultimately found to be that she could no longer hear her own footsteps. This frightened her. She had become so light, so bodiless in a manner of speaking, that she had ceased to exist. I experienced great difficulty in reacquainting her with her body. She had received a strict education in a Catholic convent, where she was brought up to believe that her body was something evil and belonged to the realm of the shadows. She had never been given the opportunity to behold her own body. On the basis of a dream, I once asked her, "Well, have you ever seen yourself naked?" She replied, "No!" To which I replied, "Well, I'm sure you take a bath from time to time." Indeed, she did take baths. I asked her how she did that. She would cover the bathtub with a linen cloth, and therefore she had never seen her body naked. I gave her the following advice: "Now run along home, stand naked in front of the mirror, and look at yourself." Which is precisely what she did. Next time she told me, "Now I have seen myself. I don't look so bad. But I wouldn't have thought that my legs were covered with so much hair."

This lady had cast all her darkness into the other person. She had been brought up to believe that if she were a good girl, then she would never be black but forever white and clean. Wherever there was anything black, it would most certainly be with other people. Consequently, she noticeably lost her shadow. The first thing to appear within her as a positive symbol of life was that she felt a black snake lying asleep deep down in her stomach.

I am not reporting this story due to a flight of ideas, which I am said to be afflicted with from time to time, but because it belongs within the larger context of dream symbols, which we'll need to discuss. You must never forget that dream symbolism is tremendously wide-ranging. We must draw on extended areas of the human mind to understand these matters. Here, in the unconscious, where we find ourselves not in the tormenting and

---

[461] A probable allusion to the French exiled aristocrat Adalbert von Chamisso's *Peter Schlemihls wundersame Geschichte* (1814; translated as *The Wonderful History of Peter Schlemihl*), a highly successful fairy tale novella—invoked in *DAS* as "The Man without a Shadow"—whose protagonist sells his shadow to the Devil. This allegory of Chamisso's own political situation as a man without a country inspired "a very good film," *The Student of Prague* (1926) directed by Henrik Galeen, praised by Jung as "a sort of second *Faust*" (*DAS*, 49 and n. 6). Jung there argued that the analysand's brother-in-law, who has succeeded him, represents his shadow ("the shadow is always the follower," 48) and remarked, "The more one turns one's eyes to the light of consciousness, the more one feels the shadow at one's back" (49).

overcrowded constriction of consciousness, we are dealing only with broad regions where we find ourselves at sea, where everything can flow toward us. Here, in the unconscious, we find ourselves in vastness. Therefore, we must gird our loins and remain ready for great distances. As mentioned, this small story is therefore not without interest for our present subject.

As far as the present question is concerned—the handwriting suggests a *female* correspondent—I must reply that in my inconsequential opinion it would be technically incorrect for us to leave this precious piece, this suffering soul, to the wife. In the first instance, that would induce an inevitable sentimentalism, an idiotic pity within him [our dreamer], and later an even more irritating resistance, because he would become slavishly dependent, by virtue of this projected aspect. For whenever we have a projection, no matter what we project, whether good or evil, a fascination, a dependency, a lack of freedom arises. This explains why we must be so cautious, indeed even over-cautious, particularly if we come across a case where a person is actually doing wrong and is behaving badly toward us, so that we quite rightly take offense. We should, then, once the first storm of affects has passed, mind our own business, and admonish ourselves to refrain from projection. Because if we do project, then we are threatened by what instills the primitive with the greatest fear: namely, a loss of soul. We must always aspire to keep the various parts of our soul neatly together and under no circumstances allow any affect to make off with the parts of our soul and to settle anywhere within another person, who subsequently becomes a *bête noire* for us, the black beast, and thus exerts a fascination over us. If we happen to have such a projection, then we find ourselves in the primitive condition of so-called *participation mystique*. One can obviously quarrel about the term. But let us not fight over words, but instead clarify the facts. *Participation mystique* simply expresses the primitive condition of consciousness where the discerning faculty is not yet developed, so that the other seems to me to be what *I* am. I cannot mean *myself* and think, cannot distinguish properly between I and you, [for I] imagine that the other has *my* psychology, and if he does not do as I think he ought to, then I am somehow offended.

Thus, the old native was angry at his disobedient son. His son, however, was already a grown man. Nevertheless, the old man exclaimed, "This fellow stands there with my body and does not obey me." He couldn't grasp this. It's his body, after all, so why shouldn't it obey him? That is primitive, and so are we to a certain extent. One doesn't say as much, but in any event *it* is there. This is *participation mystique*, a primitive state of mind. Where a primitive state of mind exists, there is no freedom, but instead the coercive rule of the stars: that fate, that is, takes automatic effect and happens of its own accord. Therefore I shun projections just as the primitive does, because I have seen that the greatest wisdom lies in not projecting. For then one has everything with one that belongs to one. *Omnia bona mea mecum porto:*[462] may this be your motto. I believe that I've answered the essence of the question.

We must now return to our dream. Yesterday I deliberately omitted an important point, which I want to discuss today in a further connection. You'll remember that after the dream the dreamer draws a picture of the highway steamroller or the macadam machine. He told me this machine had worked in a certain way, forming a kind of arabesque pattern. I asked him to draw this. One could easily have overlooked the remark that the machine had created an arabesque pattern. But one should not overlook such matters. It is always useful whenever one is dealing with forms and shapes that can't be well described in language to ask the dreamer to commit such matters to paper, so that one can see what this looks like in the person's head. Sometimes it happens that the dreamer refers to a kind of triangle or a similar shape, and one contents oneself with that, believing one knows what is meant. One should never think that one knows. Instead, we should always ask, "What was it like in reality? Please make a drawing." Thus, to our next session he brought along a drawing of this arabesque path. I've done a rough sketch of this drawing for you here [see Fig. 4]. In fact, his drawing was slightly nicer. I can hold it up for you so that you gain some impression of the original, at least from a distance. Now, one can look at this and brush it aside. I am sure you take this to be a fantasy. Ladies and gentlemen, considering that your seat, this room, this lighting, our entire culture are nothing other than products of human fantasy, then you cannot overlook this. There is nothing man-made known to you that is not fantasy. Fantasy is the alpha and omega of psychic life. Fantasy is the greatest danger of all. Do you believe that a landslide or an earthquake is as dangerous as human fantasy? Well, you haven't the faintest idea! Fantasy is a catastrophe. Imagine that you were all seized by the fantasy that suddenly a fire had broken out. In such an event a few people would have been trampled to death. Or you think you are being attacked. If you have weapons, then you'd start shooting. Several people would drop

[462] "All that is good of mine, I bring with me": Lat. saying (abbreviated in the transcription) attributed by Cicero (*Paradoxa* 1.1.8) to Biante of Priene (sixth century BCE), one of the Seven Sages. It has also been attributed to the philosopher Stilpone (Seneca, *Epistulae morales* 9.18–19), as well as to Saint Paul, to the effect that holiness is built on the human baggage carried by every human being [with thanks to Ernst Falzeder].

dead immediately. That is fantasy. In such a case, you cannot say, "Excuse me, I didn't mean it like that." That does not bring the dead back to life. Human history has witnessed the most amazing fantasies, which have exterminated whole peoples, have led to bloodbaths leaving countless people dead, and have made world history. That is fantasy. So please do not say, it is *merely* a fantasy. That's the least intelligent thing you can do. A fantasy is something incredibly important, something enormously powerful. It's the life of the psyche. For instance, your everyday psychic well-being hinges on a fantasy. If you awaken in the morning with an incorrect fantasy, then you'll feel bad for the whole day, perhaps even for weeks. If you awaken with a correct one, however, you'll feel fine. In another case, a person might even die because he has not had the right fantasy. Fantasy is what is most essential. If one ceases to have the right sort of fantasy, one is afflicted by a wrong one. Then one either feels bad or even contracts tuberculosis, everything merely on account of fantasy. Oriental physicians, for instance, know about this, and therefore they grab people by their fantasy, and not with the help of chemical molecules as they do here. Our entire science of chemistry, this whole sordid kitchen of chemicals, also has a great effect on fantasy. We shouldn't delude ourselves in this respect. So if you are looking at a shoddy scrawl, do not hold it in contempt, but instead busy yourself with observing what you see. You might be able to learn something from it. If you take it to be mere childlike scribbling, then you'll have learned nothing about life, bypassing it heedlessly, whereas life might have held something delicious under your nose.

Does anything come to your mind about this drawing?

ANSWER: It recalls a mandala.[463]

So it does. But let us develop the connections from within the dream itself. Did anything strike you earlier on?

ANSWER: Yes, about the hall.

Yes, indeed, I already drew a suspicious-looking drawing on the blackboard for you when we were considering this [see Fig. 3, p. 141]. That drawing did not particularly seem to arouse your curiosity. Let me sketch the drawing once again to refresh your memory. Here are the benches, here the dinner table. This would be an analogy. It is also an elongated quadrangle, and yet it is something utterly different. It is a *salle de pelote*. It has not been made by a machine. A ball game is played there or a dinner served, whereas here a steamroller is at work, crawling around like a snail and making all these paths. Finally, there is one other thing which

---

[463] See above [ZL], n. 358, and below, 231ff.

the dreamer would have suppressed outright. At one point, the steamroller takes a turning quite irrationally and moves toward the center, forming a spiral and coming to a halt in the middle. He did not mention this point, which transpired only when he made the drawing. Then he recalled that that's how matters had happened in the dream. But because he hadn't understood what that could mean, he didn't pay enough attention.

To begin with, let me mention that these drawings have the greatest conceivable importance for those concerned. Please do not think that such material matters only in cases of pathology. These drawings are important for whole cultures, for thousands of years of culture, for long periods about which we Western barbarians have not even the faintest notion. These drawings once boosted millennia to embark on a path of mental development. Thus, we have no reason to turn up our noses at such things. These peculiar drawings, and you must admit that you're looking at no more than meager traces, begin in dreams in this way. They already appear, just as they do in this case, in those initial dreams, often well before I have even seen the case. This idea can occur to a person without any knowledge of analysis. Ancient India or ancient China, for heaven's sake, knew nothing whatsoever about analytical psychology, nor did medieval Europe have an inkling about these Eastern things. And yet they existed, nevertheless; it was not some sick-minded simpletons who discovered these ideas, but some of our greatest minds, among others.

I must confess that when such a paltry image appears in a dream, it inspires me with awe, and I spend a lot of time keenly observing it. Because there is a great deal that I can learn from it.

In the present case, for instance, I saw with the greatest astonishment that ball games were played in this hall, or indeed meals served at table. I note this down and tell myself that here was a new aspect. Food is eaten and a ball game played. Then, I turn to the other drawing and see that a drawing of a strange steamroller is being made. A strange path is being laid out, which ultimately enters the center, where a dinner table is laid or a ball game is being played; here, moreover, a spiral shows up: what does this remind you of?

ANSWER: Of the Kundalini snake.

Now the word "Kundalini" drops in. That's the Indian technical term for this spiral snake, which lies asleep, encoiled.[464] For that reason, a short

---

[464] Sanskrit *kundalini* refers to "twisting oneself" as a snake does. It is a feminine noun, mentioned in the Upanishads, indicating the the formless aspect of the Goddess and the divine feminine energy par excellence: that is, the Shakti believed to be located at the base of the spine, in the *muladhara* chakra. In Shaktism and Tantrism it designates the energy at

while ago I mentioned the dream of the patient who had this snake lying at the bottom of her small pelvis.[465] The young lady concerned was neither a student of Sanskrit, nor an anthroposophist who had heard about the Kundalini snake at the anthroposophy school. On the contrary, she was an ordinary Swiss woman who had no idea about all these odd things. Nevertheless, she had dreamed this symbol. I'd add, moreover, that over the course of her treatment, she observed how this snake ascended step by step. Ultimately, it reached the roof of her mouth, from where the snake's head then emerged, and she discovered that the head was shining gold. She said, "Now I recognize myself." Then her body belonged to her once more.[466]

That is how these Eastern things happen among us. So, we have this symbol of the coiled snake. Only now do we have a parallel. We don't understand, as yet, what this is actually supposed to mean. Therefore let us instead try to feel our way into this curious drawing, based on what the dreamer knows.

Naturally, I asked him, "Well, what occurs to you about this?"

He replied, "Well, these could be certain kinds of symbols."

He had read books about psychoanalysis, of course, and then asserted that one could refer to these round circles as female symbols and to this elongated one as a male symbol.

In reply, I asked him, "Would you also refer to them in this way if you had not heard of this kind of interpretation?" Well, he would have barely reached such a viewpoint if he possessed no such knowledge.

"I'd have been inclined to say something else."

---

the base of the spinal column, which through meditation practice can rise through the energy centers called chakras. See Jung, *Psychology of Kundalini Yoga*, cit., Shamdasani's "Introduction" to the same, passim, and Joseph Campbell, *The Mythic Image* (Princeton, NJ: Princeton University Press, 1974), 331ff. See also above [ZL], 102–4 and n. 319.

[465] The *kleines Becken*, the pelvis minor or small pelvis in medical terminology, is the cavity in the center of the the the surrounding brim rising above it. See above, 213.

[466] For further details, see Jung, *Psychology of Kundalini Yoga*, cit., 84–85. Jung subsequently affirms, "To activate the unconscious means to awaken the divine, the *devī*, Kundalini—to begin the development of the suprapersonal within the individual in order to kindle the light of the gods. Kundalini, which is to be awakened in the sleeping *muladhara* world, is the suprapersonal, the non-ego, the totality of the psyche through which alone we can attain the higher chakras in a cosmic or metaphysical sense. For this reason Kundalini is the same principle as the *Soter*, the Saviour Serpent of the Gnostics" (68–69). Cf. in this regard the classic *The Awakening of Kundalini: The Evolutionary Energy in Man*, with a psychological commentary by James Hillman (Boulder, CO: Shambhala, 1997 [1975]), by Gopi Krishna, who famously recounts and discusses the dramatic mental and spiritual effects resulting from his own personal experience of the rise of Kundalini energy.

"Well, what might that have been?"

"I'd have said that the essence seems to be that this line always moves outward and inward, outward and then again inward, like waves, as if this thing were something that was breathing, which extends outward, contracts, and then extends again, as if it were a wave rhythm emanating from some unknown center."

To this I replied, "Well, from this you can gather, for instance, that this sexual meaning is not that far off. The male would be the upward movement, the female the drawing-into-itself." One can, of course, apply the primitive language here. One can express this in sexual terms, just as primitive cultures do uninhibitedly. It's quite self-evident that this upward movement represents the phallus. In the Indian language, as is well known, the phallus is the hieratic name for the male member. Thus the lingam is a symbol of the creative male force, simply the Yang principle in Chinese philosophy.[467] If one thinks it is the penis, some people look at one with a pitying smile and consider us stupid. You know that these round forms are symbols of the Yoni (of the female organ).[468] And the longish Padma symbols.[469] That's the hieratic name, and it is in fact called the lotus. This is equivalent to the phallus. Phallus is never an anatomical designation

---

[467] *Lingam* (Skr.): "mark" or "sign": an aniconic symbol for Shiva which, though commonly seen as the male sexual organ, actually refers to both that and the female yoni, whose union with the lingam composes the visible sign of universal creativity. Cf. Zimmer: "As symbol of male creative energy, the lingam is frequently combined with the primary symbol of female creative energy, the yoni, the latter forming the base of the image with the former rising from its center. This serves as a representation of the creative union that procreates and sustains the life of the universe. Lingam and yoni, Shiva and his goddess, symbolize the antagonistic yet cooperating forces of the sexes. Their Sacred Marriage (Greek: *hieros-gamos*) is multifariously figured in the various traditions of world mythology" (Heinrich Zimmer, *Myths and Symbols in Indian Art and Civilization*, ed. by Joseph Campbell [Bollingen Series VI] [Princeton, NJ: Princeton University Press, 1946], 127).

[468] *Yoni* (Skr.): "maternal abdomen" and "vagina," therefore the origin of the life. Among its many meanings are "place," "receptacle," "temple," "source," and "origin." As a sacred symbol of female power, the yoni is considered "the gateway between life and death, as well as the generative force behind all existence"; its "quintessential symbol" has become the lotus. Furthermore, the "inner sanctum of Hindu temples, regardless of religious sect, is called the *garbhagriha*, which means 'womb-house'" (entry "Yoni," in Constance A. Jones and James D. Ryan, eds, *Encyclopedia of Hinduism* (New York: Infobase, 2007), 516.

[469] For this reason the throne of many Buddhist divinities is represented as a blossoming lotus (*padma*), symbolizing purity, beauty, perfection, and transcendence. Also indicating an aquatic plant (*Nelumbo nucifera*), the padma (feminine name of Persian and Sanskrit origin) is associated with the *sahasrara*, the "thousand-petaled chakra" (see above [ZL], 113 and n. 345). It also names the female sexual organ in the *Vajrayana*, a Buddhist manual of techniques and doctrines for salvation dating from the fourth century, as well as the womb in Tantrism (see "Concerning Mandala Symbolism" [1950], CW 9.1, § 630 and § 652). In his comment on the padma linked with Fig. 5 (ibid., 363), Jung points out that

for the male member, but for the god who, as you might not know, even rescued himself into the Church, in the shape of a phallus.[470] The Church, in its all-embracing love, ultimately declared these indecent gods saints. The Yoni also wandered cheerfully across into the Church, only to appear in the Litany of Loreto[471]—as what exactly?

ANSWER: As a baptismal font.

That doesn't turn up there.

ANSWER: As an ivory tower.[472]

No.

"the significance of the rose as the maternal womb was nothing strange to our Western mystics" (ibid., § 652).

[470] Similarly in DAS Jung deplored the fact that "the old phallic cult, for instance, taken over from paganism by the early Christian church, is never mentioned; a remnant of it appears in one of the forms of the cross, but people avert their eyes" (26). On the symbol of the phallus, see Alain Daniélou, The Phallus: Sacred Symbol of Male Creative Power (Rochester, VT: Inner Traditions, 1995).

[471] The Litany [sic]. i.e., Litanies of Loreto ("Litaniae Lauretanae," or Litanies of the Blessed Virgin Mary) are antiphonal chants sung since the early sixteenth century at the Sanctuary of the Holy House of Loreto in central Italy (Le Marche). The fame of the Marian sanctuary caused the Lauretan litanies to become one of the most popular forms of oratio fidelium. The epithets that coalesced in these litanies, drawing on devotional formulas from the Old Testament and other early sources, include "Mater Creatoris" (Mother of our Creator), "Virgo fidelis" (Virgin most faithful), "Speculum justitiae" (Mirror of justice), "Sedes sapientiae" (Seat of wisdom), "Causa nostrae laetitiae" (Cause of our joy), "Turris davidica" (Tower of David), "Domus aurea" (House of gold), "Foederis arca" (Ark of the Covenant), "Ianua coeli" (Gate of Heaven), "Stella matutina" (Morning star), "Salus infirmorum" (Health of the sick), "Refugium peccatorum" (Refuge of sinners), "Consolatrix afflictorum" (Comfort of the afflicted), and "Regina angelorum" (Queen of angels). Before being approved and codified by Pope Sixtus V in 1587, the Litanies were attested as dating from a twelfth-century Magoza manuscript; later popes made further additions. See Giorgio Basadonna and Giuseppe Santarelli, eds, Litanie lauretane: Guida storica e commento teologico (Vatican City: Libreria Editrice Vaticana, 1997). Jung had discussed the Litany, for example in Psychological Types [1921] (CW 6, §§ 379–80, 392, 399, 406), while focusing on the second-century Christian visionary tract the Shepherd of Hermas. His exploration of the Loreto Litanies supports his major contention with regard to the non-Christian source symbol of the feminine Grail vessel, as being "a genuine relic of Gnosticism, which either survived the extermination of heresies because of a secret tradition, or owed its revival to an unconscious reaction against the domination of official Christianity" (§ 401). The abundance of Lauretan epithets testifies to the power of the late medieval "soul-image [. . .] as a vessel of devotion, a source of wisdom and renewal." Such Gnostic elements, however, "nipped in the bud the psychic culture of the man; for his soul, previously reflected in the image of the chosen mistress, lost its individual form of expression through this absorption" (§ 399).

[472] Turris eburnea: one of the epithets of the Litany's derivations "directly from the Song of Songs, as in the case of the tower symbol" (Psychological Types [1921], CW 6, § 392).

ANSWER: As a *vas*, that is, a vessel.

Yes, as a *vas*.[473] It is known as a *vas insigne devotionis*;[474] that is, a superlative vessel of devotion.[475] Yet another form occurs, indeed three others. The *hortus conclusus*, the enclosed garden,[476] the sealed spring,[477] and lastly, the most beautiful of all, the *rosa mystica*,[478] the rose of the Cathedral of Chartres.[479] It is a padma. And it's also the rose in Dante's paradise, the mystic rose of heaven.[480] It is a lotus.[481] One cannot say that

---

[473] See the Hymn of Konrad von Würzburg which describes Mary as a flower of the sea that hides Christ (linked to Fig. 5 in "On Mandala Symbolism" [1950], CW 9.1, § 652–55, and the *Terry Lectures* [1937], CW 11, § 123). On the alchemical vessel, see also *Psychology and Alchemy* [1944], CW 12, part 3, ch. 1.

[474] Lat.: "superlative vessel of devotion" (in the Vatican translation, "consecrated abode of God"), a Lauretan epithet expressing Mary's status as the first person entirely devoted to God.

[475] "Our German word for vessel," Jung points out in the *The Tavistock Lectures*, "is *Gefäss*, which is the noun of *fassen*, that is, to set, to contain, to take hold of. The word *Fassung* means the setting, and also, metaphorically, composure, to remain collected" ("Fundamental Psychological Conceptions: A Report of Five Lectures" [1935], CW 18, § 407).

[476] A medieval Latin locution ("walled garden") drawing on descriptions in Genesis, the Gospels, the Apocalypse of John, and the Song of Songs. Traditionally it was a secret, fantastic, and protected place, later meaning a garden set aside for recollection and devotion. In the Middle Ages *hortus conclusus* referred to gardens whose walls secured flowers and medicinal herbs from human and animal depredation. As a place "where Nature is subdued, ordered, selected and enclosed," the garden "is a symbol of consciousness as opposed to the forest, which is the unconscious, in the same way as the island is opposed to the ocean. At the same time, it is a feminine attribute because of its character as a precinct" (entry "garden," in J. E. Cirlot, *A Dictionary of Symbols*, trans. by Jack Sage [London: Routledge & Kegan Paul, 1962 (1958)], 110).

[477] An epithet already used in the Song of Songs (4:12) for Mary's virginity. Jung elsewhere observes that such an "inviolable source and Feminine quality" is an attribute "of the Mater Dei and of the spiritualized earth," arguing that "the quaternity is entirely absent from the dogma, though it appears in early ecclesiastical symbolism. I refer to the cross with equal arms enclosed in the circle, the triumphant Christ with the four evangelists, the tetramorph, and so on," adding in a note, "for the Gnostics the quaternity was decidedly feminine" ("Psychology and Religion" [(1937) 1938/1940], CW 11, § 126 and n. 18). Saint Ambrose (333–397) first identified "the bride in the Song of Songs as the Virgin Mary, an element that would exert an enormous influence on subsequent Christian liturgy and spirituality" (Gianni Barbiero, ed., *Cantico dei cantici* [Milan: Edizioni Paoline, 2004], 464).

[478] Lat.: "mystical rose." This association of the Holy Virgin awaiting the Holy Spirit with the rose—preferably white, but also pink or red—has a wealth of symbolic implications. The rose blooms in May, the month sacred to the Virgin Mary.

[479] The cathedral of Notre-Dame de Chartres, including its rose window, epitomizes the High Gothic. Begun in the eleventh century on the foundations of a pagan temple, it preserves as a relic a smock traditionally attributed to Mary at the Annunciation.

[480] Near the end of Dante's *Paradiso* (canto 31) the host of the blessed takes the form of a white rose.

[481] Cf. Zimmer: "The earthly sphere itself also came to be seen in Indian mystical cosmography as the lotus blossom. And down among the stems of its petals lay the Indian

these things are[482] due to migration—and we'll return to this hypothesis—which in fact are local, autochthonous notions.[483]

Thus, the dreamer's interpretation is the unprejudiced form. It gives us a much better picture of what is alive in this matter than if he had clung to a one-sided sexual interpretation, which, after all, is no more than symbolic language. When ancient man referred to the phallus, he obviously meant divine creative force. Thus in ancient India this would have also meant a highly abstract concept.

So let us bear in mind that this is a strange living thing, a thing that breathes, expanding and contracting, like a vacuole inside an amoeba, or an amoeba that thrusts out and retracts pseudopodia,[484] and thereby constitutes something that is alive.

---

subcontinent—which came to be known as 'the Little Girl Lotus'" (Heinrich Zimmer, *Indische Sphären* [Munich: Oldenburg, 1935], 127 [2nd edn: Zurich: Rascher, 1963, 123]); "The cup of the opened lotus symbolizes the pure, blooming castle to which the world give birth: the castle of the primordial ocean, the world-manifesting god Brahma enthroned on a lotus blossom. The unfolding world, as seen from above by the divine gaze, is the spreading lotus blossom. Thus this lotus is the entire world of life, which comes into being and passes away, the entirety of all becoming, all Buddhas, all created and uncreated forms of world and life" (Zimmer, *Ewiges Indien: Leitmotive indischen Daseins* [Potsdam: Müller & Kiepenheuer, 1930], 55 n. 27. On lotus symbolism in Hinduism, see Zimmer, *Myths and Symbols*, cit., 5, 90ff. and Zimmer, *The Art of Indian Asia: Its Mythology and Transformations*, 2 vols (Bollingen Series XXXIX) (New York: Pantheon, 1955), 1: ch. 6. See also Zimmer's parallelism, cited above [I], 65–66, between Dante's white rose of the *Paradiso* and the Hindu lotus (Heinrich Zimmer, *Artistic Form and Yoga in the Sacred Images of India*, trans. and ed. by Gerald Chapple and James B. Lawson with J. M. McNight [Princeton, NJ: Princeton University Press, 1984], 239).

[482] Ger.: *sind*; Jung replaced this in his personal copy with the subjunctive *seien*.

[483] Here, as usual, Jung contests the general validity of the migration theory of symbolism, which accounts for similarities among symbolic patterns from different cultural and religious traditions as being the result of cultural evolution, interchange, and transmission. While not excluding evidence in favour of the diffusionist paradigm, Jung predominantly advocated—in line with the nineteenth- and early twentieth-century quest for a common origin in comparative linguistics, anthropology, and mythology—for a comparative, thus a tendentially essentialist model, attributing to archaic, common levels of the collective unconscious the spontaneous generation of symbols. The diffusionist paradigm became established in the German-speaking areas at the beginning of the twentieth especially through the work of Friedrich Ratzel, his pupil Leo Frobenius (who introduced the concept of *Kulturkreis*, the cultural circle or field), and Wilhelm Schmidt, spreading in both Europe and America in contraposition to the evolutionists (notably Edward B. Tylor, Herbert Spencer, and James Frazer), who argued for the unilinear evolutionary progression of every society.

[484] The pseudopodium is the extruded cytoplasm from certain unicellular organisms (e.g., amoebas) and isolated cells of multicellular organisms, which englobes the prey, or a particle of food, in a vacuol. As is evident from differing pseudopodia, the same cell can

I can now reveal that this kind of drawing is common among people who dance. I have seen a whole number of cases who did not dream of such forms, but instead dreamed of dancing. I requested these patients to demonstrate this dance, and I made a mental note of the figures their dance described on the floor. Thereafter, these patterns emerged. I asked them to remember their dance figures and to set them down on paper. The patients went away and made drawings: they had danced mandalas.[485] And it was easy to see what they more or less meant. Such a dance expresses a concentration on the center. In the most proper sense it is a hieratic *circumambulatio*,[486] as it's called in legal history; namely, a conversion [transformation].

I must explain this term. When the Romans founded a city, they would undertake a circumambulation, a conversion, by drawing a furrow around the city with a plough yoked to a team of oxen, the so-called *sulcus primigenius*,[487] in a ceremonial procession. Within lay sacred land, a taboo

---

change its form. Jung also discusses the pulsing movement of the vacuol in relation to Hatha yoga's "rhythmic performance of breathing" in *DAS* (467).

[485] On the *mandala-nritya* see Jung's "Commentary on *The Secret of the Golden Flower*" ([1929/1957], CW 13, § 32, also quoted above, n. 451.

[486] Lat.: circumambulation, the archaic and universally attested ritual of circling around a person, a plot of ground, or a sacred structure with the aim of propitiation, sacralization, or purification. Jung first uses the term in the "Commentary on *The Secret of the Golden Flower*" (CW 13) when describing the process of "enclosure" in a circulation that becomes "the 'movement in a circle around oneself,' so that all sides of the personality become involved" (§ 38). Later, in *Aion*, when dealing with the combination motif of quaternity and the circle and its derivative, the wheel, he observes that "the former motif emphasizes the I's containment in the greater dimension of the self; the latter emphasizes the rotation which also appears as a ritual circumambulation. Psychologically, it denotes concentration on and preoccupation with a center, conceived as the center of a circle and thus formulated as a point" (*Aion* [1951], CW 9.2, § 352).

[487] Lat.: "primordial furrow." In the "Commentary on *The Secret of the Golden Flower*" [1929/1957], Jung ascribes to the mandala, in both East and West, the "purpose of drawing a *sulcus primigenius*, a magical furrow around the center, the temple or *temenos* (sacred precinct), of the innermost personality, in order to prevent an 'outflowing' or to guard by apotropaic means against distracting influences from outside" (CW 13, § 36; see also C. G. Jung, *Dream Interpretation Ancient and Modern: Notes from the Seminar Given in 1936–1941*, ed. by John Peck, Lorenz Jung, and Maria Meyer-Grass, trans. by Tony Woolfson [Princeton, NJ: Princeton University Press, 2014], 213). According to classical philologist Georg Wissowa (1859–1931), the Romans performed circumambulatory ceremonies "in the course of a procession or circuit round some object—land, city, army, or instruments, such as arms and trumpets—or, again, the whole Roman people, if supposed to be in need of 'purification' from some evil influence; in this extended form the ritual was called *lustratio*: and this ceremonial was perhaps the most characteristic, not only of the Roman, but of all ancient Italian forms of worship" (Georg Wissowa,

district. The municipal area of the city was thus delineated; at its center, a pit known as the *mundus* was excavated, in which field crops were sacrificed.[488] On this site a treasury or temple designating this center, or so-called *forum*,[489] could often be found later; this was the veritable heart of the city, where its pulse beat, because here all life flowed in and out, through the four principal city-gates, on the *via decumana* and the *via principalis*.[490]

Religion and Kultus der Römer [Munich: Beck, 2nd rev. edn 1912 (1902)], 390). Jung had this book in his library. For a recent study of Wissowa's work and legacy, see Elisabeth Begemann and Jörg Rüpke, "Early Christianity in the Framework of Roman Religion: Georg Wissowa," in Begemann and Rüpke, eds, *Georg Wissowa on Roman Religion of the Imperial Period*, special issue of *Religion in the Roman Empire* (8, no. 1 [2022]): 4–22.

[488] Jung's source was probably Wissowa, who in turn quotes Ovid, *Fasti* 1.5, 821ff. (Wissowa, *Religion and Kultus*, cit., 234 n. 9). According to Plutarch, the *mundus* also received votive objects and occasionally sacrificed animals, as well as vegetables, first fruits, crops, and "good and necessary things" along with a small quantity of earth from the place of origin (Plutarch, *Caesar* 44; see also *Romulus* 12). Referring to Macrobius, *Saturnalia* 1.16.18, Eliade observes, "this mundus was the point of intersection for the three cosmic spheres" (Mircea Eliade, *Patterns in Comparative Religion*, trans. by Rosemary Sheed [New York: New American Library, 1974 (1958)], 374). The establishing of a *mundus* was common in the founding of Roman cities; in one study of culture and magic of which Jung owned a copy it is observed that "after the *mundus*, also known as the *Cereris mundus* in Festus 142, had been excavated at the center of the chosen site and filled with various field crops, according to ancient custom the city founder harnessed a bull and a cow of white color to a plough whose share was made of brass. He ploughed a furrow along the line of the future walls. He took care that the earth of the furrow should fall inward toward the city, and also to lift the plough and carry it over the places where gates were to be made" (E. F. Knuchel, *Die Umwandlung im Kult, Magie und Rechtsbrauch* [Schriften der Schweizerischen Gesellschaft für Volkskunde 15] [Basel: Gesellschaft für Volkskunde, 1919], 98–99). Cf. Joseph Rykwert, *The Idea of a Town: The Anthropology of Urban Form in Rome, Italy, and the Ancient World* (Princeton, NJ: Princeton University Press, 1976), 117–29 and 163ff. For a more recent study on Rome's origins, including its Etruscan roots between history and myth, see archaeologist Andrea Carandini's *Rome: Day One*, trans. by Stephen Sartarelli (Princeton, NJ: Princeton University Press, 2011).

[489] Lat.: "public market place." The *forum romanum* was for centuries the center of Rome's affairs, the public focus of social, political, commercial, legal, and religious life. Its famous ruins still stand between the Palatine and Capitoline hills.

[490] On a symbolic level, Eliade interprets the Roman *mundus* "at once [as] the image of the cosmos and the paradigmatic model for the human habitation. [. . .] The *mundus* was clearly assimilated to the *omphalos*, to the navel of the earth; the city (*urbs*) was situated in the middle of the *orbis terrarum*. Similar ideas have been shown to explain the structure of Germanic villages and towns" (Mircea Eliade, *The Sacred and the Profane: The Nature of Religion*, trans. by Willard R. Trask [Orlando, FL: Harcourt, 1987 (1957)], 47). Cf. Kerényi: "Whether in its primitive form as used in the founding-ceremony or as a solid bit of architecture, the mundus is the ἀρχή into which the older world of the ancestors, the subterranean storehouse of everything that will ever grow and come to birth,

This is the basic form of the Roman *castrum*,[491] and of the cities founded by the Romans.[492] Such circumambulation has survived in many German-speaking regions as the encircling of a municipal district on horseback, undertaken every year for the purpose of affirming fertility and sovereignty in an area. In Switzerland, such circumambulation is performed every year when the peasants, bearing banners and playing music, follow the parson,

---

opens. The *mundus* is at once the relative ἀρχή and the absolute ἀρχή, and from it the new world of 'Rome' radiates in all directions like a circle from its centre" (C. [*sic*] Kerényi, "Prolegomena," in C. G. Jung and C. Kerényi, *Science of Mythology: Essays on the Myth of the Divine Child and the Mysteries of Eleusis* [London: Routledge & Kegan Paul, 1985 (1949)], 11).

[491] Lat.: fortified camp serving a Roman legion (pl. *castra*). The *castrum* (cf. Ital. *castello*, Eng. "castle") formed a roughly five hundred-meter (1,600-foot) square or rectangle. Its main streets were the *cardo maximus*, corresponding to a city's *via principalis*, and the *decumanus maximus*, corresponding to the *via praetoria*. These crossed at the commander's *praetorium* or residence.

[492] One should note that the accounts of city-founding ceremonies by Polybius, Plutarch, Varro, Dionysius of Halicarnassus, and Ovid diverge at certain points, in particular presenting two geometrical forms. While one tradition refers to a circle made by a plough with the *mundus* at its center (Plutarch, *Romulus* 8), Ovid mentions a *fossa* (ditch) that was filled after the sacrifice, with an altar built over it (*Fasti* 4.819); another tradition speaks of "Roma quadrata" (Dionysius of Halicarnassus, *Roman Antiquities* 1.79) and of a rectangular "primal furrow" (*sulcus primigenius*). According to Plutarch, "Roma quadrata," which conformed to the *sulcus primigenius*, was not sited on the *mundus*. In the "Prolegomena" to his collaborative volume with Jung, Károly Kerényi refers to classical philologist and historian of religion Franz Altheim's attempt to solve the contradiction by noting that *quadrata* means "divided into four": "whether you took your stand at the mundus or at the Quadrata Roma, you could draw a circle round you and thus, according to the rules of Roman surveying, found a quadripartite city, a *Roma quadrata*. Such a division into four along two axes is described both in Roman surveying and in augury" (Kerényi, "Prolegomena," cit., 12; cf. Rykwert, *Idea of a Town*, cit., 125–26). Varro takes for granted that the ceremony of the circle was used in founding the Roman *coloniae*—originally the *urbes*, from *orbis*, "round"—and in fact "most of the coloniae show that in reality rectangular city-plans grew out of the ritual circular boundary" (ibid.). In the *Terry Lectures* Jung would expand his heuristic speculations into a wide-ranging metapsychological, cultural-historical perspective, addressing the historical process activated by Protestantism in terms of disintegration of the unity that had been sustained by the Catholic Church and a paving of the way for the advent of contemporary dictatorial states: "With the demolition of protective walls, the Protestant lost the sacred images that expressed important unconscious factors, together with the ritual which, from time immemorial, has been a safe way of dealing with the unpredictable forces of the unconscious [. . .]. This whole development is fate. But one thing is certain—that modern man, Protestant or otherwise, has lost the protection of the ecclesiastical walls erected and reinforced so carefully since Roman days, and because of this loss has approached the zone of world-destroying and world-creating fire" (*Psychology and Religion* [(1937) 1938/1940], CW 11, § 82, § 84). See also above [I], n. 79. And referring to the

mounted on horseback. The whole municipal area follows.[493] Thereby, this district is reconstituted once again as a kind of *temenos*,[494] a sacred precinct, by virtue of which its fertility and sovereign rule over the area are once again reinforced and reaffirmed. This is an ancient magical custom, which, as is well known, is also performed on sacred images. In Tibetan culture, these relics, for instance, a stupa,[495] a mound-like structure, are circumambulated in a clockwise direction. Such movement must, of course, follow the path of the sun. If this movement runs counter to the path of the sun, then that forebodes evil. It's black magic. Where that happens, the force of the sanctuary is employed to serve an evil purpose.[496]

---

overwhelming, potentially destructive character of numinous experiences he affirmed, "How thin these protective walls are can be seen from the positively terrifying suggestion-ability that lies behind all psychic mass movements, beginning with the simple folk who call themselves 'Jehovah's Witnesses,' the 'Oxford Groups' [. . .] among the upper classes, and ending with the National Socialism of a whole nation—all in search of the unifying mystical experience!" ("A Psychological Approach to the Dogma of the Trinity" [1942/1948], *CW* 11, § 275).

[493] See Eduard Hoffmann-Krayer, *Feste und Bräuche des Schweizervolkes* (Zurich: Olms, 1992 [1940]).

[494] Gk.: τέμενος, "precinct"; this originally names the ground set apart for a royal palace or a temple complex, such as parts of the Nile Valley and the Athenian or Argive acropolis. Later the term indicated the sacred boundaries of a sanctuary's enclosure (from its root τέμνω, "to cut," derives Lat. *templum*), and thus typically also the forecourt to the temple itself.

[495] Sanskrit term meaning "foundation of the offering," originally indicating in Buddhism a reliquary. Symbolically, the stupa is the Buddha's holy body-mind as a marker on the path to enlightenment. The Tibetan Buddhist stupa is called *chorten* or *ciorten*. Jung further explains the stupa as a tripartition on a macrocosmic level (purely spiritual world, intermediate world, and world of objects) and a microcosmic level (Self, anima, and body) in his ETH Lectures (C. G. Jung, *Psychology of Yoga and Meditation: Lectures delivered at ETH Zurich, Volume 6: 1938 to 1940*, ed. and introduced by Martin Liebscher, trans. by Heather McCartney and John Peck [Philemon Series] [Princeton, NJ, Princeton University Press, 2020], 52). In *MDR* he describes his visit to the stupas of Sanchi, site of the Buddha's fire sermon (278–80).

[496] In the expanded version of his first Eranos lecture, commenting on the circumambulation of the stupa and *chorten*, Jung maintained that in Tibet "the leftward-spinning eddies spin into the unconscious; the rightward-spinning ones spin out of the unconscious chaos. The rightward-moving swastika in Tibet is therefore a Buddhist emblem," while "the leftward-moving swastika is a sign of the Bön religion, of black magic" ("A Study in the Process of Individuation" [1934/1950], *CW* 9.1, § 564). "In general," he conjectured, "a leftward movement indicates movement towards the unconscious, while a rightward (clockwise) movement goes toward consciousness" (ibid.). In February 1935 he would observe, "Now, in choosing the black swastika turning to the left, the Germans have surely expressed the backward movement in many ways" (*Nietzsche's Zarathustra*, cit., 1:377. See also "Fundamental Psychological Conceptions" [1935], *CW* 18, § 411; and *DAS*, 344). The custom of circumambulating according to the direction of the sun (around the sacred fire, the teacher/guru, and then the temple) is known in India as *pradaksina*, and is attested in such ritual texts as the *Shatapatha Brahmana* and the *Grhyasutra*. Knuchel observes that all

Thus, the strange circular movement described by the figure repeats the ancient circumambulation, which thus also expresses itself under the sign of the spell ring, the magic circle, which is erected against evil spirits. In this way, an inner, sacred district is demarcated from external powers. Now, the *sulcus primigenius* can no longer be crossed by evil powers. You must clearly bear in mind how primitive consciousness feels about these things. For primitive consciousness, we live in a world that is animated in every respect. Trees, bushes, animals, mountains, rivers—everything extant in nature—is filled with life, with dark and eerie life. Everything we extract from nature is shrouded in this strange fluid spirit, which on a very primitive level is not recognized by those who want to have character. These are sparks of spirit or small material souls, sylphs, dryads, and so forth.[497] Consequently, when you are taking an object from nature, you are advised to perform a corresponding rite of demystification to avoid swallowing a natural soul that could affect your stomach. All these things have been preserved in the Catholic Church in a very nice and meaningful way, for instance, the *benedictio fontis*[498] and the

---

processions performed in ancient Indian ceremonies in honor of the gods, as well as circular progresses in marriage rituals, are performed clockwise, in the opposite direction to those in honor of demons (Knuchel, *Die Umwandlung im Kult, Magie und Rechtsbrauch*, cit., 24, referring to the *Shulba Sutras* of Katyayana, 5.9.17, and of Apastamba, 1.9.11). However, the Tibetan Bon (or Bön) religion, which antedated the advent of Buddhism in the eighth century, and was to be suppressed by the latter, assimilated several aspects of Buddhism while maintaining some details distinct: such as the counterclockwise circumambulation of sacred sites. It should be noted is that the Bon literature became accessible only fairly recently to North American and European scholars (see the entry "Bon" in Robert S. Ellwood and Gregory D. Alles, eds, *The Encyclopedia of World Religions*, rev. edn [New York: Infobase Publishing, 2008], 57–58 at 58).

[497] Jung here refers to mythological creatures or elemental beings populating nature, which abound in folklore and mythology generally. A mid-nineteenth century dictionary defines dryads as "nymphs, the offspring of Nereids and Oreads, [who] preside over woods and forests, roving through them day and night," and sylphs as "intelligences with a vaguer sexuality, sometimes take an interest in the cares, studies, and affections of human beings" (Angelo Sicca, *Dizionario di mitologia ossia Dizionario delle favole degli antichi compilato sui migliori autori* [Florence: Tipografia di Pietro Fraticelli, 1845], 105 and 292). Traditionally, the elemental beings are linked with the four classical Empedoclean elements of antiquity, so undines or nymphs preside over the waters, sylphs the air, gnomes the earth, and salamanders fire. Such a quadripartition, adopted to a great extent in Rosicrucian and hermetic literature, goes back to the Swiss physician, alchemist, and astrologist Paracelsus's (1493–1541) posthumously published *Liber de nymphis, sylphis, pygmaeis et salamandris et de caeteris spiritibus*, or *Book on Nymphs, Sylphs, Pygmies, and Salamanders, and on the Other Spirits* (1566).

[498] Liturgical Lat.: "blessing of the fountain"; that is, baptismal water, which, together with the Easter candle (thus *benedictio ignis et fontis*), is employed during the Easter (or the Pentecostal) vigil; salt added to the water makes it *aqua pontica* (sea water), or *aqua permanens* deemed to behave like baptismal water (see Jung, *Mysterium Coniunctionis* [1955],

saldo sanctus.[499] On the night before Easter, holy water is made from ordinary water by disenchanting the latter, in order to extract the *admixtio omnis diabolicae fraudis.*[500] Thus, the possession of objects by elemental forces is repealed by benediction, just as the salt added to the holy water of the Church must be freed from the admixture of the elementary spirits, just as the wax used for the candle, and just as the incense grains

---

CW 14, § 316; *Psychology and Alchemy* [1944], CW 12, § 160). Jung also notes that the sea is mythologically conceived as the "matrix of all creatures" (*Mysterium Coniunctionis*, CW 14, § 100 and § 246), thereby connecting it to the unconscious, which is "both good and evil and yet neither, the matrix of all potentialities" (§ 253). Again, the *aqua pontica* is the alchemical equivalent of the Red Sea understood as healing and transforming baptismal water, as Jung affirms (recalling Saint Augustine's equation of the the Red Sea with baptism, and Honorius of Autun, according to whom "the Red Sea is the baptism reddened by the blood of Christ, in which our enemies, namely our sins, are drowned" [*Speculum de mysteriis ecclesiae*, PL, 172, col. 936]), in *Mysterium Coniunctionis*, CW 14, § 256; he also mentions "the patristic and Gnostic interpretation of the Red Sea as the blood of Christ in which we are baptized; hence the paralleling of the tincture, salt, and aqua pontica with blood" (§ 259). Not only crossing the Red Sea, but also the Flood was considered by some Church Fathers as a prefiguration, or type, of salvation through baptism (Saint Cyprian of Carthage [200–258], Epistle 75; and Origen [ca. 185–254]; note also Saint Paul, who writes [1 Corinthians 10:2] that the Israelites were "baptized into Moses" when they crossed the Red Sea).

[499] Very possibly an incorrect transcription of *sal sanctus*, as this was used in the baptismal rite. In "The Visions of Zosimos," Jung recalls that the idea of holy water derived originally from the Hellenistic philosophy of nature, probably following Egyptian influences, and quotes from the *Missale Romanum* the priest's invocation following the quartering of the water: "Descendet in hanc plenitudinem fontis virtus Spiritus Sancti, totamque huius aquae substantiam regenerandi feconde effectu" (May the power of the Holy Ghost descend into this brimming font, and may it make the whole substance of the water fruitful in regenerative power [Jung's source is J. B. O'Connell and H.P.R. Finberg, eds, *The Missal in Latin and English* (London: Sheed & Ward, 1949), 431]) ("The Visions of Zosimos" [1938/1954], CW 13, § 104 and n. 59; see also § 89 and n. 8). In the revised *Terry Lectures*, Jung adds that the Easter Eve rite "was known as the 'lesser (or greater) blessing of salt and water' from about the 8th century" ("Psychology and Religion" [(1937) 1938/1940], CW 11, § 161 n. 70). On the symbolism of water in relation to baptism, Jung wrote to pastor Fritz Pfäffin, "Through historical development 'the way to baptism' has departed so far from its original meaning that the idea of baptism is left hanging in mid-air because the actual experience of baptism has somehow disappeared. If your unconscious now comes up with the water symbol, this means that it is trying to give you back the experience of baptism in its original form. Originally it was an immersion to the point of death and thus had the significance of rebirth" (Jung, *C. G. Jung: Letters*, cit., 1:192 [July 5, 1935]).

[500] Lat.: "the admixture of every devilish fraud" (also mentioned in *DAS*, 589 and n. 3, and *Nietzsche's Zarathustra*, cit., 1:42). This is very similar to a formula in the "Ordo ad faciendam aquam benedictam," or "Order for the Blessing of Water," in the *Rituale Romanum*. In the course of the exorcism over the salt (which will be mixed with the water at a later point in the ritual), the priest asks God to dispel the "versutia [cunning, guile, artifice] diabolicae fraudis" (*Rituale Romanum Pauli V Pontificis Maximi jussu editum aliorumque pontificium cura recognitum atque autoritate SSMI. D. N. Pii Papae XI ad normam Codex Iuris Canonici accommodatum* [Vatican City, Edito juxta Typicam Vaticanam, 1925], titulus VII, cap. 2, 188 and passim). Probably, Jung quoted from memory and substituted *admixtio* for *versutia*.

must be purged of impure spiritual elements prior to their admixture in the incense cauldron. These actions are performed on account of the age-old recognition of the fact that no natural connection may be interrupted, unless one establishes a spiritual equivalent for it.

Let us now return to our picture. It is quite obvious that the dreamer has in mind something like an animated center. I asked him, "What are your thoughts on this?"

He replied, "It's like a path into something inner, like a labyrinth garden. But I also have the feeling that this somewhere leads into an inner sphere, which one approaches slowly."

Also, however, everything seems to him like our analytical path, like this connecting path that we take to understand our dreams. You sensed this, too. It feels to you as if one were circling around the village with the church [procession], and therefore you asked, why this pussyfooting around?[501] We are beating around the bush, but for a perfectly good reason: namely, because we *cannot* even touch these things. We cannot draw closer to the center. The dreamer felt this and represents these things in a way that suggests that the center can be entered only slowly, as it were, though endless windings, and by no means directly. At the outset, it seems to him as if this were only a circular path. But at some point, precisely at an unexpected, irrational point, the inner way all of sudden begins to move downward. Everything that seemed confusing, erroneous, and sterile is now given meaning as an outer surrounding, an outer manifestation. He notices that the way leads inward unequivocally, and that it reaches a definite end, precisely here, at the center. What this was he did not know. He simply said that perhaps the purpose of the whole analysis was to get onto this path. But what for? What were his thoughts in this respect? He couldn't grasp it. Where did the way lead? Did it lead to him, somehow into his interior? But he had no notion whatsoever of the inner sphere.

If we knew nothing else about such a drawing than what the dreamer contributed to it, then we'd find ourselves in a somewhat impaired situation. Fortunately, however, we have a parallel at our disposal. Nowadays, we know—obviously such knowledge has not always existed—that it is not only dreamers who dream such drawings and images, but also quite a few other people who dream these things, who have neither had analysis with Doctor Jung nor to whom he has made suggestions, as is kindly claimed about me. To this day I am credited with no more than a minimum of scientific character. I have observed these things in all possible corners of

[501] The German expression used here, "Wir gehen wie die Katze um den heissen Brei herum," alludes to dodging something (disagreeable or even dangerous) to avoid getting burnt.

the world, and I know perfectly well that quite a few people have not allowed me to put ideas into their heads. You will find such drawings among
the Pueblo Indians in North America, among the Navajos.[502] A pot has
even been found inside a pyramid.[503] Not even my patients would be capable of creating the turquoise mosaic found inside this pot, which is at
once so beautiful and so simple. You will find this in India as the epitome of
wisdom, in classical Chinese philosophy, and in medieval philosophy, in the
works of Jacob Böhme among others. We shall return to these matters.

These parallels demonstrate that this has to be an archetypal symbol,
and that it must correspond to a natural instinct that such things in fact
are done under circumstances that, shall we say, are born out of necessity.
Thus, for instance, fortified cities are built out of a fear of enemies. To
begin with, these enemies are real enemies. But the enemies are not only
real. The worst are the unreal ones that lurk at crossroads at night, against
whom peculiar apotropaic images must be erected, and who in the first
instance are thwarted with the help of the *sulcus primigenius*. Other people
simply string a thread across the path, to which a small bundle of straw
is tied. This suffices to keep the ghosts away. Or they build a small gate
out of three bamboo sticks, to which *Tschuktschukbeutel* are attached to
medical kits,[504] which contain bones and a medicine man's other bric-a-
brac. One thus prevents the ghosts, which come out of the forest or the

---

[502] US photographer Edward S. Curtis, who for over thirty years studied—and
portrayed—Native North Americans, wrote early in the twentieth century, "Navaho life is
particularly rich in ceremony and ritual, second only to some Pueblo groups. Note is made
of nine of their great nine-day ceremonies for the treatment of ills, mental and physical.
There are so many less important ceremonies occupying four days, two days, and one day
in their performance. In these ceremonies many dry-paintings, or 'sand altars,' are made,
depicting the characters and incidents of myths. Almost every act of their life—the building
of the hogán, the planting of crops, etc.—is ceremonial in nature, each being attended with
songs and prayers" (Edward S. Curtis, *The North American Indian: Being a Series of Volumes Picturing and Describing the Indians of the United States and Alaska*, ed. by Frederick Webb Hodge, vol. 1: *The Apache. The Jicarillas. The Navaho* [Cambridge, MA: University Press, 1907], 138).

[503] Reference to a circular limestone jar about a foot high found by Earl Morris (leader
of a Mayan expedition by the Carnegie Institution of Washington) in the so-called Temple
of the Warrior at Chichén Itzá (as added in Jung's hand in his copy of the Seminar). It contained "a wooden plate on which was fixed a mosaic design. It was a mandala based on the
principle of eight, a circle inside of green and turquoise-blue fields. These fields were filled
with reptile heads, lizard claws, etc." (*DAS*, 115 and n. 2).

[504] Austrian dialect and folklore. *Tschuktschukbeutel*: apotropaic sacks filled with objects intended to repel strangers and interlopers. [T.N.] In his personal copy, Jung made the
following correction: "to which kits are attached" (und hängen ~~Tschuktschuk~~beutel [over-
writing "zu sie [*sic*] daran ~~Medizinbeutel~~" with "Beutel" (pouches)] and deleting the German for "medicine bags").

desert every night, from entering the kraal or village. Primitives believe that the dead are always aggrieved at having died. They are foreover impatient until the living have joined them. And if the living hesitate, then the dead return to fetch them. This must be prevented, because if the ghosts enter the kraal, someone will fall ill and the person afflicted can be cured only if the ghosts are driven out of the kraal. To prevent that from happening, a circular fortification is built, as it were. Thus, ghost traps, magical fences, and magic circles are erected, whose purpose is to keep the spirits outside.

These are, of course, only psychological projections. However, they are very real on that level, and nothing whatsoever is gained from referring to them only as projections. All this must be enacted in reality. If it's not actually performed, it has no effect for primitives. But what the primitive *does* is what the *thought* or *drawing* is supposed to bring about within the dreamer; that is, it's meant to create an inner *temenos*, an inner sacred precinct that liberates one from all *admictio* [sic] *diabolicae fraudis.*[505] Nothing should come near this precinct from outside, nothing that somehow injures and soils the interior; instead, this inner zone should be a pure receptacle, which is protected against everything without. Nothing should be brought to these things from outside that doesn't already belong to them from within. That's the meaning of such a *circumambulatio.*

I have presented a number of terms designating the contents of such drawings, such as a sacred precinct. There's a great deal of evidence for this, such as Eastern mandalas. I must explain this term. "Mandala" is Sanskrit for circle, magic circle, or ring; the word is used in various ways. In this usage, it's a so-called circle yantra, a cultic instrument that supports the projection process. Such mandalas are produced on a day-to-day basis in many places in India—for instance on a sheet of paper, or on a plate made of leaves, or from woven straw, which is offered to the gods. This is how the day begins. Then the mandala is either thrown into the river or burned. That's the morning prayer, and with it one declares, "I have cleansed my receptacle. Nothing has entered. I have freed myself from every impure mixture. Thus this day may begin under this sign."[506] The

---

[505] See n. 500 above.

[506] See above [I], 37–38 and 62ff., and [ZL], nn. 357 and 358. In *Psychology of Kundalini Yoga*, cit., Jung (speaking of Tantrism, and of the role of mantra in the magic rituals of sadhana practice), described the yantra as "a cult image, in the center of which the god is depicted. Through intensive contemplation of the god [in the yantra], it comes alive. The viewer enters into the god, and the god is in the viewer (identification with the god, deification of the viewer). This method may be used to attain unity with the All, but also for sheer

mandalas that we know stem chiefly from Lamaist Tantrism and usually have a Tibetan origin. They are drawings of circles that always contain a specific symbol, namely, the symbol of this temenos, that is, a sacred precinct, a wall that zones off a plot of land. Inside stands a temple. This walled property is the temenos, that is, a forecourt. In the Tibetan or Lamaist mandala, the area within is a cloister, a quadrangular walled space with four gates, precisely like the Roman castrum. These four gates are marked by four qualities. This is the temenos itself: quadrangular, but encircled by fire, which in turn is encompassed by the circle of suffering, the torments of hell and the burial ground, where the souls and bodies of the dead are torn to pieces by demons—in accordance with the Buddhist idea that fire augments suffering and brings forth death and all the torments of hell. The circle of fire, the protective magic circle, has been furnished with many Tantric signs. It has the classic shape of either the thunderbolt, lightning rod, or diamond wedge. It all probably comes down to the same meaning, which simply expresses gathered and focused energy. It forms a magic circle and protects the yogi against the fire of desire, which introduces all contaminating admixtures and must therefore in particular be shut out or cut off.[507]

*   *   *

magic with wordly aims (fakir trickery)" (72). See also Martin Brauen, *Mandala: Sacred Circle in Tibetan Buddhism* (Boston, MA: Shambala, 1997); Georg Feuerstein, *The Deeper Dimension of Yoga: Theory and Practice* (Boston, MA: Shambhala, 2003), ch. 68.

[507] Cf. *DAS*, 479; see also ibid., 465.

LADIES AND GENTLEMEN,

Let us continue with these parallels, so that you may build up a sense of the nature of such an archetype. As already noted, this can be found in all possible regions of the world. Discussing the Lamaist mandala, I mentioned that it is the inner realm, and so we come to the part where the spiral way begins off in some corner. What's most important seems to happen here, at the center.[508] As mentioned in several ways, the patient's dreams have provided us with some hints about that: for instance, that a dinner table stands at the center, a table of communion, or that a ball game is being played, or indeed that a path winds itself like a spiral, a path with a head at its end and which thus resembles a coiled snake. It's very difficult to readily establish connections between these hints. In actual fact, we have to consider each one in turn.

I have already begun doing so by discussing the ritual game of pelota. Perhaps you thought that that was yet one more of those wicked, elusive byways. I referred to the game to adduce, at least temporarily, some notion of the fact that this inner zone is a place where rituals take place, in fact strange things, which also preserved themselves in the Christian Church during the Middle Ages, but which somehow have also become extinct. For that reason I also mentioned the touching story of the burial of "Halleluja," who is buried as a woeful lump of soil in the middle of a cloister garden. The cloister, as you know, is usually a quadrangle inside which lies a cloister garden. That's where the "mournful Halleluja" is interred. So it's also a burial site.

The ball game, the communion, the coiled snake: of course, it's impossible to interpret all these things with such ignorance. Thus, we must possess quite some knowledge to come to grips with these things. For instance, we need to establish some kind of concept for the possible meaning of the sacred ball game. In that connection, I must supply a few references, or better only one in particular. The early Christians were accused, for instance, of playing such ball games using small children. Christians accused the Gnostics of the same misdemeanor, namely, that they would toss a child around among themselves until the child was dead. Similar claims have been made about the Jews, extending as far as the accusation

---

[508] In "Concerning Mandala Symbolism" [1950], CW 9.1, § 647ff., he deals with a woman patient's painting of a mandala in which a snake's head and tongue begin to penetrate a structure surrounding the center, at a juncture like this one ("off in a corner"), in that respect paralleling the drawing by the male patient here. The "corner" turned in each case precipitates the inward development (picture 4 being, in the woman's series of twenty-four pictures, the "turning point of the whole process" [§ 562]). [T.N.]

of ritual murder.[509] Such claims were always advanced whenever it concerned a matter that the ordinary person could simply not imagine. If something happens that eludes his understanding, the ordinary person asks himself, "Well, what could this be?" It always has to be some obscenity or even a crime. Because it is dark. What is dark, moreover, is evil. When he no longer grasps anything at all, that is a catastrophe. Then something strange definitely must be at work. Consequently, everyone standing around outside assumed that the Christians played such games where their mysteries took place: that is, in their baptisteries and circular temples, in such a way that a small child was brutally murdered. What does such a projection mean?

ANSWER: Hatred.

Well, people certainly weren't on good terms.

ANSWER: Well, they have this need.

Precisely not. That explains why they assumed others did such things. However, they were getting their bearings on the matter in one way or

---

[509] In *DAS* this is tied to the *jeu de paume*: "The idea of a ritual game survived until about the thirteenth century. They really used to play ball in the churches, the jeu de paume, and this gave rise to the rumor that Christians killed a child by tossing it to one another like a ball till it died. The Gnostics accused the Christians of this, and the Christians in turn accused the Jews. There was a rumor in Bohemia only thirty years ago that the Jews had killed a child, a ritual murder. The jeu de paume had a ritual meaning just as carnival had. In the monasteries during the spring carnival, they used to reverse the position of the abbot and the young lay brothers, the youngest lay brother became the abbot, and vice versa" (25–26). No source for these accusations, or for the rumor of the Bohemian ritual murder, has been found. Jung would discuss them further in the Visions Seminar, in relation to child sacrifices and the Black Mass and, again, the story so widely read by Christians about Gnostic heretics (forming a circle and tossing the child until its death), adding, "These rumors are really the first inklings of that idea of the *jeu de paume* which was played later in the church" (*Visions*, cit., 1:223–24). He then connects the magical meaning and purpose inherent in the *jeu de paume* ritual (and its sacrificial original nature, as Jung seems to maintain here) to the "preoccupation of the primitive mind" with regard to helping the sun to rise (224). He briefly mentions the accusations in *Psychology and Alchemy* [1944], CW 12, § 182, and "Psychological Aspects of the Kore [1941], CW 9.1, § 324. Indeed, the early Christians, accused of ritual infanticides and cannibalism by pagans, reacted by pinning the same allegations on the Jews. The Christian charge of ritual infanticide by Jews with the purpose of using Christian blood in the matzoh and wine used to celebrate Passover, thereby gruesomely parodying the Eucharist, was consonant with several anti-Judaic stereotypes, such as the stabbing of the Mass communion wafer and the poisoning of wells. These false accusations dating back even to medieval Christianity were exploited as pretexts to justify persecutions and pogroms even into the twentieth century: see, for example, Stephen Benko, *Pagan Rome and the Early Christians* (Bloomington, IN: Indiana University Press, 1984), and Alan Dundes, ed., *The Blood Libel Legend: A Casebook in Anti-Semitic Folklore* (Madison, WI: University of Wisconsin Press, 1991).

another. These people, the Gnostics and early Christians, grappled with a mystery that simply existed at the time, and had to emerge: namely, a higher level of consciousness. Like the Christians themselves, the pre-Christian Gnostics sought to attain a spiritual construct which, however, was completely misapprehended by their surroundings. Consequently, their attempts necessarily resulted in *mysteria*.[510] So the sacraments were not referred to as *sacramenta* in the early Church, but as *mysteria*. For instance, baptism and the Lord's Supper were *mysteria*, and they were addressed and celebrated as such.[511] Hence the separation of the bapistery from the principal church building, into which the commoners could flock.[512] Only allegorical

---

[510] In "Transformation Symbolism in the Mass" [(1942)/1954], CW 11, after mentioning the gradual transformation of the originally esoteric mysteries into a broadened experience, in ancient Egypt as well in Greece, to the point that "at the time of the Caesars it was considered a regular sport for Roman tourists to get themselves initiated into foreign mysteries," Jung wrote, "Christianity, after some hesitation, [. . .] made celebration of the mysteries a public institution, for, as we know, it was especially concerned to introduce as many people as possible to the experience of the mystery [. . .]. The ground was prepared from the realization that, in the mystery of transubstantiation, it was not so much a question of magical influence as of psychological processes" (§ 448). "What we now call the sacraments of the Church"—Jung recalls elsewhere—"were the mysteria of early Christianity" (*The Tavistock Lectures* [1935], CW 18, § 254). See also Jung, *Nietzsche's Zarathustra*, cit., 1:79–80.

[511] The use of the term *mysterium* was linked, as illustrated here, to the initiatiatory nature of Christian ritual: the doctrinal teaching or catechism in this connection was an initiatic introduction to the *mysterium* of baptism, inaugurating the sacramental life that culminates in the Eucharist. Apart from a single instance in the Gospels and three in the Apocalypse, *mysterium* occurs frequently in the Pauline letters. Paul's precedent for this treatment of *mysterium* were the pagan mysteries, according to Richard Reitzenstein, *Die hellenistischen Mysterien-religionen* (Leipzig: Teubner, 1910). The term *sacramentum* was first used by Tertullian (see above, n. 442) with reference to the baptism, as entrance into the so-called *militia Christi*. Only later in the West did it take on its present meaning, referring exclusively, that is, to rituals pertaining to salvation and sanctification (which the Greek Orthodox Church still calls "mysteries"), while "mystery" meanwhile assumed other meanings in Church usage such as in the liturgy and in catechism. We speak, for example, of "the chief mysteries of the faith" in the sense of the truth of dogma believed by the faithful. The *mysteria* are also called *mysticae figurae*, *intima sacramenta*, etc. In this regard, the *sacramenta* are the *corporalia mysteria* and the *mysteria* are the *spiritualia sacramenta*. For more on this subject, see Louis Bouyer, *The Christian Mystery: From Pagan Myth to Christian Mysticism*, trans. by Illtyd Trethowan (Petersham, MA: Saint Bede's Publications, 1990); Henry de Lubac, *Corpus Mysticum: The Eucharist and the Church in the Middle Ages* (Notre Dame, IN: University of Notre Dame Press, 2007).

[512] The baptistery (Lat. *baptisterium*) is the building adjacent to a (generally Italian) church, where baptism is performed. The first baptisteries appeared when immersion in the "living water" (in springs, rivers, or the sea) was abandoned, and also because ritual blessing of the water required situating the sacrament in a specific place. The term goes back to Rome's pre-Christian thermal baths. Separation of the site took place in the first centuries CE, when unbaptized neophytes were not admitted to liturgical celebrations inside the sanctum. The octagonal structure of most baptisteries draws on the symbolism of the eighth day, representing the reversal of death and prefiguring resurrection in the grace of Christ.

language was used to allude to these customs. For this reason, the Lord was often not mentioned by name, but instead was referred to mystically as a fish, Greek *ichthys*,[513] or *poimen*,[514] or indeed as a shepherd.[515] He was also compared to Orpheus,[516] or expressed by other symbols, which were sometimes necessary to shroud the secret and to prevent profane hands from soiling what was most sacred. However, there were also other reasons for this curious symbolizing, such as the fact that the perfectly ordinary, historical, matter-of-fact designation was not solid enough to express everything performed in the sacred [baptismal] space. The full force of images was called upon to supply this rich array of manifestations. At the same time, these early Gnostics and Christians were modern people who strove for a higher level of consciousness, and who needed the mystery to protect that higher level against the misunderstanding of the blind mass of people. The latter had no knowledge of what these moderns were doing, nor would they have understood. Therefore they could project their own fantasy into this darkness, and so they projected this strange idea that a child was toyed with, cruelly, until it died. The little-known difference between the mystery initiates and the profane multitude is that the latter are childish, whereas those who attain a certain higher level of consciousness have overcome that childishness. Consequently, it must have seemed to the multitude as if Christians had subjected a child to fatal torture. That is what was projected, and attributed to these people as a crime, just as the intelligent or more conscious person is mostly seen as frightening, a source of doubt, dangerous or immoral.

Bourget's novel *L'Étape* includes a charming episode in this respect.[517] It tells the story of a married couple, from the Parisian petite bourgeoisie,

---

[513] *Ichthys* spells in acrostic form "Jesus Christ, son of God and Savior" (Ἰησοῦς Χριστός Θεοῦ Υἱός Σωτήρ, romanized as "Iesous Christos Theou Huios Soter" (see Augustine, *The City of God* 18.23). This symbolism goes back to Titus Flavius Clements or Clement of Alexandria and Tertullian (*De pudicitia* 7.1.10, 12ff.) In a famous passage, the latter compares Christians to "little fishes" who follow the image of *ichthys*-Jesus Christ (*De baptismo*, ch. 1). During the persecutions in ancient Rome, Christians apparently used this symbol as secret identifying mark. Images showing the symbol of the fish are still visible in the catacombs of Saint Callisto in Rome, dating from around the middle of the second century.

[514] See *Aion* [1951], *CW* 9.2, particularly section 6, "The Sign of the Fishes."

[515] This is among the oldest symbolic representations of Christ (Psalm 23 begins, "The Lord is my shepherd"; John 10:11 has Jesus proclaim, "I am the good shepherd").

[516] For instance, the epithet "our Orpheus" for Jesus—referring to the ancient Greek hero, son of Apollo and the muse Calliope, endowed with extraordinary poetic and musical skills—has long appeared in Christian hymnody.

[517] Paul Charles Joseph Bourget (1852–1935), psychological novelist, conservative intellectual, monarchist, anti-modernist, and critic of democracy, converted to Catholicism in 1901; *L'Étape* (1902) was the first of four so-called "Catholic novels." In his *Essais de psychologie contemporaine* (1883) and *Nouveaux essais de psychologie contemporaine*

who have been summoned to the Ministry of Culture. Lacking a proper invitation, they must bow and scrape since they have come to petition the minister himself. They are asked to wait in the outer office, amid an incessant to-and-fro of all kinds of people entering and leaving. Gesturing toward these visitors, the lady finds these proceedings immensely important: "The gentleman over there is a member of the high diplomacy. That one over there is from this and that department," and so forth. Then a man enters, the novel's protagonist, a philosopher who leads a life completely removed from public affairs. He, too, has been been summoned to see the minister, who wishes to consult him on a matter concerning education. He enters the outer office, the lady sees him, nudges her husband, and says, "Vois-tu cet homme là? Il est de la police secrète! Il a l'air si méchant."[518]

Now, that's what happens all too easily to the conscious person, more developed and no longer childish, who arouses suspicion precisely because his actions are neither understood nor seen straight. That, of course, is the Promethean sin of higher consciousness: namely, that it affronts all lower consciousness and also offends nature. Consequently, higher consciousness casts a deep shadow. The brighter the Promethean fire burns, the deeper the shadow it casts, and the more dangerous the envy of the gods.[519]

It follows, quite obviously, that the strangest ideas are entertained about the sacred mysteries taking place within this center. These ideas are incredibly diverse. I'm referring to symbolic ideas, because we are concerned here with the plainly unknowable. It's extremely interesting to observe how the unknowable, which is nevertheless thought to be effectual, is symbolized by the human mind. Earlier on, I mentioned some of these symbols from the Litany of Loreto. Now there are other symbols in Lamaist Tantrism, which all sound quite different to us, but which nevertheless designate the inscrutable nature of these matters. I feel somewhat embarrassed discussing these things, since we have in our midst an exquisite expert on mandala symbolism, Professor Zimmer. So may I ask you,

---

(1885), he subjected the customs and protagonists of nineteenth-century intellectual and literary history to psychological investigation.

[518] "See that man over there? He's from the secret police! He has such a malicious air about him!"

[519] In Greek mythology, Prometheus—whose name means "forethought"—is the Titan known for stealing fire from Olympus and giving it to humanity in the form of development of knowledge, technology, and civilized arts, then serving as a champion for humankind. For his hubris towards the gods, as narrated in Hesiod's *Theogony* and other tellings of myths, he was condemned by Zeus to be eternally bound to a a rock with an eagle eating his liver. Jung famously wrote, "To be 'unhistorical' is the Promethean sin, and in this sense the modern man is sinful. A higher level of consciousness is like a burden of guilt!" ("The Spiritual Problem of Modern Man" [1928/1931], CW 10, § 152).

Professor, to take the floor to say a few words about what Tantrism perceives at this center? Please do take my chair.

PROFESSOR ZIMMER: I shall do so, although reluctantly. However, this field obviously holds an extreme fascination for all of us, though it perhaps fails to afford instant gratification and, if I might add, will by no means unravel itself even after prolonged contemplation. Fortunately, Lamaist literature does not explain matters, but merely observes them as they are. There again, all this material inside the mandala presents these people with no problem whatsoever. These matters are practiced without being pondered. Practice and contemplation mutually exclude one another. At the moment when people start thinking about things, they would already no longer be involved and might just as well desist. I must say that I enjoyed looking at these things, but did not think about them, because I told myself that nothing would come of my contemplation. But instead things must happen so that one suddenly grasps a mandala on the basis of connections of some kind that have established themselves underground. Nothing can be more tedious or desolate, however, than someone reading the literature in the hope of seeing daylight. In fact, nothing will dawn on him; while he can obviously acquire general knowledge and come to a psychological conception of the sphere in which these things stand, he should rather leave off trying to understand them all.

PROFESSOR JUNG: Please tell us what kind of figures can be found at the center.

PROFESSOR ZIMMER: Simply every possible kind of figure can be found there. In actual fact, that is quite obvious, since in the Tantrist cult the individual does nothing other than create a deity from within himself, in general only the congenial sort, of course, not the evil kind. But he can also conjure up evil ones if he wants to harm others. What he sees at the center, however, can turn into everything. This corresponds exactly to the Hathayoga saying that Shiva has 21,600 shapes.[520] It doesn't matter how many shapes one assumes. The human being can assume only a few dozen shapes. Shiva, however, sits in the middle and can metamorphose into every animal.[521]

---

[520] On Shiva and Shakti, see above [ZL], 121–22, and passim.

[521] Elsewhere Zimmer writes, "Shiva is apparently, thus, two opposite things: archetypal ascetic, and archetypal dancer," being, viewed from one side, total calm, and from the other, total activity; his "aspects are the dual manifestations of an absolutely non-dual, ultimate reality" (*Myths and Symbols*, cit., 167; cf. also 151). Besides the *smarta* (traditional) cult of Shiva, this god's attributes find expression in various other sects, the oldest of which is the *pashupata*, where Shiva is called "Lord of the domestic animals."

PROFESSOR JUNG: You will gather from this demonstration how useful it is if one doesn't know too much, because then one cannot surmise that one knows anything, whereas if one is burdened by as much knowledge as Professor Zimmer has, it is less than obvious where to begin. There are so many motifs that he can't let them pass. From his very interesting presentation, we may gather that at this center sits Shiva, who can create and destroy everything, who can turn into 21,600 figures, all of which he can ensnare, and moreover that the Tantrist mandala can contain any kind of deity. But the inherent deity is Shiva. Therefore, the matter is both utterly intangible and inexpressible. That is why it expresses itself in millions of forms and is inexhaustible. Whatever human fantasy can conceive is pressed into service, as it were, to articulate this center, so that not only a ball game or banquet takes place at the center, but an infinite number of other things. Everything that is life, everything alive, happens there. Because just there is the inception of all creation. In ordinary mandalas, Shiva is therefore depicted in the embrace of his wife Shakti, the many-armed Shiva dissolving into countless duplications of himself and gradually filling the infinite spaces of the universe with himself and his divine wife. Such is Eastern fantasy.

As a rule, our Western mandalas are not blessed with such abundance, but instead are always something specific and always seek to grasp something specific, without, of course, quite attaining it. One can, however, at least say that this center, which is reached by a laboriously entwined, mysterious path, expresses an inner experience, which from time immemorial has been branded with the image of God.[522] For this reason, a divine symbol fills this center, a symbol of God in medieval representations, such as those produced by Hildegard von Bingen,[523] or Jacob Böhme,[524] or those

---

[522] On this topic, cf. "Psychology and Religion" [(1937) 1938/1940], CW 11, § 157.

[523] Hildegard von Bingen (1098–1179), German Benedictine nun, abbess, and mystic, canonized by the Catholic Church. She was an innovator in the language of mystical experience and was among the few women to be influential in medieval philosophy. Her work embraced poetry, music, philosophy, the natural sciences, and medicine. She was in correspondence with several luminaries of her own day, some of whom she counseled spiritually, and took vigorous political stands, challenging the emperor Frederick Barbarossa, who had been her protector, when he backed two antipopes against Pope Alexander III. The impressive character of her visions, as depicted by her scribe Volmar, was quickly recognized by Bernard of Chiaravalle, among others. In her *Liber divinorom operum* (1174) the cosmos is shown in round form growing in the breast of a divine figure that represents the macrocosm. See also Avis Clendenen, "Encounter with the Unconscious: Hildegard in Jung," *Jung Journal* 3, no. 1 (2009): 39–56.

[524] Although the German philosopher and theologian Jakob Böhme (1575–1624), a preeminent modern mystic, is poorly represented in histories of philosophy, his writings on mystical theology, his cosmology, and his labyrinthine prose have been a storehouse of

found in the glass windows of old cathedrals. As you know, Saint Augustine conceived of God in terms of the circle, which is the symbol of the most perfect figure.[525] Thus here we can think of a strange experience, which is proposed quite unexpectedly by the unconscious.

So if you recall why this man consulted me, and observe what his unconscious has planned for him, you will probably ask yourself whether we are not completely and utterly mistaken. Is this not an enormous deception to which we are abandoning ourselves? And should we not retrace our entire progression so as to establish where we began, whether we have taken the correct path, and whether somewhere or other we have abandoned ourselves to a deception and backhandedly smuggled something into the equation? But if we come across such symbols where such

---

insights for European, particularly German, metaphysics, beginning with Spinoza and including Hegel, Schelling, Schopenhauer, Nietzsche, and Heidegger. He also deeply influenced the visionary work of William Blake. Böhme's first book, commonly known as *Aurora: The Dayspring, or Dawning of the Day in the East*, became a source of local theological scandal and remained incomplete. He went on to author, among other works, *The Threefold Life of Man* (1620), *The Treatise of the Incarnations* (1620), *The Six Theosophical Points* (1620), and *Mysterium Magnum* (1623). For further detail, see below, 246–47 and n. 534.

[525] Elsewhere Jung writes, "The Christian mandalas probably date back to St. Augustine and his definition of God as a circle" ("Brother Klaus" [1933], *CW* 11, § 484; see also *Psychology of Yoga and Meditation*, cit., 202, and n. 467). See too "The Philosophical Tree" [(1945) 1954], where he states that the circle, apart "from the the point itself, is the simplest symbol of wholeness and therefore the simplest God-image" (*CW* 11, § 557). The commonly cited phrase "Deus est circulus cuius centrum est ubique, circumferentia vero nusquam" (God is a circle whose center is everywhere and circumference nowhere [ibid., § 457 n. 4]), derives from the twelfth-century pseudo-hermetic text the *Liber XXIV philosophorum*. Cf. "Psychology and Religion" [(1937) 1938/1940] *CW* 11, § 92 and "A Psychological Approach to the Dogma of the Trinity" [1942/1948], *CW* 11, § 229, and ibid., n. 6, which also refers to the form "'Deus est sphaera infinita' (God is an infinite sphere) [. . .] supposed to have from the *Liber Hermetis, Liber Termegisti*, Cod. Paris 6319 (14th cent.); Cod. Vat. 3060 (1315)." The image was commonly invoked by medieval theology and has been also attribuited to theologians such as Saint Augustine and Alain of Lille, and the hermetic tradition. A variant attributed to Saint Bonaventure begins, "Deus est figura intellectualis cuius centrum" (God is an intelligible sphere whose centre) (Bonaventure, *Itinerarium mentis in Deum*, 5.8; cited by Jung, *Psychological Types* [1921], *CW* 6, § 791 and n. 74). Cf. Angelus Silesius: "God is my center when I close him in; My circumference when I melt in him" (Angelus Silesius, *Cherubinischer Wandersmann* [1657], trans. by Campbell, *The Mythic Image*, cit., 64). In this vein Jung affirms in *Psychology and Alchemy* [1944], "The Self is not only the centre but also the whole circumference which embraces both conscious ad unconscious" (*CW* 12, § 44). See also "A Psychological Approach to the Dogma of the Trinity" [1942/1948], *CW* 11, § 229 n. 6, and *Mysterium Coniunctionis* [1955], *CW* 14, § 41 n. 42; and cf. the related discussion in *Nietzsche's Zarathustra*, 2:1048–49 and n. 7 at 1049. Note too that the English poet and cleric John Donne (1572–1631) wrote, "One of the most convenient hieroglyphics of God is a circle, and a circle is

thoughts emerge from the innocent fantasy of a human being unencumbered by any knowledge whatsoever, simply growing, like a flower that opens at night, then one can probably not but be convinced, and at least advance the initial working hypothesis that the unconscious is indeed aimed at something of the sort. I can assure you that if my experience were limited to this case, I would never dare suggest this parallel. But over many years, I have seen these things repeat themselves quite often. For a very long time, I had doubts. I waited for seventeen years until I uttered a word to anyone about this, because I was far from certain whether it was actually true.[526] Only when I was absolutely certain did I claim that such a thing as the collective unconscious, which contained archetypes, exists.

As you will probably realize, however, it's extraordinarily difficult to gather evidence, because one depends on material things, and these matters are simply unknown. When I speak to an audience of doctors about such matters, I do so differently than if I were addressing an audience of historians, philologists, or philosophers. What do they know about the human soul? In this respect, one is forever caught between stools, and one must, as it were, wait things out, until there are enough people who possess enough inner experience to know that these things actually mean the psyche, and until the same people have at least some notion both of an archetype and of a collective image. Reading the works of Lévy-Brühl, for

---

endless. Whom God loves, He loves to the end" (Donne, *The Sermons of John Donne*, ed. by Evelyn M. Simpson and George R. Porter, 10 vols [Berkeley, CA: University of California Press, 1953–62], vol. 6 [1953], 173). The circle or disk is very frequently an emblem of the sun, while the point "signifies unity, the Origin and the Centre. It also represents the principles of manifestation and emanation, and hence in some mandalas the centre is not actually shown but must be imagined by the initiate. There are two kinds of points to be considered: that which has no magnitude and is symbolic of creative virtue, and that which—as suggested by Raymond Lull in his *Nova Geometria*—has the smallest conceivable or practicable magnitude and is a symbol of the principle of manifestation" (entry "point," in Cirlot, *Dictionary of Symbols*, cit., 248).

[526] Jung refers to the period beginning 1912–13, which led to the *Liber Novus* (*The Red Book*). In a similar way, commenting on the unconscious and hypnagogic images with striking symbolic content produced by one of his patients, he affirmed—rather apodictically—in 1931 that "[o]f course, we need certain terms today for the inner things if we want to talk about them. For this reason, but also out of the motive of *captatio benevolentiae*, people often use the terminology coined by myself. This is why I hesitated for seventeen years before publishing something about mandala psychology. I knew that the moment I did that I would no longer obtain any genuine mandala drawings" (*On Active Imagination: Jung's 1931 German Seminar*, ed. by Ernst Falzeder, trans. by Ernst Falzeder with Tony Woolfson [Philemon Series] [Princeton, NJ: Princeton University Press, forthcoming]).

instance, who has established the notion of *participation mystique*,[527] you will come across parallel terms—*représentations*; collective mythological images, general images of universal validity—which have a collective effect and are also recognized. These are archetypes. It wasn't me who invented the term, but I merely employ it in this special psychological application. In actual fact, the term turns up in Augustine's writings, for what according to Plato can be called primordial images.[528] The basis of our nature is a realm of images, and these images are ineffable; they are empty, as if they did not exist. And yet our experiences express themselves in solid forms. If we had no individual experience whatsoever, none of these archetypal images could appear. If we have individual experience, a preformed image fulfills itself. Our individual experience does not preserve the individual shape in which we have experienced it, but pours into the performed shape and fills this with an invisible drawing, which is given to everyone in the soul, which, if these drawings could be copied, would entail a world of inexhaustible vividness. You will find ample notions of this world of images both in Eastern philosophy and in the ideas of the Pleroma,[529] and among the Gnostics and in the works of Plato, who observes that these images are stored in a transcendent place, in the realm of fullness, which it is impossible, however, for the human being to see, and which becomes vivid only through individual experience.

Now, here we have such an instance: my dreamer's confusion, that is, about what the machine is doing, which runs in parallel to the confusion that you experienced yourself. This confusion is an impression, an

---

[527] On this concept, used by Jung from 1912 onwards and borrowed from the French anthropologist Lucien Lévy-Bruhl (1857–1939), author of *Les Fonctions mentales dans les sociétiés inférieures* (1910) (*How Natives Think* [1912]), *La Mentalité primitive* (1923), *L'Âme primitive* (1927), and *L'Experience mystique et des symboles chez les primitives* (1938), see Sonu Shamdasani, *Jung and the Making of Modern Psychology: The Dream of a Science* (Cambridge: Cambridge University Press, 2003), 290–93, 296–97, 314–16; and Robert A. Segal, "Jung and Lévy-Bruhl," *Journal of Analytical Psychology* 52, no. 5 (2007): 635–58.

[528] For Jung's in-depth treatment of this vast theme, see, for instance, *The Relations between the I and the Unconscious* [1928], CW 7, and "The Archetypes of the Collective Unconscious" [1934], CW 9.1. For a close examination of the historical, philosophical, and epistemological roots of Jung's medical-psychological constructs, see Shamdasani, *Jung and the Making of Modern Psychology*, cit., passim.

[529] The Greek term πληρωμα generally refers to the complete array of divine powers. It means "fullness" or "totality," and was used both in Gnostic and in Christian contexts (e.g., Paul in Colossians 2:9). A fundamental element of Gnostic cosmology, the pleroma designates the light beyond the earthly realm, whose emanations are the Aeons or Archons. It enters into Philemon's teaching in Jung's *Red Book* (C. G. Jung, *The Red Book [Liber Novus]*, ed. by Sonu Shamdasani, trans. by Mark Kyburz, John Peck, and Sonu Shamdasani [New York: Norton, 2009]): see "Scrutinies," 347–55.

experience. It is something that has somehow seized him, since from the outset he had the idea that I was not completely mad, but instead a very balanced individual, who does not necessarily utter confusing things. And now I had led him onto an eerie arabesque path that he could not understand. This had caused him such problems that in the second dream he seriously considered whether or not he should continue. The thought definitely occurred to him—but he didn't actually utter this thought—whether I was in my right mind, because *he* was confused. This experience is the individual experience that fills the primordial image—the primordial image of his paradoxical, incomprehensible, living central creature or nature that inhabits him. The dreams themselves help him to somehow express this peculiar creature. They already indicate the primordial forms, which of course become more differentiated over the course of the subsequent dreams.

I have deliberately omitted something, although I hinted at it earlier. These are the four paths that issue from the intersection of the *via decumana* and the *via principalis*, whereby either four or eight or indeed twelve gates are formed. Multiplications by four exist as well. These things, which are standard matters in theory and are related to the padma or lotus or mandala, are elementary factors for every educated person. For this reason, they turn up precisely in Western mandalas with the greatest regularity. Ordinary Lamaist padmas are also formed out of four, eight, twelve, or sixteen rays, and so forth. These vivid mandalas as a rule possess such fourness. There are exceptions, of course, particularly among linear mandalas. This fourfold quality coincides in a strange way with the early Greek intuition of the great Pythagoras—namely, that everything is based on the tetractys,[530] the set of four. This structure before us contains the

---

[530] For Pythagoras of Samos, philosopher and mathematician of the sixth century BCE and his followers, called *mathematikoi*, the *tetraktys* (Gk. τετρακτύς), a triangular representation of the sum of the first four natural numbers (positive whole numbers), informs cosmic structure, symbolically joining earth, sky, and the human realm. A ternary law governs the constitution of living beings, and a septenary law their evolution. Jung deals with the tetraktys in *The Relations between the I and the Unconscious* [1928], CW 7, ch. 2.4; "Psychology and Religion" [(1937) 1938/1940], CW 11, § 61, 90, and 92, and "A Psychological Approach to the Dogma of the Trinity" (1942–48), CW 11, § 246. Already in *Psychological Types* [1921], he included the Pythagorean tetraktys along with the quaternity among the "empirical symbols" of the Self (CW 6, § 791). As has been noted, "The model of the tetrad, ten dots or *alphas* arranged into an equilateral triangle, makes it visually obvious that the addition of four independent loci of elements results not in a fifth locus, but rather in the trascendent number ten, the number that Pythagoras believed to be 'the very nature of number'" (Lawrence Durrell, *Comprehending the Whole*, ed. by Julius Rowan Raper, Melody L. Enscore, Paige Mattey Bynum [Columbia, MO: University of Missuouri Press,

number four. Apart from its quadrangular shape, this mandala contains various other things; in actual fact, fourfoldness is not clearly indicated in it. In Tantrist and other mandalas, it moves increasingly into the foreground with ever greater differentiation, and is then indicated regularly in the four complementary colors. The colors are mostly complementary, but not always. These four are symbolized in a strange way: namely, as the four cardinal directions, the four dimensions of the horizon. In the sacred ceremonies of the Navajo Indians, the four historical points recur in the so-called sweat lodges, the little houses in which the sick lie sweating.[531] The number four is also represented by four different gods. Very often the center is represented as an equilateral cross, just as in the early Christian Church, whose quintessential symbol is the old Alemannic sun-wheel.[532] This is a Bronze Age sun-wheel, which at the same time is the early Christian symbol. It's a mandala. It expresses the deepest meaning. I mention all this because I shall be speaking about the number four in the next dream.

1995], 169). For more on the tetractys in Jung's metapsychology, see Marie-Louise von Franz, *Number and Time: Reflections Leading towards a Unification of Psychology and Physics* (London: Rider & Company, 1974).

[531] Simply built, typically circular hut made from natural materials, used by Native American tribes for healing ceremonies and the realignment with the four sacred elements. Anthropologist, ethnologist and naturalist Colonel James D. Stevenson (1840–1888), also an executive officer of the US Geological Survey, extensively researched religious ceremonies and sand paintings of the Navajo Indians: see his "Ceremonial of Hasjelti Dailyis and Mythical Sand Painting of the Navajo Indians," in *Eighth Annual Report of the US Bureau of American Ethnology to the Secretary of the Smithsonian Institution, 1886–87* (Washington, DC: Government Printing Office, 1891), 229–85. Stevenson was allowed to assist in a ceremony of healing, which had been underway for nine days and nights, drawing "some 1200 Navajo" (235). Using native images, he described the rituals that accompanied the construction of the "Medicine Lodge" and "sweat houses," which were oriented on the cardinal compass points (237 and 239ff.). For further details, see Ivan A. Lopatin, "Origin of the Native American Steam Bath," *American Anthropologist*, n.s. 62, no. 6 (1960): 977–93. Jung refers to "sweat lodges" in *Mysterium Coniunctionis*, CW 14, § 365.

[532] In *DAS* Jung reported that "quite a number of prehistoric mandalas from the Bronze Age have been excavated and are in the Swiss National Museum. They are called sun-wheels and have four spokes like old Christian crosses. This is also the design of the Host in the Catholic Church and on the bread used in the Mithraic cult, a sort of 'mandala bread' as shown on a monument" (120; see also ibid, 354 and his "Commentary on *The Secret of the Golden Flower*" [1929/1957], CW 13, § 45). The diffusion of the solar wheel (a circle around a cross) goes back at least to the Bronze Age, leading archaeologists to hypothesize the existence of a religion focused on a solar god who drove across the sky drawn by horses. As for the cross, the Byzantine Monophysite emperor Anastasius I (491–518) replaced the monogram of Christ with the cross; under Justinian it became the symbol of Christ and Christianity. During the 1930s, the Germanic sun-wheel was popularized from sources in archaeology and ethnology as a cryptic swastika, a broken solar wheel.

Lecture 5

Friday, 30 June 1933

Ladies and gentlemen,

One member of the audience wants to know why the Old Woman says nothing, only the Old Man. In life it's usually the case that old women do not speak, only young ones. Naturally, every dream is not only a message from the Old Man, but also one from the Old Woman. Now, of course, there is nothing that we can establish in exact scientific terms, but instead we can approach the nature of the psyche only through allegories, parables, and symbols. With regard to the question of whether every dream contains what might be called a message, I might say that you shouldn't take that too literally. Nature has no message. Nature tells you nothing at all. Not even the compass tells you that it is pointing north. Instead, we project our conceptions and interpretations into such a dream. If this were a real message in the human sense, the dream would use a decent language and express itself more clearly. Only seldom does a so-called unequivocal dream occur, where one really must assume it is actually meant that way, that it's a kind of exhortation.

I recall one patient, a representative of ultramodern natural science, whose misfortune it was to be afflicted by a nervous disorder. I was assigned to treat him. This man, of utmost intelligence, by the way, had dreams like those of a primitive. In his dreams, a voice appeared that can be described as nothing other than the voice of the Old Man. This voice told him, quite simply, what to do, what the meaning of the dream was, where he made mistakes, that is, what he had to do differently. Everything fit in wonderfully with his rationalism. If something bothered him, he immediately had a dream in which things were said that suited his intellectual conception. Only gradually did I manage to persuade him that there were no rational dreams, but instead that dreams had to be understood almost always in symbolic terms. Ultimately, he became so limp that when a wasp flew into the consulting room recently—he is afraid of wasps—he asked me quite fearfully, "But you don't mean, do you, that the wasp came into the room accidentally?" From a rational standpoint, one could say that I had simply looked on calmly while the unconscious had driven this man insane. Alas, he is now much better. He's less insane than earlier. In this sense, there are in actual fact dreams that are messages, which one can simply not explain otherwise than as the voice of what can only be referred to as symbolism.

There are other dreams, of course, which grow like weeds or flowers, without intention or deliberation. They speak only to those who listen into them. Thus, we would need to listen into the vast majority of cases. For instance, does a flower, or a dew drop hanging from a leaf, tell us something, whereas in fact it is only H_2O that has condensed as a droplet, or that hangs there by accident? If you observe this, if you listen to it, then you might experience what Jacob Böhme, our *philosophus teutonicus*,[533] experienced upon entering his cellar workshop one morning. The sun was shining through the window onto a copper plate. The plate reflected the ray of sunlight, which struck Böhme in the eye. He wrote, "Then I was enchanted into the center of nature."[534] For us, this would merely be a

---

[533] The epithet *philosophus teutonicus* was already in common use among Böhme's contemporaries.

[534] Jung refers here to the crucial illumination experienced by Böhme as a twenty-five-year-old shoemaker in Görlitz, Saxony, in 1600. He was enraptured, his entire being seized by a supernatural sense of joy, as reported in *Morgenröte im Aufgang*, or *Dawn of the Day in the East*, better known as *Aurora* (1612). The book, which wasn't intended for publication, circulated privately, and was harshly condemned as heretical by the pastor of Görlitz when it eventually came to his attention. Böhme's description of his mystical rapture did not fit with the classical *topoi* of the loss of individuality in the divine abyss of the blossom—a Dantean contemplation of the Heavenly Rose in union with the Godhead. He wrote, "I did not climb up into the Godhead, neither can so mean a man as I am do it; but the Godhead climbed up into me, and revealed such to me out of his Love as otherwise I would have had to leave quite alone in my half-dead fleshly birth" (*Aurora*, viii, 7, quoted in John Joseph Stoudt, *Jacob Boehme: His Life and Thought* [Eugene, OR: Wipf and Stock, 2004], 61; translation modified). He also wrote about the light which surrounded him and the heavenly knowledge he achieved: for example, "in this light my spirit directly saw through all things, and knew God in and by all creatures, even in herbs and grass" (*Aurora*, xix, 13, quoted by Stoudt, *Jacob Boehme*, cit., 59). Jung owned Böhme's book *Answers to Forty Questions on the Soul* (1647), which deals with the spiritual paths to enlightenment and God and includes a supplement on the "reversed eye" ("Psychologiae supplementum, *Das umgewandte Auge*," 4.7, 181). In *Psychology of Kundalini Yoga*, cit., Jung speculated that Böhme could have experienced the attainment of a higher chakra (58); cf. the bestselling book by Lee Sannella, *The Kundalini Experience: Psychosis or Transcendence?* (Lower Lake, CA: Integral Publishing 1987 [1976]). Jung also dealt with the light of nature in "Paracelsus as a Spiritual Phenomenon" [1942], *CW* 13, § 148ff., 229. In the Visions Seminar, he mentioned Böhme's "reversed eye of the philosophical globe, or the mirror of wisdom," crediting him with being "one of the first to create a mandala consciously," and

reflection in the plain, ordinary sense: that is, a ray of light that struck his retina, and nothing else. But instead he was prepared to conceive of this as a message, and the corresponding effects occurred.

Things are as meaningful as we take them to be. If we consider them to be utterly insignificant, as they also are in a certain sense, then, of course, nothing is meaningful. We can explain everything that exists in the smallest possible terms, or we can explain it in the largest possible ones, just as suits our character. Another question that has been raised by a member of our seminar is whether right and wrong exist in the domain of human fantasy. Yesterday, I spoke about correct or incorrect fantasies.[535] That was a mistake pure and simple. Fantasies are neither correct nor incorrect; rather there are only facts. Whether we speak of correct or incorrect fantasies depends essentially on how we grasp the matter. So, for instance, it can be quite correct that someone has a fantasy that is not right—for instance, a fantasy that has particularly impaired a sick person. The question is always the standpoint from which one considers these things. Such a dream should be taken literally. Why has the dear Lord given us intelligence if we fail to use it?

There is yet another question from the floor:

Could you draw somewhat clearer parallels between the *rites d'entrée* among primitives and yoga exercises? As far as I understand, the purpose of the *rites d'entrée* is to consciously and deliberately concentrate libido on a specific point, in order to have on hand the amount of libido thus obtained for the fulfillment of the outer task. It is thus a kind of "yoking" just as yoga is, which seeks to lead one toward the fulfillment of an inner necessity. With respect to yoga, it is very difficult for the European, who suffers precisely from too much willing, to find the right way and the correct measure of will; that is, of conscious practice. If he has just learned with difficulty "to do nothing" and must now seek the transition to a necessary and healthy exertion of the will, he is threatened by the danger of falling back into cramps of consciousness and will. It seems to me important to establish even more clearly the difference between the

---

knowing the necessity to keep the eyes open both inwardly and outwardly: "[s]ince there are those two forms of existence, one must have two eyes in order to see them, the one turned out to this world, and the other turned in," and Böhme knew that "[w]ith the one eye, one sees the circumference of the horizon here, and with the other eye the mandala" (*Visions*, cit., 1:557 and 792).

[535] See above, 215ff.

mistaken exertion of the will which must be overcome, and the employment of will intended by *rites d'entrée* and yoga exercises.

This is a very complicated question, which should be discussed in a yoga seminar.[536] For these problems are closely related to practical analysis in various ways. You must not forget, in particular, that analytical work is done with individuals. What is an utterly correct truth in one case could be utterly wrong in another. In the one, we are compelled to establish principles that are absolutely binding. In another we would say something utterly different, precisely because something else is necessary. Psychological truths shimmer, and shimmer they must, as soon aswe wish to represent them somehow in generally valid terms. It is therefore hard to make generally valid statements in psychology, because matters are differentiated individually, and because differentiated individuals exist in quite different states, so that the individual acts according to this principle on one occasion and according to another on another occasion. That's simply the way it is.

Consequently, one can't say that one shouldn't do either this or that. In other words, it's a question of, as it were, relaxing to the point of simply allowing things to happen, and sheer determination should be avoided. Exactly that tight grip can be quite wrong, just as exerting sheer determination can be what is needed. Certain things must be undertaken with great effort, others precisely not. The same applies to the connection between the *rites d'entrée* and yoga. Yoga originally arose from the *rite d'entrée*. Yoga is a complicated system of the *rite d'entrée*. In some cases, it is important to strike out on such a path; in others, doing so would be utter folly. In principle, such strange aids as the *rites d'entrée* are valid

---

[536] This question concerns the modalities of will-power in relation to the practices necessary for activating libido in both yoga disciplines and initiation rites. Jung claimed that the chakra system in Kundalini yoga symbolizes the development of "suprapersonal" psychic life beyond the I, "the light of the gods" (*Psychology of Kundalini Yoga*, cit., 68); it is "at the same time an initiation symbolism, and it is the cosmogonic myth [. . . ;] this is the symbolism of all initiation cults: the awakening out of *mūlādhāra*, and the going into the water, the baptismal fount with the danger of the makara, the devouring quality or attribute of the sea" (30). On *rites d'entrée*, see also "On Psychic Energy" [1928], CW 8, § 86, where the *atninga* ceremony of the Arunta is mentioned, following Lévi-Bruhl. In this regard, see Shamdasani, *Jung and the Making of Modern Psychology*, cit., esp. 290–97 and 328–38. On the matter of volition in primitive psychology in relation to *rites d'entrée et de sortie*, one can turn to the episode from Jung's East African expedition of 1925–26 as reported in "A Radio Talk in Munich" [1930], CW 18, § 1288ff., and "Questions and Answers at the Oxford Congress, 1938," in McGuire and Hull, *C G. Jung Speaking*, cit., 99–115 at 100.

only where the task in question cannot be accomplished with reasonable concentration; that is, if something must be done that cannot possibly be undertaken with one's normal best effort. Thus, it would be extremely uneconomical if we were to perform a *rite d'entrée*, a dance, before writing a letter. And so we also say that we made a song and dance because someone refused to do something. We can in fact do all those things with a quite definite and specific measure of energy. We do not need *rites d'entrée*.

But as soon as we are concerned with matters that are extremely hard for us to achieve, such as inner self-development or the transformation of inner fantasy life, we cannot rely on the will. As you can see, the unconscious itself places yoga or the *rites d'entrée* directly at our disposal. Therefore I would say that we cannot decide with the will or with reason if instead the indirect path must be taken. This always presents itself as a *necessity*. It must strike one as a matter of necessity. If you take my comments as a recipe—that this procedure must be followed, or that such is the correct way to achieve some kind of goal—then you'd be colossally mistaken. Here [in the case of our dreamer] the path refers to the path that has happened, not to one that I was looking for. I wouldn't have dreamed that this path from the unconscious would already emerge in the first fourteen days. Under no circumstances can I imagine what will come up next. On this occasion, I might have waited helplessly for a whole year, without anything like mandalas or some other clue showing up during that time. In the other case I referred to, that of the learned young man, the first obvious and unmistakable traces of mandalas became evident only after a year. So in such a case, for example, matters are quite different.

Thus, it cannot be said *a priori* that in general one must unlearn the cramping of the will. There are very many people who do nothing but "allow things to happen," and who suffer from a complete absence of focused determination. They must learn will-power. In this case, the unconscious gives people all possible kinds of guidance about how to proceed with intent and volition. But it is not done for one, one must do it oneself. People who suffer from an absence both of will and of reasonable intent are served other fare. The relationship between the conscious and the unconscious is compensatory, and therefore it is so very difficult to establish the ultimate nature, that is, the essence, of the unconscious. I am convinced that we cannot determine this [nature] in principle; that our consciousness and the acts of the unconscious form part of it and work together, that the entire psychic phenomenon of the conscious and the unconscious, that is, the entire psychic disposition, puts all of these things into play simultaneously, and that we stand in their midst. This is something in which we are completely contained. We represent consciousness, and are surrounded by the unconscious. We are the ball that is thrown back and forth. How, for instance, is the tennis ball supposed to decide the outcome of the game? That makes no sense whatsoever.

Let us return to the subject of our session yesterday. First I would like to come back to the spiral path or way. You will remember that this spiral way appears at the center of the mandala. It expresses itself quite evidently in an Eastern mandala, the *muladhara* chakra,[537] where the Kundalini snake lies coiled around the lingam of Shiva, the creator god. It surrounds the lingam in three-and-a-half turns. That is the dormant state. One text tells us that the sweet murmuring of the Kundalini snake resembles the humming of a besotted beehive.[538] This spiral is an initial state, the condition before the awakening of libido, which is represented precisely by the Kundalini snake. This center is the *muladhara* chakra, or the lowermost of the six or seven chakras in the Tantric system. These chakras are superimposed mandalas that represent different developmental states of the soul.[539]

Thus, if you come across this spiral motif somewhere, you can almost certainly expect to discover a certain static, sleeping state, that is, a way that has yet to be undertaken. One must first reach the center. The spiral is the summons to see this center, and it therefore appears not only in

---

[537] See above, n. 362 and [ZL], 113 and n. 344.

[538] Likely reference to the *Shiva Samhita*, a celebrated text of Hatha yoga probably written in the seventeenth century and including elements of Kundalini yoga, Jnana yoga and Vedanta. The text enumerates the mystical sounds *(nada)* (listing them in a slightly different order from the *Gheranda Samhita*) heard by the practitioner: "The first sound is like the hum of the honey-intoxicated bee [*matta-bhrnga*], next that of a flute [*venu*], then of a harp [*vina*]; after this, by the gradual practice of Yoga, the destroyer of the darkness of the world, he hears the sounds of ringing bells [*ghanta*]; then sounds like roar of thunder [*megha*]" (cited in Guy Beck, *Sonic Theology: Hinduism and Sacred Sound* [Delhi: Motilal Banarsidass, 1995], 103). Jung similarly mentions "a humming sound like a swarm of erotically excited bees" with regard to the Kundalini's initial sound in his letter to Elined Kotschnig of July 23, 1934, in Jung, C. G. *Jung: Letters*, cit., 1:170).

[539] Cf. Jung, *Psychology of Kundalini Yoga*, cit., 60ff. Jung considers the chakras first and foremost as "symbols" which "afford us a standpoint that extends beyond the conscious. They are intuitions about the psyche as a whole, about it various conditions and possibility. They symbolize the psyche from a cosmic standpoint. It is as if a superconsciousness, an all-embracing divine consciousness, surveyed the psyche from above" (67). So the chakra system represents "the standpoint of a fourth dimension, unlimited by space or time" and "is created from a standpoint . . . that transcends time and the individual" (65). On the symbolism of the spiral in its various transgeographical variations, see Jill Purce's beautifully illustrated *The Mystic Spiral: Journey of the Soul* (London: Thames & Hudson, 1974).

Indian but also in Eastern symbolism more generally where the necessity to reach the center occurs and where nothing else happens, other than to establish this concentration of libido on the center; that is, in practical terms, contemplation or observation.[540] "To observe [*betrachten*]" derives etymologically from "to make pregnant [*trächtigmachen*]."

Observing a thing involves concentrating your libido on it until it becomes pregnant. *Observation is the act of filling something with libido* until birth occurs, that is, until suddenly movement occurs. You can fill any starting point with concentration until there is movement. Imagine a painting that depicts the surface of the sea, and on it a sailing ship. If you look at the picture intently, frescoed onto the wall, the sailing ship gradually begins to move. All kinds of things happen. The people aboard move because observation has made this pregnancy result in birth. This spiral motif is also an architectonic principle. As you know, it was used for the lovers' labyrinths in medieval gardens.[541]

It also appears as a deeper symbol, in Francesco Colonna's book, for instance, where it is represented by a system of canals, and where the island of the blessed is gradually reached by ships undertaking a spiral passage. The book is entitled *Hypnerotomachia*. It is an amusing humanist account written in Italian, Latin, and Greek, and couched in a peculiar jargon. A bibliographical rarity, it features a language composed of Latin, Greek, and Italian, and interspersed with Arabic, as was common practice at the time.[542] It was written in the mid-fifteenth century, by a monk, and concerns his inner path.[543] Highly acclaimed at the time, it was no

---

[540] In the ETH Lectures, vol. 6 (*Psychology of Yoga and Meditation*, cit.), Jung will say, "The Latin word contemplation comes from *templum*—a zone for living encounter is defined, a specific field of vision in which observation takes place" (7).

[541] See above, n. 454.

[542] The work is also studded with neologisms, and exhibits a preoccupation with gods and cults of late Antiquity, and certain symbolic motifs which have led to the hypothesis that the author was an alchemist. Its complete title is *Hypnerotomachia Poliphili, ubi humana omnia non nisi somnium esse ostendit, atque obiter plurima scitu sane quam digna commemorat* (Love's strife in Polia's lover's dream, where he teaches that all human things are nothing but a dream. And where he gives account of many other things worthy to be known [*Polia*: "many things"]).

[543] As the last incunabulum printed by the renowned printer and humanist scholar Aldus Manutius, in 1499, and among the handful of non-classical texts published in his famous Venetian workshop in addition to Dante's *Divina Commedia*, this anonymous, enigmatic, and esoteric masterwork incorporates a heterogeneous iconographic corpus of 172 woodcuts variously attributed to painters such as Andrea Mantegna, Giovanni Bellini, Cima da Conegliano, and Sandro Botticelli, with a significant array of architectural images that also led to its being regarded as the first illustrated printed architectural book in history. Widely discussed, though sometimes dismissed because of its lighthearted character, it

longer understood later. While referred to again in the Renaissance, it was no longer read, nor would it have been understood, because such things were no longer accepted in those days. The title comprises three terms: *hypnos*, which means something like "sleep" or "dream"; *eros*, "love"; and *machia*, "struggle." Thus, a modern translation would be "the conflict between sleeping or dreaming and love." The story is about the monk's quest for Lady Polia. There is an early translation into Old French, entitled *Songe de Poliphile*. In other words, [the dream of] Polia's lover.[544] He calls her Lady Polia. His soul seeks the anima everywhere. He is deprived of his anima by regal Venus on a blessed island. Ultimately, he is promised that he will reach this island, where he shall be united with Lady Polia. And so it comes to pass. After many, many turns in the plot, the soul's *unio mystica* occurs. At that moment he wakes and hears the tolling of bells. It is the first day of May, and thus the book ends.[545]

exibits an unparalleled encyclopedic culture comprising ethological, architectonic, mathematical, and botanic elements, as well as an array of analogies, mythemes, deities, and cults of late Antiquity and certain symbolic motifs which raised speculation that the anonymous author was an alchemist. It has been attributed to the Venetian Dominican monk Francesco Colonna (ca. 1433–1527). The core plot is an initiatiory journey of love inspired by the *Metamorphoseon* of Apuleius, and with borrowing from Dante's *Divina Commedia*, such as the tripartition of Inferno, Purgatory, and Paradise. The story begins with the classic topos of the dream or hypnagogic vision that comes to the protagonist when he falls asleep, fatigued by not finding a solution to his problem. The erudite conception of love of this extraordinary work displays an eroticism, as has been noted, "so intense and so exaggerated [. . .] that at a certain point it ceases to be just a piece of erotic writing and becomes writing about eroticism" (Liane Lefaivre, *Leon Battista Alberti's "Hypnerotomachia Poliphili": Re-Cognizing the Architectural Body in the Early Italian Renaissance* [Cambridge, MA: MIT Press, 1997], 61).

[544] The attribution of the work to Colonna largely rests on a concealed acrostic made up of the thirty-eight capital letters which begin the chapters, spelling "Poliam frater Franciscus Columna peramavit" (Brother Francesco Colonna greatly loves Polia). See Maurizio Calvesi, *Il sogno di Polifilo prenestino* (Rome: Officina, 1980); and Calvesi, "*Hypnerotomachia Poliphili*: Nuovi riscontri e nuove evidenze documentarie per Francesco Colonna signore di Preneste," *Storia dell'Arte* 60 (1987): 85–136. Lorenzo de Medici, Giovanni Pico della Mirandola, Pomponio Leto and, later, Leon Battista Alberti are the other principal candidates for authorship. For the argument against Colonna's authorship, see Lefaivre, *Leon Battista Alberti's 'Hypnerotomachia Poliphili,'* cit.

[545] The *Hypnerotomachia*, Jung writes elsewhere, "takes the form of a dream which depicts the apotheosis of love. It does not tell the story of a human passion, but describes a relationship to the anima, man's subjective image of woman, incarnated in the fictitious figure of Polia" ("Psychology and Literature" [1930/1950], *CW* 15, § 154). In philologist and Jungian analyst Linda Fierz-David's view, "the secret aspect of the ancient world," the life in its mystery cults, constitutes the background "elixir of life" that symbolizes "the living process of growth which had been set going, obscurely and incomprehensibly, in the men of his [Colonna's] time, and had made of the Renaissance the beginning of a new era" (*The Dream of Poliphilo: The Soul in Love* [1947], related and interpreted by Linda

It is worth noting that the book was written by a monk. Over the next two centuries it was revered as an extraordinarily mystical book containing the most profound wisdom. Indeed, it contains the most interesting symbolism, partly very difficult, but which forms an absolutely original parallel to the *Divine Comedy*.[546] It was never reprinted.[547] There is a two-volume commentary in French, and the original is available only in the older French translation, which costs several hundred marks.[548] The Italian edition is a bibliographical rarity of inestimable value. Only a few copies exist, for which one would obviously have to spend a very handsome sum. So here we find this spiral motif.

Moreover, it's an architectonic principle, which you'll also find in the Buddhist temple at Borobudur,[549] where a circumambulation of all the bas-reliefs occurs, until the pilgrimage ends at the level of the imageless

---

Fierz-David, trans. by Mary Hottinger, with foreword by C. G. Jung [Bollingen Series XXV] [Dallas, TX: Spring, 1987 (1950)], 24–27).

[546] Cf. *Psychology of Kundalini Yoga*, cit., where Jung compares the beginning of the *Hypnerotomachia* with that of Dante's *Inferno* "but expressed in entirely different terms. He [the author of *Hypnerotomachia*] depicts himself as traveling in the Black Forest, which in those days, especially to Italians, was still the ultima Thule where the unicorn still lived, as wild and unknown as the forest of central Africa to us. And there he loses his way, and then a wolf appears. At first he is afraid, but afterward he follows the wolf to a spring where he drinks of the water—an allusion to baptism. Then he comes to the ruins of an ancient Roman town, and he goes in through the gate and sees statues and peculiar symbolic inscriptions, which he quotes, and which are most interesting from a psychological point of view. Then suddenly he is afraid; it becomes uncanny. He wants to go back, and he turns to go out through the gate again, but now there is a dragon sitting behind him that bars the way, and he cannot go back; he simply must go forward. The dragon is Kundalini. You see, the Kundalini in psychological terms is that which makes you on the greatest adventures. [. . .] Kundalini [. . .] is the divine urge" (21).

[547] Today anastatic reprints as well as critical editions of the *Hypnerotomachia* are available: see, for instance, *Hypnerotomachia Poliphili: The Strife of Love in a Dream*, trans. by Joscelyn Godwin, (London: Thames & Hudson, 1999). An updated bibliography of the secondary literature can be found in Linda Barcaioli, "Una banca dati per il *Polifilo*: La catalogazione dei contributi," *Bibliothecae.it* 4, no. 1 (2015): 123–47.

[548] This would probably be the two-volume translation by Claudius Popelin, *Le Songe de Poliphile ou Hypnerotomachie, littéralement traduit pour la première fois, avec une introduction et des notes* (Paris, 1883; present in Jung's library).

[549] Borobudur is an eighth–ninth-century Mahayana Buddhist temple in central Java (recent studies place it between 760 and 830 CE). Lost for centuries to volcanic ash and the jungle, it was cleared and restored beginning in the early nineteenth century. With a base 123 meters (400 feet) square and a height of 30 meters (100 feet), it is the world's largest Buddhist temple. Six nearly square levels, surmounted by three nearly round ones, lead the pilgrim on a winding five-kilometer (three-mile) ascent through paradises and hells, ending with scenes from the lives of the Buddha and bodhisattvas who attained Liberation. The whole construction is based on the Buddhist concept of the three spheres, climaxing in the empty dome-like stupa. See also above [I], 68.

dome. For all the images are visions. The final truth is unrepresentable, empty. Other sacred buildings also share the same spiral shape. This circular motif, the mandala motif, can also be found in the sacred buildings of Islam.[550] One such building is the mosque in Old Cairo, which represents a very interesting mandala. It forms a perfect square, with great colonnades featuring magnificent columns on all four sides. At the center stands the hall of purification, where ritual ablutions are performed. There we find fountains and the bath of rejuvenation, of spiritual rebirth.[551] If you enter such a vast space from a dusty, busy street, all of a sudden you enter a heavenly realm. It holds nothing but heaven. The wind stirs through this vast, open space, and you really have the impression of perfected concentration, of acceptance into the immense emptiness of the sky. Then you will understand this cult in which God is in fact a summons. When you hear that inimitable call—"Allah," so it rings through this vast space— you'll have the impression that this call penetrates the heavens. Thus, you have the full picture. Nothing else remains to be said. All one can say is, this is how it is.

The Sultan Hassan mosque has the same structure, except that it features an interesting innovation, a concession to Christian ideas: namely, that it's no longer shaped like a quadrangle, but instead a cross.[552] It has a courtyard, and towering niches. Naturally, the scale is enormous. Here we find another form of the mandala. Then, of course, there are the Byzantine

---

[550] *Tafa*, the Arabic verb referring to the circumambulation of a sacred object means originally "to attain the summit of a thing by spiraling round it": Purce, *Mystic Spiral*, cit., 31. Notably, the sevenfold circumambulation of the sacred stone the Kaaba, the representation of the *axis mundi* and the goal of the Islamic pilgrimage to Mecca, must be performed counterclockwise, so that the Black Stone remains to the pilgrim's left (Paul B. Fenton, "The Symbolism of Ritual Circumambulation in Judaism and Islam. A Comparative Study," *The Journal of Jewish Thought and Philosophy* 6 [1997]: 345–69 at 352ff.).

[551] The description relates to the Mohammed Ali mosque, built between 1824 and 1848 by the Greek architect Youssef Bochna and known as the "Mosque of Alabaster." It is grandly situated and imposing (the central dome rests on four pillars 52 meters [170 feet] high and 21 meters [70 feet] in diameter), and has a central courtyard dominated by a fountain for ritual ablutions. Jung had visited it before continuing towards the Giza pyramids and Luxor on his cruise to the eastern Mediterranean with Hans Fierz in spring 1933 (see Andreas Jung, "Carl Jung and Hans Fierz in Palestine and Egypt: Journey from 13 March to 6 April 1933," in Erel Shalit and Murray Stein, eds, *Turbulent Times, Creative Minds: Erich Neumann and C. G. Jung in Relationship (1933–1960)* [Asheville, NC: Chiron, 2016], 131–34 at 133).

[552] The Sultan Hassan mosque in Cairo, built by Ibn Bilik el-Mohsini between 1356 and 1363, was in its time the greatest in the world. Its unusual cruciform ground plan arranges the four halls around a central courtyard. Jung visited the it on the same trip with Fierz in 1933—as well as the Hagia Sophia and the Sultan Ahmed mosque on their earlier stop in Istanbul.

round churches like the Hagia Sophia,[553] which is an axial structure. We find the same in baptisteries, particularly those in Ravenna.[554] Their interiors are clearly divided into four sections, just like secular structures—for instance the quadrangular Diwan-i-Khas,[555] or the audience hall of the maharaja, that is the emperor or sultan, of Agra in India. He had an audience hall built in accordance with the mandala. At the center of this hall stands a pillar, on which rests a kind of flat basin in which the emperor sits. He is surrounded by the diwan, the assembly, whom he addresses, or who address him. He sits at the very center of the mandala, elevated above the floor, as it were in the lotus position. In actual fact he is a Mohammedan. I have cited these examples to indicate how widespread this

[553] Hagia Sophia or "Basilica of Holy Wisdom" (Gk.: Αγία Σοφία; Turkish: *Ayasofya*) represents the apex of Byzantine architecture. Begun by Constantine and consecrated in 360 after his death, it burned down twice and was twice rebuilt, being reconsecrated anew in 537 under Justinian. Its gigantic dome of thirty-one meters (100 feet) in diameter is the fourth biggest in the world after those of St. Peter's Basilica in Rome, St. Paul's Cathedral in London, and the Duomo in Florence. The Basilica has been an imperial church, a mosque, and from 1935 a museum. In 2020 it became once more a mosque.

[554] Reference to the fifth century Neonian baptistery (baptistery of the Orthodox) and Aryan baptistery, built by Theodoric with the aim of promoting peace between the Aryan Goths and the Orthodox Latins. Both baptisteries are erected on square ground plans which support octagonal drums and domes decorated with mosaics above octagonal baptismal fonts. During Jung's (apparently third) visit to Ravenna (the former two being, it seems, in 1914—not in 1913, as mistakenly recorded in *MDR*, 284—and 1928), which took place in 1932 before the Kundalini Seminar (see Jung, *Psychology of Kundalini Yoga*, cit., 16 and n. 27), he had, together with Toni Wolff, a visionary experience. Both believed that they saw in the Neonian baptistery four great mosaics: Jesus's baptism in the Jordan, the Red Sea crossing, Namaan's cleansing, and Jesus's rescue of Peter in the storm on Lake Genasereth; only later did Jung learn that none of these exists, and speculated that his anima had probably been projected onto Empress Galla, who commissioned similar mosaics (in a church now destroyed) after surviving a storm at sea (see *MDR*, 284–86; Bennett, *What Jung Really Said*, cit., 80–81; Daniel C. Noel, "A Viewpoint on Jung's Ravenna Vision," *Harvest* 39 [1993]: 159–63; Ronald V. Huggins, "What Really Happened in Ravenna? C. G. Jung and Toni Wolff's Mosaic Vision," *Phanês* 3 [2020]: 76–115; and Daniela Boccassini, "Jung and Ravenna: *The Red Book*, Visioning, Rebirth," *Psychological Perspectives* 65, no. 1 [2022]: 43–81).

[555] The private audience chamber of the Indian Mughal emperor Akbar (1542–1605) at Fatehpur Sikri (the City of Victory), near Agra, the capital of the Mughal Empire for about ten years before being abandoned in favor of Lahore in 1585. In *DAS*, Jung connected the building's structure with the design of the third dream and further commented, "Akbar used to sit in a sort of saucer in the middle of it, while learned men from all parts of the world told him about all sorts of religions and philosophies and discussed them with him. There he tried to make an integration for himself. The red sandstone saucer is supported on a pillar with a jet-black stem, in the middle of the square hall [where four gangways converge on it from a raised gallery]. When one looks up at the saucer the black stem of the pillar is practically invisible so that the whole thing seems to be suspended in mid-air"

archetypal image is, particularly in ecclesiastical architecture, and that it always seems to have been granted symbolic importance.

Let me turn to another aspect of the mandala. You will remember that our dreamer felt that the steamroller work was tedious and obscure, and that he compared it to the course of analysis, which seemed to him to be wearisome and fussy. From this we can conclude that analysis actually involves a circumambulation. One could say without further ado that the technique, as we understand it, and if one can speak of a technique at all, is observation, that is, contemplation. For this reason, I believe that one should not undo and destroy a dream that one does not understand by tinkering with it. Instead, it should be contemplated from all sides, and filled with libido so that it begins to live and finally gives birth to its inner meaning and tendency. In that way, one performs what the mandala and circumambulation mean. That is why the dream also selected the symbol of the machine, because the machine is concretized working capacity. It's a mechanized capacity for effort that we must exert in the *circumambulatio* so to speak, to perform the conversion or transformation. The transformation changes everything that can be changed. That's why the ancient Israelites long ago marched around the walls of Jericho when their siege tactics failed. Finally the walls of Jericho gave way.[556]

Therefore this *circumambulatio* or rather mandala symbolism is used as a healing symbolism; for instance, among the Navajo Indians. Thus, if a rich man falls ill, a large mandala is made, whose diameter can measure up to two hundred meters [650 feet]. Certain kinds of wood (pine shrubs) are heaped up at the four cardinal points, and a temenos, a circular area,

---

(115–16). Akbar, whose empire grew to include much of the Indian subcontinent, promoted a syncretic creed (*Din-i Ilahi*) encompassing Zoroastrian and Christian elements, along with Islamic and Hindu influences, and encouraged a philosophical and interreligious dialogue. Jung would visit the fortified city during his 1937–38 travels in India (see "The Dreamlike World of India" [1929], *CW* 10, § 983; and Sulagna Sengupta, *Jung in India* [New Orleans: Spring, 2013], 112–14). A plan and photos of the palace can be found in Christian Gaillard, *Le Musée imaginaire de Carl Gustav Jung* (Paris: Stock, 1998), 100–101.

[556] Jung refers to the siege of Jericho ("City of the Palms") narrated in the Book of Joshua (6:11). On the seventh day of siege, Joshua's troops, after marching once round the city on six consecutive days, with in their midst the Ark of the Covenant preceded by seven priests blowing ram's horns, marched seven times around the city walls, which then, upon a shout from the whole army, finally collapsed. This episode is paradigmatic for any ancient Jewish ritual circumambulation (*hakafot* or *haqqafot*; *haqqafah*), the classic example being Sukkot, the Feast of the Booths (or Tabernacles), a seven-day ritual starting at the full moon of the first month of the year and involving a counterclockwise procession (Fenton, "Symbolism of Ritual Circumambulation," cit., 346ff.).

is marked out. Mets in legs are erected both at the power points and at the center,[557] where the sick person is then treated by the medicine man. Inside the hut [medicine lodge], another mandala is drawn, using colored sand. Interestingly, this mandala also contains the Indian motif: namely, the motif of the snake united with God. It is a cross at which two figures stand.[558] It is a swastika, as you will gather, and these figures are regularly shaped Hostioken,[559] the male and female god. The thing itself is a circular movement, to which small sun symbols are often attached with eight swastikas, which indicate the rotational character. This ceremony lasts about eight or nine days. It is called the Mountain Chant.[560] During the ceremony, the sick person is subject to every conceivable symbolic situation.

These things are, of course, established just as they are in the Indian technique of yoga. Ancient traditions are followed, and they should not be confused with what happens in this dream. All these things have followed from the primordial phenomenon, which the ancient primitives have recognized as significant and healing for thousands of years. Consequently, it was systematized after a period of long use, and further differentiated and codified by secret tradition. The European never finds out anything about these matters. He might be allowed to attend a ceremony, but the primitives refuse to give any explanations. In principle, there's nothing for them to explain. Interestingly, primitives have no knowledge

---

[557] The transcript at this point appears to be conjectural in several respects. "Mets in legs" (as in the protocol: "An der Braccialpunkten werden mets in legs aufgestellt, ebenso im Zentrum") might conceivably correspond to the medicine bags in Matthews's summary of the nine-day ceremony (cf. below, n. 560), assembled by the shaman for the healing ceremony's outdoor phase ("erected both at the power points and at the center" of the large circle for the climactic fire-dance after the medicine-lodge healing). Jung's summary then moves on, taking up "another mandala." This is one of the sand paintings done inside the lodge; and roughly corresponds to the one that he later described in his 1950 study of mandala symbolism, also from the Mountain Chant (see below, n. 579; cf. CW 9.1, Fig. 45, and § 700), where some of the features highlighted in the lecture also appear. This lecture identifies one pair of male and female gods, while the later essay refers to four pairs, and four large male gods; the protocol identifes a snake deity, while the essay's Mountain Chant painting shows the rainbow goddess. [T.N.]

[558] Apparently in error for "four figures," the large outer figures (see "Concerning Mandala Symbolism" [1950], CW 9.1, § 700).

[559] Unidentified term.

[560] See Washington Matthews, "The Mountain Chant: A Navajo Ceremony," in Fifth Annual Report of the Bureau of Ethnology to the Secretary of the Smithsonian Institution, 1883–84 (Washington, DC: Government Printing Office, 1887), 379–446, and Jung's comment on a mandala pertaining to this chant in "Concerning Mandala Symbolism" [1950], CW 9.1, Fig. 45, § 700.

whatsoever of what they're doing. They simply *do* it. The explanations that they do give are precisely such symbolic events. If you ask them, "Why? What does this mean? Why does one do this?" then they'll simply tell you another strange story, which would also require explanation. Ultimately, you will be forced to lay down your arms. These people strike us as so primitive, because they don't even know what they're doing. In turn, we consider ourselves so immensely superior because we know what we're doing. I have already told people, "Well, if you want to know what we're doing, then here is what we do: at Christmas, we ask Mr. Jones or Mr. Smith, 'What are you doing? What is this strange ritual? You said you were not religious and had no particular rites. But what does this tree mean, and what about the candle?'" In reply we're told, "This is how it's always been done." If we inquire, "Why? What does this mean?" then we get no answer—for this is the symbolic. Such stories exist, in effect, believe you me. That is when the Lord Jesus was born, but what does his birth have do with a Christmas tree? Nobody knows.

At Easter, we visit another Mr. Jones and ask, "Well, you are so enlightened and have given your word of honor that old symbolic rites are no longer practiced in this country. And now, lo and behold, I see that you have bunnies everywhere, fertility customs, with differently colored eggs. Well, what does that mean?" In response, we are told that "this is how it's always been done." [561]

You see, then, that this is just as cuckoo as the primitives who do not know what they're doing. But we are supposed to know what it is we're doing? I ask you, who knows what a Christmas tree is! In this respect we think just like the primitives: yes, our fathers knew, and people once knew, too. This is precisely as stupid as if the primitives claimed, "Our medicine man is not that good, but there's a good one over there." On the island they'll tell you, "We don't have a good medicine man, but there is one back there." Finally, you are sent back to wherever you came from.

For that matter, people have never known. We have far greater knowledge of the Christmas tree than all our ancestors. We know much more

---

[561] Cf. *DAS*, 221 and 331. The Christmas tree is one of the most widespread Christmas traditions: a fir or other evergreen, it represents the rebirth of light or "the tree of life," with a probable pagan derivation from European myths of renewal, notably pre-Christian conceptions of magical powers in evergreen trees. They were used in year-end festivals such as the Roman Saturnalia and the January Calends for adorning homes and temples. The first references to the modern Christmas tree come from mid-sixteenth-century Germany, specifically Alsace, but the "oldest evidence for decorated trees at Christmastime comes from Riga, Latvia (1510), and from Reval, Estonia (1514)" ("Christmas Trees," in William D. Crump, *The Christmas Encyclopedia*, 3rd edn [Jefferson, NC: McFarland, 2013], 139).

about these symbolic things than people did in the previous era. Our ancestors *did* these things, but we *know* about them, because we are beginning to account for them. Mostly, we do not account for our own symbolism, just as little as the primitives did. The primitives that I observed in East Africa were members of a tribe that lives in the jungle, in the savannah. I watched them for a long time, and finally I asked, "Isn't there anything that you do?"—"No, nothing at all." It was precisely as if I were to ask someone on the street in Berlin, "Tell me, where do you perform your symbolic customs?" You can imagine how your typical snotty Berliner would respond.[562]

You can also imagine how these natives answered my question. They considered it complete and utter nonsense, until one day, while on an excursion, by accident I came across a small house. It was a little kraal, with puppets, shards of pots, native millet flour, some rotten bananas and a small pot of water, some simulacra of instruments, digging sticks, hacks, and other such things. I discovered a sleeping figure made of clay, a god or a dead person, stretched out on a kind of bed. It was, of course, a ghost house. Triumphantly, I returned to my people, and said, "But you build ghost houses." They looked at me as if I had said something rude in polite company. It was most embarrassing, a minor obscenity, and socially impossible. Later on, the chief came to me and seriously reprimanded me for uttering such matters in public. In my stupidity, I did not understand this at first, but since I was obviously interested in the matter, I asked in a roundabout way, "Tell me, why do you build these small houses?" I was told that children built these small houses. I knew that this was an absolute lie. The houses were far too well constructed. I thought that they were built to heal the sick. This assumption was confirmed. In the kraal there lay a sick person, without any doubt whatsoever. For the huts were always situated by a path, moreover in a highly sophisticated manner.[563]

These bush trails are small paths. They traverse the whole of Africa, from the Cape of Good Hope to Algiers. You can cross the whole continent on such trails, from ocean to ocean. They form a fine network, and all have the same shape. They meander, and their radius measures six to seven meters [ca. twenty feet]. Our farm trails, dating from the Middle Ages, are similar, except that their radius is much greater. Man's original

---

[562] The German phrase *Berliner Schnauze* or "Berliner snout" refers to Berlin dialect but also characterizes an attitude of rough irony, sarcasm, and brusque, even aggressive behavior.

[563] The episode occurred during the Kenya expedition, on the southern slopes of Mount Elgon: see "Psychology and Religion" [(1937) 1938/1940], CW 11, § 30.

path was a snake's path, in that he veered to the left and right in his un-conscious perfection. But we want to follow the straight path. Conse-quently, the farm trail gradually became straighter, so that its radius now measures between a hundred and two hundred meters [325–650 feet], while the radius in Africa is very short, so that you if you advance at a native pace, then you will sway from one side to another. But you will see that you can very easily maintain the speed, because you will walk far more easily, for you will walk not only on the right-hand path but also the one on your left. You'll make good progress, in spite of the tropical sun, because it's easier to walk that way. Thus, at night the ghosts of the departed come. Here is the bamboo forest. This is a somber term. One wanders there silently in a greenish twilight. Deadly silence prevails in such a forest, and one's feet are covered ankle-deep in bamboo leaves. You will not hear a sound. This is terribly eerie. The ghosts live here. And at night they descend. Down here is the kraal, and here is a person whom the ghosts are supposed to fetch. To prevent this from happening, a ghost trap is built, which is prepared by laying out a small, well-tended path, paved with small stones down that extension. Eventually, after much work, they discovered the best place for such a ghost house.[564]

Then they told me how the ghost descends, and what it says: "Here it continues. Here is a nice path. Here is a hut. This is a beautiful bed. Oh, this is beautiful. One can sleep here. I shall sleep here. Here there is water. Here I shall live." The ghost then lies on the bed and sleeps. Next morn-ing, the sun rises. "Alas, the sun. I must be on my way, right now." But he hasn't noticed that he wasn't in the kraal. He leaves quickly and vanishes into the bamboo forest. The sick man, meanwhile, recovers.

This illustrates how the primitives deal with such matters; not only do they say it's the work of children, but they even believe this. I'm inclined to assume that they believe this. For instance, we also maintain that Father Christmas is for children, and so, too, are the bunnies. Ultimately, how-ever, such matters are in the hands of adults. So they are the silly-billies really. But they claim that it's children who love such things. But this is our childlike side. This is the primitive. Moreover, it is true when they claim that children do it. In reality, it is the ancients within us whom one calls children, and they want to play. On the one hand, we have no knowl-edge whatsoever about them; on the other, there is nothing about them that we want to know. It's indecent to speak about these things; for in-stance, to mention the word for ghosts. This is the greatest social gaffe that

---

[564] For details of these episodes, see *MDR*, 253ff.

you can commit. People will scatter in all directions if that is mentioned, because they forever turn their backs on those things, and I mean literally. When at a palaver for the first time I uttered the word *selelteni*[565]—that is, the native word for ghosts, the taboo word that should cross no one's lips—they literally turned their backs on me and began discussing the weather. We behave the same way in such cases.

I mention this because, as you'll recall, we already encountered this motif on the occasion of the first attempt at a mandala. One abstains from it. Our consciousness is always so constructed that it prefers not to return to these childlike, primitive games, but instead strives forward, outward. We do so to prolong our youthfulness. We assume that the young person must always look outward, must not see these things, and must devalue them to a certain extent. Otherwise, he will remain stuck in the past if he believes for too long that these matters are true or important. For this reason, the natives quite correctly maintain that it was the children who did this, while they themselves had no such customs. All that, they maintained, had no meaning whatsoever.

This continued until one of the tribal elders said, after long and painstaking efforts on my part, "Well, Sir, there is something we do in the mornings when the first ray of sunlight descends. We leave our huts, turn toward the sun, and do this: we spit into our hands and turn our palms outward." When I asked him what this meant, he said, "Our fathers always did this. One has always done it this way."[566] In reply, the young members of the tribe claimed that they had never done this. They do not want anything like that to be done; it is stupid. They laughed when the elder uttered these words. The elder and his fellow elders retorted, "But you do it, too." In reply, however, the youngsters were having none of it, they didn't want to admit it, they simply would not look at things, and

---

[565] Elgonyi term for ghosts or spirits, not found in *MDR* but again in "Psychology and Religion" [(1937) 1938/1940]: "I wanted to inquire about the ghost houses I frequently found in the woods, and during a palaver I mentioned the word *selelteni*, meaning 'ghost.' Instantly everybody was silent and painfully embarrassed" (*CW* 11, § 30).

[566] The elder believed that veneration of the sunrise was common to all peoples; see Jung's nearly identical report of this "offering [. . .] to the sun divinity" (*MDR*, 267) in "Archaic Man" [1931], *CW* 10, § 144, 146. Cf. "The Structure of the Psyche" [1927], *CW* 8, § 329, and *DAS*, 220–22, where McGuire notes (221, nn. 7 and 8) that according to A. C. Madan's *Swahili–English Dictionary* (Oxford: Oxford University Press, 1903), used by Jung, *roho* (death rattle) also means "soul, spirit, life," and refers to Jung's "Spirit and Life" [1926], where the latter contends that "it can scarcely be an accident that onomatopoeic words like *ruach*, *ruch*, *roho* (Hebrew, Arabic, Swahili) mean 'spirit' no less clearly than the Greek πνεῦμα and the Latin *spiritus*" (*CW* 8, § 601).

instead made a joke of it. Finally I said, "Quiet now! Let me listen to your elder."

I asked him about the reason for this. Of course, he didn't know the reason. But we can know what spitting into one's hands means. This is *ruh*. This is Swahili for breathing or rattling. It is our German word *röcheln*. The Swahili even pronounce it as "roho." It is onomatopoeic, a word that brings itself into being. It is probably connected with the Arabic words *ruh* and *rich*. This is the word *pneuma*.[567] It is the rattling breath of the dying, when they exhale their soul. This blowing is referred to as *roho*. The soul is blown out. Most primitives also consider saliva to be magical. In the New Testament, you will recall, when Christ heals the blind man,[568] he makes a dough out of saliva and earth, because saliva is mana. Spitting or blowing into one's hands is mana, a living force, which is given into the hand. The force of the soul is given into the hand and offered to the god. From this I concluded that they actually worship a deity, because they also denied this, and would have none of it. I assumed that the sun was the god, and so I said, "Here, the sun is *mungu*"—that's probably a loan word and denotes *mana*, their word for sacred. *Mana* is used in Swahili to express God. They use the word *mana* in Swahili.[569] *Mana jadoga*[570] means the meaning of the dream or the sense of the dream. This is important for us, isn't it? Then I said, "The word for sun is *adhîsta*, and here

---

[567] Gk. πνεύμα (*pneuma*), "air in motion," or "breath/breath of life," indicating among the Presocratics the vital principle or *arché*. The Stoics connected it with spirit and soul regarding it as "an agent which generates all the physical qualities of matter" (Samuel Sambursky, *Physics of the Stoics* [Westport, CT: Greenwood Press, 1973], 7). Both material and immaterial, it differs from the Platonic and Aristotelian *nous* or mind. In Christianity, *pneuma* translates Hebrew *ruah* (חור, spirit), a noun of female gender also meaning "wind" and "breath." Spirit becomes a *locus classicus* in the history of the philosophy, especially in such Renaissance thinkers as Agrippa von Nettesheim, Paracelsus, and Giordano Bruno. See also *DAS*, 220–21; *Psychology of Kundalini Yoga*, cit., 37; "Spirit and Life" [1926], *CW* 8, § 319; and *MDR*, 267.

[568] Mark 8:22–26.

[569] This appears to be a transcription error. *Mana* commonly refers in Polynesian and Melanesian cultures to the supernatural energy and/or healing power ascribed to, or emanating from, specific objects, venues, persons, and spiritual beings. The term went on to be widely studied, interpreted, reinterpreted, or even misunderstood in the social sciences and entered the Western imagination of the early twentieth century. Cf. "On the Nature of the Psyche" [1946], *CW* 8, § 411 and n. 120, and "On Psychic Energy" [1928], *CW* 8, § 177, on *mana* and Swahili *mungu/mulungu*. See also Shamdasani, *Jung and the Making of Modern Psychology*, cit., 293–95, and Nicholas Meylan's critical account, *Mana: A History of a Western Category* (Leiden: Brill 2017).

[570] Unidentified term.

this adhîsta is mungu." The sun was standing in the position of half past nine in the morning. They erupted into laughter, and I had no idea why. They exclaimed, of course, that it was utterly stupid to believe that this was mungu. In response, I said, "But if you hold out your hands to the sun, you do mean the sun." "Yes, yes, of course." "So, then, the sun is mungu." "Yes," they replied, "it is mungu." "But if it is there, then it is not mungu?" "No, of course not."[571]

Helios greatly exercised the early Church.[572] For this reason, Saint Augustine preached in the fifth century that "Non est hic sol . . ."—"Not

---

[571] Elsewhere Jung refers to the discrepancy between designating God as the rising sun (Swahili *adhîsta*) and the sun itself as God: "To the primitive mind it is immaterial which of these two versions is correct. Sunrise and his own feeling of deliverance are for him the same divine experience, just as night and his fear are the same thing. His emotions are more important to him than physics; therefore what he registers is his emotional fantasies. For him night means snakes and the cold breath of spirits, whereas morning means the birth of a beautiful god" ("On the Structure of the Psyche" [1927/1931], CW 8, § 329; cf. "On the Nature of the Psyche" [1946], CW 8, § 411). In *MDR* Jung adds, "Besides 'adhîsta' the Elgonyi also venerate 'ayîk,' the spirit who dwells in the earth and is a *sheitan* (devil). He is the creator of fear, a cold wind who lies in wait for the nocturnal traveler" (267). Therefore, "from the sunset on, it was a different world—the dark world of *ayik*, of evil, danger, fear. The optimistic philosophy gave way to fear of ghosts and magical practices intended to secure protection from evil. Without any inner contradiction the optimism returned at dawn" (267–68). See also Hannah, *Jung: His Life and Work*, cit., 173–75 and Burleson, *Jung in Africa*, cit., 160ff. (as well as Burleson's comment [n. 65] regarding the inaccuracy at times of Jung's recollection in *MDR* of Elgonyi terms related to their funerary ceremonies).

[572] In *Symbols of Transformation* [1912/1952], CW 5, in particular ch. 1.5, on the vast theme of solar symbolism, Jung attended closely to the links between Christianity and Mithraism. The philosophical aspects of this mystery cult, masculine in character, Zoroastrian in affinity, include the identification of Mithras with the reborn sun (as documented by inscriptions from the second century CE, in which the god is named *sol invictus Mithras*, *Deus sol invictus*, *Sol Mithras*, and so on). In fact, the centrality of the sun was a main point of connection between Christianity and Mithraism, and other mysteries. In this regard, Rahner notes that "the greater the religious appeal during the third century of ancient heliolatry to Neo-Platonic Theosophy and to what had become the solar mysteries of Isis and Mithras, the harder-pressed were the churches to lend cultic form to their own supranatural mystery of the sun" (Hugo Rahner, *Griechische Mythen in christlicher Deutung* [Basel: Herder, 1984 (repr. "mit 11 Abbildungen und einem Geleit- und Schlüsselwort von A. Rosenberg")], 132). Early Christian apologists such as Tertullian and Hieronymus already saw Mithraism as an imitation of Christianity. Justin Martyr (ca. 100–165), the first Christian apologist to deal with both Christian and pagan mysteries, saw them as independent (and thus avoided using the term *mysterium* to refer to both in the same context). Justin regarded the offering of bread and water in the Mithraic mysteries as a diabolical caricature of the Christian Eucharist (*First Apology* 66): see Bouyer, *Christian Mystery*, cit., 133–34 and passim. Since the 1970s, the scholarly research on Mithraism has largely superseded the dominant paradigm set by Franz

this sun is our Lord Christ, but He who made this sun."[573] "Das ist der Helios Natolos,[574] die Sonne des Aufgangs. Das ist Gott, wo die Sonne aufgeht." (This is the sun Natolos [*sic*], the sun of the rising. Where the sun rises, there is God.) The primitives said that mungu was not the sun, but instead the *moment*. I replied, "If that's right, then the new moon must also be mungu." That is not the case at all. But when the new moon appears and the first bit of the rim slowly becomes visible, then they all performed the same action. They greeted the new moon because it was mungu.[575]

Cumont (see also above, n. 410), who derived Roman Mithraism from Iranian Mazdeism and stressed its parallels with Christianity. Recent studies indeed tend to doubt or even deny that Mithraism was genuinely a competitor to Christianity, downplaying Ernest Renan's famous assertion that "if Christianity had been struck down in its cradle by some fatal illness, the world would have become Mithraistic" (Renan, *Marc-Aurèle et la fin du monde antique* [Paris: Levy, 1882], 579). For recent perspectives, see Richard Gordon, "Mystery Metaphor and Doctrine in the Mysteries of Mithras," in John R. Hinnells, ed., *Studies in Mithraism* (Rome: "L'Erma" di Bretschneider, 1994); Roger Beck, *The Religion of the Mithras Cult in the Roman Empire: Mysteries of the Unconquered Sun* (Oxford: Oxford University Press, 2006).

[573] "Non est dominus christus sol factus, sed per quem sol factus est" (Augustinus Hipponensis, *Tractatus in Iohannis Evangelium*, 34.2, in *PL* 35, col. 1652). Augustine often connects Christ with light and the sun, thus developing the frequent biblical image of the messianic age as one of light: references to the eschatological light as well as the Latin epithet *sol justitiae* (sun of righteousness) are commonly found in the prophets as well as the Gospels (cf. Isaiah 9:1; 30:26; 62:1; Malachi 4:2; and Luke 1:78–79; 2:32); the Gospel of John in particular presents Christ as the new light meant to illuminate the world (see John 8:12). It was Augustine who effectively introduced the theological linkage of two of the major Roman feast days (the Fors Fortuna on 24 June, and the Dies Solis Invictus on 25 December) with the births of Saint John the Baptist and Christ. Also, Augustine polemicized against the sun worshippers among the Mithraists and even more often among the Manichees, who held the sun to be an emanation of the principle of the good. Yet sun worship was still diffuse among both pagans and Christians in the fifth century: see the Leo the Great's fascinating allusion in his Sermon 27, "On the Feast of the Nativity" (Philip Schaff and Henry Wace, eds, *A Select Library of the Nicene and Post-Nicene Fathers of the Christian Church* [NCNF], 2nd series, vol. 12 [New York: The Christian Literature Company, 1895], 140). It might be recalled, finally, that Max Müller considered the solar myth primeval for human civilization: "I look upon the sunrise and sunset, on the daily return of day and night, on the battle between light and darkness, on the whole solar drama in all its details that acted every day, every month, every year, in heaven and in earth, as the principal subject of early mythology" (Max Müller, *Lectures on the Science of Language, Delivered at the Royal Institution of Great Britain*, 2 vols [New York: Charles Scribner, 1872], 2:537).

[574] Unidentified term, which does not appear in the *Tractatus in Iohannis*; but presumably an incorrect transcription for Lat. *neonatus* (newborn).

[575] See also Jung's lecture "Archaic Man" [1931], CW 10, § 138.

If you cast a critical eye on these customs, then you'll reach the peculiar conclusion that things happen here that these people do not even *think*. They simply *occur*. We would express this in the form of a prayer, such as the following well-known formula, for instance: "Father, into your hands I commit my spirit"[576]—the sacrifice of the soul—that is simply an event on this primitive level.[577] We believe that this prayer is conscious action. Only the primitive does not know what this is. In principle, however, this is exactly what it's like among us. It simply happens. It's exactly the same, and therefore we must ask, who is actually thinking this? Who, for instance, thinks this mandala in the case of my dreamer? He himself most certainly does not. But *someone* does. It happens. And why can something like that happen? This problem has already caused me a considerable headache.

* * *

[576] "In manus tuas, Domine, commendo spiritum meum" (Psalm 31:5 [Vulgate Psalm 30:6]). According to Luke's Gospel (23:46), after crying out on the cross, Jesus uttered his last words in paraphrase of this psalm saying, rather than "Domine," "Abba": "Father, into thy hands I commend my spirit" (Matthew 27:46 and Mark 15:34 report instead, "My God, my God, why hast thou forsaken me?"). Burleson aptly comments, "This empatic evaluation of an indigenous African ritual as spiritually authentic was unusual in an age when European explorers and missionaries constantly and fallaciously viewed Africans as godless savages who worshipped the Devil" (*Jung in Africa*, cit., 163).

[577] Regarding the devout Elgonyi practices around *mungu*, cf. *MDR*: "If the gift was spittle, it was the substance which in the view of primitives contains the personal mana, the power of healing, magic, and life. If it was breath, then it was *roho*—Arabic, *ruch*, Hebrew, *ruach*, Greek, *pneuma*—wind and spirit; the act was therefore saying: I offer to God my living soul. It was a wordless, acted-out prayer which might equally well be rendered: 'Lord, into thy hands I commend my spirit'" (267). The deep impression made on Jung by this "wordless, acted prayer" that could be set alongside the words of the Psalm led him wonder in the final passage of "Archaic Man" [1931], "Does this merely happen so, or was this thought brooded and willed even before man existed? I must leave this question unanswered" (*CW* 10, § 146).

LADIES AND GENTLEMEN,

To convey a sense of the original atmosphere of these things, I'd like to present some samples of a primitive text. They are the memoirs of a Navajo Indian,[578] which discuss, among other matters, these mandalas. Later

---

[578] The source of the text cited here, which is not in the JA or JFA, has not been found, notwithstanding inquiries and archival searches elsewhere with the assistance of John Peck. So one can only hypothesize that such testimonies reached Jung through one of the following: a) William Morgan, husband of Christiana and author of "Navaho Treatment of Sickness: Diagnosticians," *American Anthropologist* 33, no. 3 (1931): 390–402, and "Navaho Dreams," *American Anthropologist* 34, no. 3 (1932): 390–405, in which he adopts Piaget's and Jung's theories and refers to unpublished material based on a Navajo informant of his (see also his later collection of Navajo stories, *Coyote Tales*, translated by Morgan and Robert W. Young [Washington, DC: US Department of the Interior, 1949]); b) Jung's friend the Native American chief of the Taos Pueblo Tribal Council, impresario, and medicine man Ochwiay Biano (also known as Antonio Mirabal and Mountain Lake), with whom he had engaged in intensive conversations in Taos, New Mexico in 1925 (see *MDR*, 247ff.; for an assessment of Jung's encounters with Native Americans, see David G. Barton, "C. G. Jung and the Indigenous Psyche: Two Encounters," *International Journal of Jungian Studies* 8, no. 2 [2016]: 75–84); c) anthropologist and ethnomusicologist Jaime de Angulo (1887–1950), the husband of psychologist and translator Cary Fink Baynes; de Angulo had studied the Taos and after first meeting Jung in 1923 stirred his interest in the Pueblos and carefully prepared his pivotal 1925 encounter with Mountain Lake (Shamdasani, *Jung and the Making of Modern Psychology*, cit., 318–21). A list of de Angulo's publications is included in *Jaime in Taos: The Taos Papers of Jaime de Angulo*, compiled with a biographical introduction by Gui de Angulo (City Lights Books: San Francisco 1985); or d) Margaret Erwin Schevill (or Shevill Link; 1887–1962), an American artist and self-taught ethnologist who took a deep interest in Navajo sand paintings in the early 1940s and collected a wide array of copies. Jung would quote her book *Beautiful on the Earth* as the source for the Navajo myth of Estsánatlehi (Self-renewing Woman) and Yolkaíestsan (White Shell Woman) (see below, respectively, nn. 582 and 583), similar to the account offered in the pages that follow here, in the revised 1954 version of the "Visions of Zosimos" lecture at Eranos in 1937 (*CW* 13, § 130). Schevill was equally the source for the two sand mandalas (respectively from the "Mountain Chant" ceremony and the "Male Shooting Chant") commented on by Jung in "On Mandala Symbolism" [1950], *CW* 9.1, Figs. 45 and 46, and § 699–700). Schevill recalls, "Twenty-five years ago [so ca. 1920] on the Navajo Reservation one could still see how gods were born and reborn [. . .]. I was not an ethnologist. [. . .] Made conscious of the dynamism of the art forms on a day when I saw the first sand painting, I cherished for many years the desire to have other Americans appreciate these symbolic paintings" (xiii); these she found "stimulating in the sense that modern art" was, but also impressive for their "timeless quality" (Margaret Erwin Schevill, *Beautiful on the Earth*, with illustrations reproduced from the author's drawings by Louie Ewing [Santa Fe, NM: Hazel Dreis, 1947], xiii, 8–9; see also 130). She wrote that because of the Navajos' devout respect for healing processes, "the power of divine designs was too great to remain in the permanent possession of the earth people, so they decreed that no reproductions were to be made" (129), whence the ritual required that sand paintings be destroyed. Lastly, it is worth recalling the 1932 essay by the Swiss Lutheran pastor Oskar Pfister, "Instinktive Psychoanalyse unter den Navaho-Indianern," *Zeitschrift für Anwendung der Psychoanalyse auf die Natur- und Geisteswissenschaften* 18 (1932): 80–109. This psychologist and early associate of Freud provided a psychoanalytic interpretation of Navajo healing rituals, recognizing in them a "deeper knowledge of true knowledge than in all of pre-analytical psychotherapy" (103).

on, he also talks about the so-called Night Chant and Mountain Chant.[579]
This mandala is, of course, also associated with dance and song. He writes,
"These chants"—I would like to read the English version, but perhaps not
all of you understand enough English, so I shall translate:

> These chants are intoned by someone who is sick or who has been
> sick. Sometimes they are also intoned for a dream. Certain dreams
> are bad and must be sung away. When people have colds, it is also
> good for them to dance. The whites have medicines for all sicknesses.

---

[579] The Night Chant and the Mountain Chant are two major Navajo healing ceremonies,
designed to preserve, promote, and restore the psychophysical integrity of the individual and
the community. They are among the ancient Native American religious traditions first recorded, at the beginning of the sixteenth century, by Europeans arriving on the continent.
Subsequently, the first major dictionary, *A Vocabulary of the Navaho Language*, would be
published by the Franciscan Fathers only in 1912, and *A Manual of Navaho Grammar*, compiled by Fr. Berard Haile, in 1926. The first full account of a Native American ceremony ever
published was probably that of Washington Matthews in 1887, as noted by Paul Zolbrod in
his foreword to Washington Matthews, *The Mountain Chant: A Navajo Ceremony* (Salt Lake
City: University of Utah, 1997), who also observes that "[e]ach Navajo ceremony builds on a
specific story, which in turn contributes to a network of interlocking narratives as poetically
rich as the Homeric epics or the Arthurian cycle"; while Navajo healing ceremonies revolve
around a form of spiritual communication which also includes "a nonhuman community that
today's conventional thinking cannot yet fully accommodate" (vii). Washington Matthews
(1843–1905) was a US Army surgeon during the American Civil War who undertook ethno-
graphic research in the Dakota Territory and the Navajo reservation (1880–84, 1890–94)
supported by Army Medical Museum in Washington, earning the trust of various medicine
men. The Bureau of American Ethnology (BAE), created in 1879 under the umbrella of the
Smithsonian Institution to collect archive testimonies and other materials from Native North
American tribes, supported research by Washington, Franz Boas, Frances Densmore, and Paul
Radin, among others. See Matthews, "Mountain Chant," cit.; Stevenson, "Ceremonial of
Hasjelti Dailjis," cit. Jung, who mentions Matthews's and Stevenson's texts in "Commentary
on *The Secret of the Golden Flower*" ([1929/1957], CW 13, § 31 n. 10) and in "Concerning
Rebirth" [(1939) 1940/1950] (CW 9.1, § 240 and n. 4), owned the entire collection of the
annual reports of the Bureau of American Ethnology along with other studies such as Pliny
Earle Goddard, *Indians of the Southwest* (New York: American Museum of Natural History,
1921) and Gladys A. Reichard, *Navaho Religion: A Study of Symbolism*, 2 vols (Bollingen
Series XVIII) (New York: Pantheon, 1950). Reichard is among the most important students of
languages and cultures of Native Americans in the first half of the twentieth century, paying
special attention to their social life (Gary Witherspoon, *Navajo Kinship and Marriage* [Chi-
cago, University of Chicago Press, 1975], ix). According to her, "the number of Navaho songs
is incalculable. The 576 songs Matthews mentioned as actually sung at performances of the
Night Chant he witnessed are the very minimum and do not include those for the various
branches or phases of the chant" (Reichard, *Navaho Religion*, cit., 1:286). See also Paul G.
Zolbrod, "On the Multicultural Frontier with Washington Matthews," *Journal of the Southwest* 40, no. 1 (1998): 67–86, and his vivid introduction to Navajo culture in Dine Bahane:
The Navajo Creation Story (Albuquerque, NM: University of New Mexico Press, 1987
[1984]). For a phenomenological analysis of the psychophysical and social and healing process
of the Navajo ceremonies, see Gregory W. Schneider and Mark J. DeHaven, "Revisiting the
Navajo Way: Lessons for Contemporary Healing," *Perspectives in Biology and Medicine* 46,
no. 3 (2003): 413–27.

We savages, so to speak, use them, too, if we can obtain them. In most ceremonies, the man for whom the ceremonies are sung will take some of the hard, white medicine that the doctor working for the American agency has given him. The white medicine is taken, but within the context of a chant. Occasionally, these medicines are either taken first and then the dance is performed, or the medicines are taken during the dance. But even if one seems to have been healed by the white man's medicine, nevertheless it is not good not to dance for a longer period. Because the sickness was inside you, and *one must cleanse one's own life.* One is really healed because one remembers that one was sick. But after chanting one feels as if one has been reborn. Then the path of life extends before us in a straight line. If a man only seems to be sick, or if he might get sick, or if nothing has happened to him for some time, or if he has sold sheep and holds the money in his hand, then he must be healed.

You see, this is the psychology that I mentioned earlier. Here, the connection between man and nature is sacred if we draw sacred spiritual force from nature. If we get something out of this sacred nature and gain an advantage, we must pay for it; that is, there must occur, once again, a purification or reconciliation.

Then he must be given relief again. For this sin has burdened him.[580] He will then have the Night Chant or the Mountain Chant sung around him, and he will be reborn. The singer comes and chants for five days. But it is better to let the singer chant for nine days.

And so forth.

Here, he tells us what mandalas are good for. It's particularly interesting to observe that these primitives also undergo the analytical procedure if they are not sick, or also if nothing has happened to them over a longer stretch of time. The absence of incident is felt to be uncanny. Something must always happen. That's the way it is with us—namely, that from time to time fate seems to stand curiously still. Nothing seems to want to happen. It is a condition of lostness, which the Indians say can be healed in this way. I have already mentioned that Indian mandalas make a great fuss about the number four. These texts contain ample references to it, and I'd like to quote a relevant passage:

---

[580] For the Navajo, "sin" in the sense of the classic Greek *hamartia* or flaw (with thanks to Prof. Paul G. Zolbrod).

One day my grandfather heard me chant. He asked me where I had learned the chant. But I couldn't tell him. I didn't know. It seemed to me as if I had always known the chant. But I believe I learned it when I was sitting at such a place, that is, a mana place, a sacred place. At the time, my grandfather began to take an interest in me. He told me a lot about the creation of man. And from him I learned about the ways of many things. First, he told me about the colors of the four parts of the world. He said, look south, and you will see blue. You will see the color white on your left hand, the color yellow on your right, and behind you it will be black. These colors were in the beginning, and the whole world was like that at the time.[581] Then he said, Estsánatlehi lived in the west,[582] and rain comes from the west. But all

---

[581] In Navajo cosmogonic myths, two principal color systems symbolize the four quarters, which vary only in the placement of black and white (indicating east and north, respectively). See Jeff King, Maud Oakes, and Joseph Campbell, *Where the Two Came to Their Father* (Bollingen Series I) (Princeton, NJ: Princeton University Press, 1991), 71–72 and 89ff.; Washington Matthews, *Navaho Legends* (Memoirs of the American Folk-Lore Society 5) (Boston, MA: Houghton, Mifflin, 1897), 215–16; and Margaret Schevill: "Four colors are used over and over again. These are the ceremonial colors of the four directions of the earth. White is generally in the east, blue in the south, yellow in the west, and black in the north. They are used in that order, also clockwise, - although I have heard that in some paintings used in witchcraft rituals, these colors are laid on in the reversed direction. Black and white are sometimes interchangeable. White is used at times in the north and black in the east. When this is done it is said the painting usually presents a situation which happened in one of the former worlds. Black and yellow are said to be masculine colors, and blue and white feminine ones, but I have seen them used in opposite connotations. Objects which are white are surrounded with black lines and *vice versa*. Those which are blue or yellow are outlined in the opposite color. This is to suggest a harmony in the inclusion of the contrasting male and female colors" (*Beautiful on the Earth*, cit., 131–32). For a full description of the Navajos' color symbolism, see Reichard, *Navaho Religion*, cit., vol. 1, ch. 12, esp. 187–97, and ch. 13.

[582] The name Estsánatlehi (deriving from *estsán*, "woman," and *natléhi*, "to change") translates as "Changing Woman" or "Self-renewing Woman" (Schevill, *Beautiful on the Earth*, cit., xi), or "A Woman She Becomes Time and Again" (Berard Haile, *Origin Legend of the Navaho Enemy Way* [New Haven, CT: Yale University Department of Anthropology, 1938], 85, cited in Oakes and Campbell, *Where the Two Came*, cit., 62). Estsánatlehi is the Navajos' "most powerful and revered deity," whose attributes are "plants of all kinds," which "is further evidence that she is an apotheosis of Mother Nature" (Matthews, "Mountain Chant," cit., 31 and 32). Margaret Schevill maintains that Estsánatlehi "encompasses all the mystery and paradox of the coming of eternity into the net of change. She is the transmuter, changing timelessness into time. She is fluidity of life, the Heraclitean flux, the ever-rippling Veil of Becoming, changing the white light of Unchanging Being into the colors of the living world" (*Beautiful on the Earth*, cit., 63), and points out that "[t]he Turquoise Goddess is so revered it is forbidden to make any representation of her. Her home is sometimes depicted in the center of a sand painting, but her actual presence would seem to be implied through the presence of her sons" (25–26). Jung writes on Estsánatlehi, "She was the mother of the twin gods who slew the primordial monsters, and was called the

prosperity is in the west. Hard substances abound there. But Esdan is not alone. Yolkaíestsan,[583] her sister, lives with her. One is white, the other blue. They are two. All things in the world have two ways, my son. All things are male and female, and Esta Nadi is both, male and female. From her come the rains, which are both male and female. The summer rains are male. They come with thunder and lightning.

And so it goes on.

If you make this primordial idea specific, you will see into the manifestation of things. The nature of all things is that they are one. The oppositions are united in them. The whole thing is a so-called unifying symbol, into which all oppositions are incorporated. In the Navajo system, this mystery is thus a tray or pot containing dew, which is animated and confers fertility. It is indicated by turquoise sand, because that is the sacred, greenish-blue color of water.

There is a further passage in which he speaks about the number four, in a very interesting psychological way:

Say your prayers often, and your face, that is, your gaze, will become more beautiful. When people offer incense, then the whole world will be open to them; but he who offers no incense and who takes on a cunning look, is a sorcerer and his heart blossoms under the ashes. Knowing the ways of the gods is especially good. They, too, follow their paths and trails. They, too, are among what is beautiful, and thus one can lay a blue glass pearl on their path. All things are arranged in beauty, and the gods, too, stand in beauty. There is beauty in everything that is four. If you do something four times, then you will have done it well. Pray four times. He who understands the beauty of the world is a wise man. No hairball, no knot will prove too complicated for him to unravel. His path is long. He understands things, and he doesn't ask the birds to guide him through the cliffs. But when he reaches them, he asks Bighorn, who is a special bird and who has pity on him, and who then opens the cliffs for him. Because

---

mother or grandmother of the gods (*yèi*)." She is "immortal, for although she grows into a withered old woman she rises up again as a young girl—a true Dea Natura" who also represents "an anima figure who at the same time symbolizes the Self" ("The Visions of Zosimos" [1938/1954], *CW* 13, § 130 and 131).

[583] "Yólkai Ěstsán," or twin sister, also called "White Shell Woman whom the gods created by ceremonially giving life to a white shell image" (Matthews, "Mountain Chant," cit., 32). See Jung, "The Visions of Zosimos" [1938/1954] *CW* 13, § 130, and Schevill, *Beautiful on the Earth*, cit., xi and passim.

the birds in the sky are wise, like the four-footed animals on earth. And he who preserves the grasshopper, that is, spares it, is also a wise man. When it is cold in winter, he will place the grasshopper beside the fire; the room will become warm and the grasshopper will be able to sweat. Outside it is winter, but all of summer is contained in the grasshopper's tiny little body. The wise man also understands the ways of the wind. From the wind comes life, and the wind has blown into him and left him again by the tips of his fingers. Lo and behold, those are the traces of the wind, where it came out of his skin, because here you can see these small waves, which you can also see in the desert when a storm has blown across the sands. [Thus, he compares these whorls on the fingertips with the rippled desert sands.] But the wind makes no sound. Above the wind there is darkness, and the darkness is silence, and in silence and in stillness a man's heart makes beauty.

Compare this account with a missionary's sermon, which I've witnessed and heard often enough in Africa. How could we object to it? That's simply ludicrous. Indeed, these awovals are often filled with inner beauty and an empathy for nature, and they reveal such refined observation that we can do no more than cite them as exemplary. I'm convinced that if we could evoke such a mood within ourselves, then much evil would be remedied. However, the way that might lead one back to or onward toward such an experience of reality is not plainly obvious, unless we follow the allusions of the dream just as the primitive follows the flights of birds. Then, it seems to me, it might just be possible once again to attain the inner state that regards the world as an experience meant for us. But precisely this requires that we get to know the secret of this design. One cannot turn one's back on what lies within and gaze out the window, thinking of this as something indecent. We shouldn't assume that we have discovered a cesspool if we discover the soul. Nature is no cesspool. We must turn our eyes toward these inner things, and we must take to heart what this primitive tells us: namely, that it is wisdom to see the beauty of things and not their abysmal ugliness. That is sheer pathological folly, nothing else, the stupid sickness of our time.

Let us turn to the next dream.

At this juncture, I'd like to take the opportunity to note that a member of the audience has asked about how the dreams are connected: whether, that is, they form a sequence. They are indeed a sequence. The sequence I shall refer to in what follows was dreamed over the course of four days.

Nothing else intervened, no other dreams came along, and so these dreams are attuned to one another. I have mentioned everything I know about this man. Nothing has been trimmed, and so all the darkness, all the imperfection, all the doubt is there, and I've given you everything so as to enable you to account for matters. It makes no sense to play games and present incorrect details. For then you wouldn't be able to account for how matters actually proceed.

In the next dream, he discovers that he owns a kind of cage that consists of various compartments. The cage contains four small hens, which, so he says, he must guard carefully, since they keep trying to escape through a hole.[584]

> *In spite of my frantic efforts they do escape near the hind wheel. I catch them in my hand and put them in another compartment of the cage, the one I believe to be the safest. It has a window but it is secured by a fly-screen. The lower end of the screen is not properly fastened, so I make up my mind to get some stones and put them on the lower edge of the screen to keep the animals from escaping. Then I put the chickens in a basin with smooth, high sides, assuming that they will find it difficult to get out. They are down at the bottom of the basin, and I see that one does not move and I think that it is because I have pressed it too hard. I think that if the chicken is dead it cannot be eaten. While I watch it, it begins to move, and I smell an aroma of roast chicken.*

It was rather difficult to get the dreamer to produce associations. Eventually, quite a few associations emerged. Curiously, the fly-screen irritated him most, and he described it at great length. It was made of cords in a wide mesh fastened to the window of the cage. The cage itself was made of iron rods, so that the flies could enter freely on all sides, circulation was unhindered, and the hens could breathe fresh air. So the window was quite unnecessary, just like the fly-screen attached to it. At least it served to keep the animals caged, provided the screen was fastened, which he decided to do.

The window, which exists quite illogically, might well be my tics.

Here he refers to what I haven't mentioned so far, namely, that he easily loses his temper, particularly when irritated; like many people, he gets crotchety.

---

[584] Cf. the initial account in *DAS*: "I possess a sort of cage on a wagon, a cage which might be forlions or tigers. The cage consists of different compartments. In one of them I have four small chickens. I must watch them carefully because they are trying to escape" (105).

At this point, he mentions the brief conversation we had about this. He asked me how such a matter should be understood. I told him that in any event some sensory motor matter was not under his control, that there was a leak in his system, a hole through which something emerges that he cannot stop from happening, a kind of autonomous complex.[585] Tics cannot be suppressed. If one tries to suppress them nevertheless, so much tension rises that matters must be called off for a while. I explained it to him in terms of a leak. And he mentions this.

He goes on,

Now the window is secured by the fly-net, but down below, toward the ground, it isn't fastened. And I must fasten it there if the cage is to do its job properly, and actually bring the functions under my control.

For he had thought that these animals were psychic functions.

So I had to plug the hole or opening facing the ground.

He asks himself,

Would these tics also go away if I managed to close the window securely?

In actual fact, those tics did go away after a while.[586] But for now we are confronted with a new situation, namely, a cage in which four hens

---

[585] Reference to a cornerstone of Jung's system, the "gefühlsbetonter Komplex" (emotionally charged complex), following the word-association experiments carried out with Franz Riklin at the Burghölzli in 1903–7; see the essays in *CW* 2—and cf. Freud's crediting of the "Zurich school (Bleuler, Jung and others)" for the "very useful" definition (Sigmund Freud, "The Origin and Development of Psychoanalysis," *The American Journal of Psychology* 21, no. 2 [1910]: 181–218 at 199). The existence of these complexes, Jung claims, "throws serious doubt on the naive assumption of the unity of consciousness, which is equated with 'psyche,' and on the supremacy of the will. [. . .] The complex must therefore be a psychic factor that, in terms of energy, possesses a value that sometimes exceeds that of our conscious intentions, otherwise, such disruptions of the conscious order would not be possible at all" ("A Review of the Complex Theory" [1948], *CW* 8, 2nd edn, § 200). "The complex," Jung pointed out in 1934, "is a very unruly animal, quite independent of our I. Complexes are disobedient, autonomous beings" (Jung, *History of Modern Psychology*, cit., 116). See also Jung's "General Considerations on the Theory of Complexes" [1934], *CW* 8, and cf. his exemplification of the (Christ-)complex related to Saint Paul's conversion in "The Psychological Foundations of Belief in Spirits" [1948], *CW* 8, § 582ff.

[586] *DAS* does not mention "tics" as such. The discussion of this dream in that seminar initially took a different tack, emphasizing the wish expressed by the dreamer's unconscious to keep the chickens confined in order to "hold his individuality together." Jung later noted that "the running away of the chickens may mean the blind escapades of his life," reiterated that "[t]his man is not naive, he is a hard-boiled business man, yet he is something of an idealist. [. . .] The chicken-catcher is not the conventional man. It is the

are intent on escaping. If we compare this to the earlier dreams, then we can guess what this might refer to. What would you assume with regard to the cage?

PROFESSOR ZIMMER: His correctness which controls everything, his self-consciousness,[587] tells him that something inside is bound to escape him at some point. Conspicuously, there are four hens, and it's strange that he suppresses precisely what seems to slip away most easily.

PROFESSOR JUNG: This dream is unmistakably a very fitting expression of his psychology, which has grown tiresome to him. His self-consciousness behaves like a cage from which he is always trying to break out. Precisely that is the unpleasant situation. So, of course, he crushes much that were better not suppressed. What comes to your mind in this respect? What does one usually suppress if one is self-conscious?

ANSWER: Feeling.

Feeling! It is characteristic, is it not, that such an inhibited person, who is self-conscious is incapable of giving natural and warm expression to his feeling. This frightens him most of all, because he thinks he might forget himself. He could misbehave, and that would be awful! That fear is evoked by his self-consciousness, which suppresses natural, plain, warm, and affectionate feeling. We can assume that this little hen is probably the feeling that will be hurt most of all, and which evidently, because he treats it somewhat harshly, is most inclined to run away. Because if one is correct, then one runs away from oneself. One would like to jump out of one's skin. That is why self-conscious people suffer terribly from constantly having to watch themselves, why they never enter a function room or a large gathering of people, because, after all, one cannot show oneself. For in actual fact one wants to jump out of one's skin.

This dream speaks his language. Such is his conflict on the surface. But we have noticed that the previous dreams don't move quite so much on the surface, although this surface language very often contains what belongs in the depths.

---

conventional man who chases prostitutes. His conventional outside has gone with prostitutes, this is convention. The chickens are fragmentary unconscious souls which organize escapades" (112).

[587] As previously, both here and below, the English term "self-consciousness" is used by both Zimmer and Jung.

Here is what an old Latin proverb says about dreams: "Canis panem somniat, piscator piscem": "Dogs dream of bread, fishermen of fish."[588] In other words, what we ourselves are, what our day-to-day occupation is. We speak the language of our profession, of our complexes, of our everyday existence, and express in that language what moves us most profoundly. We can therefore assert that such a dream, which represents his specific knowledge, shows in another sense something even much more profound, which carries the healing meaning. Just recall that image: a moving cage, from which four hens are constantly trying to escape. What crosses your mind? What would be the next parallel? It's not that difficult. We have discussed this cage at great length. Has nothing occurred to you?

ANSWER: The sacred number four.

Right, you mean this zone. This temenos, or rather this container, would be the prison, a cloistered courtyard. The way to prison is like the way to the cloister. One is also locked in the cloister, after all, and wants to leave. In the previous sessions, moreover, you got acquainted with that curious arrangement of benches on which people sit with their backs to each other and as it were gaze out the window. Likewise, the hens too want to get out. It's an explosive situation, where he wants out and precisely not back in. These hens all want out, and he should keep them inside, that is, in the prison or cage that the unconscious has prepared for him. This feels rather eerie to the hens, and they'd like to fly off cackling.[589]

---

[588] Source unknown.

[589] In the corresponding discussion in *DAS*, Jung invoked hexagram 50 of the *I Ching*, "Ting" ("cauldron" in the Wilhelm/Baynes version, or sacrificial vessel containing spiritual nourishment. Incidentally, this is the same hexagram Jung received when he interrogated the oracle itself with regard to his introduction to the Taoist text: see his "Foreword to the I Ching" [1950], CW 11, § 976ff. Jung argued there that the dream might indicate a dissociation of individuality and emphasized the four functions, along with the way in which fate elicits encounters with aspects of individuality that can then be integrated ("If we experience it, it is caged, we taste that chicken," *DAS*, 109). The dream, he remarked, does not have to do with "the problem of his consciousness," but "with the centre outside of consciousness" (107). Later he observed, "The chapter of the *I Ching* dealing with [the] pot is one of those chapters which contain Yoga procedure, and our analytical procedure produces Western forms of what in the East is Yoga. The terminology is different, but the symbolism is the same" (117). Moreover he insisted on the idea that one should "hunt down [one's] instincts, get them together and transform them," for they are "most contradictory, [. . .] like animals in a zoo, they do not love each other at all, they bite each other and try to run away" (108). In German, he added, "*heilig* is connected with *heil*, or 'being whole'; *geheilt* means cured. Healing is making whole, and the condition into which one gets is a whole or complete condition, while before it was only fragments held together. So collecting and roasting the chickens means curing or making new" (128). Finally, "my idea in summing up this dream is that it gives the ingredients for the making of the new man.

Well, could you explain this in some way? Of course, I asked him what he made of this. But I'd prefer not to divulge his observations for the moment. Let us see whether you are able to unearth this yourselves. What we do know with certainty is that what he is suppressing or crushing is, shall we say, first and foremost his feeling, the natural expression of the fact that he is suppressing and damaging this inhibition. He kills his feeling. He does not allow it to flow out naturally, but instead his morbidity and over-harsh treatment destroy the expression. Thus, we have more or less recognized one of these hens. So what then are the other hens?

ANSWER: Functions.

Indeed, here we are contemplating functions.[590] Prior to this dream, the dreamer attended one of my public lectures, in which I discussed the theory of functions. He heard me mention the existence of four functions. For many people, it is very odd and mysterious to know that we have four functions. However, this isn't so strange, just as little as the horizon has graduated degrees of arc. Thus, we also speak of so and so many degrees north and northwest, and thus determine our position. By way of comparison, we have an orientation function. This is our consciousness. Consciousness permits you to account for where you are: that is, in what kind of human surroundings or company you find yourself. You use consciousness to do this. Consciousness is an organ of perception and orientation. To establish our bearings, we have four points, at the east, south, north, and west. We need approximately four criteria to orient ourselves. For instance, we must at first see, then hear, and so forth. Thus, we require sensations to know *that* something exists at all. Therewith, we obviously do not yet know *what* it is. For this, we need a process of apperception, which is the elementary process of thought and judgment. This process indicates to us what we are perceiving. We must know, furthermore, what *meaning* and value things have. For instance, if I tell someone that a fire has broken out in the house, that there is a fire in the cellar, he'll be able to picture and think this in utterly cold-blooded terms: well, indeed, fire is hot and bright, and something is burning, namely, the cellar. But it doesn't spur him into action. Only if it upsets him, if it gets evaluated and judged by feeling, does it have a value for us. If a person is suffering from melancholy, if everything is only gray on gray, like a photograph, it makes not a jot of difference to this person if a fire has

---

Therefore we have the parallel to the *I Ching*. Whether he lets them run or whether he kills them and roasts them is practically the same thing" (112–13).

   [590] Jung's psychological function system is explained in his *Psychological Types* [1921], CW 6.

broken out; that will leave him completely indifferent. To orient ourselves properly, we thus also need feeling, which serves as a criterion.

Now there is another criterion that is very important: namely, the one concerning the movement of things in time. Everything comes from somewhere and goes somewhere. The three functions about which I have spoken so far do not yet permit us to determine the movement of things in time. Instead, they give us the present moment. However, there is a before and after, for which we must somehow account. If a person were suddenly transferred to this place from elsewhere without any preparation whatsoever, he would say, "There is a gathering of people, and this room is an amphitheater. A voice can be heard. Someone is evidently delivering a lecture." Well, this has a before and after. The before is that those present have come from somewhere to listen to this lecture. The after is that everyone leaves again, each returning to his sacred herd. Thus, he has a notion of the story, he guesses where it comes from and where it is going. This is known as intuition. It is simply the faculty of presentiment, which we must put into play under all circumstances. There are people, of course, who can't imagine where things come from and where they are going. But most people can. They have the feeling that things might contain quite unheard-of possibilities. Thus, the intuitive procedure could also be defined as a faculty allowing for the *recognition of possibilities* present in things and situations, and concerning their possible origin and destination.

If we apply these criteria to the present moment, then we would seem to be sufficiently well oriented. Thus, we could say that we're aware of this particular situation. Now, these four points can of course be multiplied by three, four, or five without difficulty, and matters will still hold true. They do so because you will always be using approximately these criteria. As for these criteria, we have grown accustomed to a minimum. Thought allows us to make out possibilities to a certain extent, but not to *perceive* them. Intuition is a function of perception. Intuition thus has a nose, and can smell possibilities. You will see this chiefly among primitive people, who are not hampered by prejudices as we are. One dare barely say anything if one has an intuition, because everyone will pounce, exclaiming, "That is mystical." Having certain premonitions, just as animals do, is one of the most natural capabilities in the world. You can also frequently observe this among primitives, who live in a continuum. They do not live in conscious uniqueness, but instead in a stream of natural before-and-after, with everything proceeding in the right sequence; if one observes this order attentively, then one will reach the goal.

Now, my dreamer had heard this lecture and concluded that these four animals were his functions, which he was unable to keep together properly. I agreed with him, because our notion of the fourfold nature of things, which we identify with our orientation of consciousness, is, of course, a modern psychological term, which rests upon the ancient foundations of a primordial human intuition, according to which some kind of being exists that contains this sacred tetractys—the oldest notion of Greek philosophy, expressed in Pythagoreanism—the most ancient idea in Chinese thinking, where you will also find this fourfold foundation of the cosmos. Why this is so, of course, I do not know. One can speculate about this, and I always find it very interesting and useful to speculate about it.

Lecture 6

Ladies and gentlemen,

It is the last evening of our seminar, and various themes and issues remain on our program. I fear that if I answer all the questions that have been submitted, we shall lose too much time. If I can manage to return to them at the end of our proceedings, I shall gladly do so. But I believe we are better advised to continue exploring our dream.

You will recall the dream about the hens imprisoned in the cage, or rather that should have remained imprisoned. Regarding the cage, the dreamer offers the following remarks: it is either the kind of cage found in menageries, in which animals are transported or kept for exhibition, or a barred wagon in a circus, in which animals are displayed during performances. He then begins to contemplate and speculate about the cage. Thus he says,

> We are ourselves the cages of our thoughts, and must ensure that they do not escape us. Because, once they have gone, they are very difficult to capture again. Are the four hens meant to express thoughts and feelings that are seeking to free themselves and that we restrain? I have made every effort possible to hold them captive until they become unpalatable. Since these are symbols, I suppose they are meant to express something instinctual.

We have already discussed the cage. The dreamer pictures it as menagerie cage, fitted with wheels for the transportation of animals. This thought is very peculiar. This cage is, as you will gather from his comments, what the dreamer associates with himself. He himself is a cage. He contains the instinctual and animal-like, as expressed by this image, and, as I have mentioned, he feels like a receptacle for instinctual contents. While this thought might strike you as strange, we must accept it at face value, since it corresponds more or less to his feeling. It is not for us to consider it any differently from how it seems to him. We can therefore safely assume that his psychology feels to him like a cage, wherein the instinctual is contained. We must now begin to apply this hypothesis to the dream. The contents expressed by these small, agile hens are evidently somewhat excited; they are seeking to flee the cage on all sides, apparently because they are supposed to be penned in. You know how hens behave. If you try to drive hens into a narrow space, for instance, a hectic chase and endless cackling ensue. The animals scatter in all directions, and they panic easily. If we pause over this, we have to acknowledge that we know this from the inside: if we find ourselves in a situation from which we seek to escape, something inside us breaks into a panic and a fluttering, and looks anywhere for an out. We are keen to avoid entrapment in a given situation under all circumstances. We seek a way out on all sides. All our excuses, for instance, are such ways out. Everywhere, we look for a hole through which we can escape. I'm sure that you're familiar with such things.

Except that we haven't yet grasped what the actual issue is with these four hens, particularly why there are precisely four. You will recall, as I mentioned earlier on, that before this dream the dreamer had heard me deliver a lecture on the four functions of consciousness. I discussed these functions in our previous session. He now mentions that these four hens probably refer to these functions. That doesn't fit too badly with the first association that I have mentioned: namely, that the thoughts and feelings about some psychological instincts are supposed to represent that function. He has psychic instincts, of course, and the dream evidently challenges him to prevent these subterfuges, this fluttering away, and precisely to what end? Why do you think it should be prevented?

ANSWER: He wants to come to grips with himself.

Well, do you believe that one can do that with hens?

ANSWER: That's why his thoughts shouldn't fly off.

Why should his thoughts be prevented from breaking out?

ANSWER: He should wrestle with himself in the forum of his consciousness.

In simpler terms, he should concentrate and keep these animals together. Whether he should wrestle with himself is another question, which we could discuss perhaps later. But to do so, these instincts would have to have assumed other forms. You can no more wrestle with the instinctual than you can with hens. One cannot grapple with animals. Animal instincts can be domesticated, surmounted, fastened, or locked up in a cage. But one cannot pin down these instincts as soon as they appear in the form of animals. What we have here is an indication of the necessity for concentration, so that the libido does not flutter in all directions and dissipate—so that something is there when it's needed. You must always view such dreams in terms of the whole path undertaken up to that point. It is not a new situation, but a matter of change, an alteration of the already existing line. Besides, we have noted that the previous dream contained this mandala, which is the epitome of concentration, circumambulation, and contemplation. In this dream, we evidently face the question of how that should be done, how such concentration might be enacted, or which difficulties confront such concentration. As we have seen, the difficulties at first come from the four hens, which are trying to escape all the time. Therefore, the difficulty here appears to reside in the instinctual, which doesn't want to snuggle up to this concentration very readily. Over the course of this dream, we also observe how these instincts try to escape. Did you notice this?

ANSWER: At the rear and at the bottom.

Yes, indeed, at the rear wheel. You observed that the fly-net was open at the bottom and needs to be weighted down, with stones, so that the hens cannot escape. What do you think this means? Let me first mention some more associations about this rear wheel. The dreamer is an enthusiastic motorist, and the word "wheel," particularly "rear wheel," obviously leads him to think of the wheel of a motor car. He mentions that the rear wheel is a very important part of a car, since it sets the vehicle in motion. It's a drive wheel. A rear wheel engages with the drive train. Why do you think that these drives, these instincts, these hens, can escape most easily at that very point? He says that the dream text mentions a rear wheel, and not a driving wheel. That's imaginative.

ANSWER: It's not easy to steer the rear wheel. One doesn't steer at the back.

ANSWER: They can escape more easily there.

But why not at the front, where the engine is?

ANSWER: He has that under control.

The engine? He can control the rear wheel and the engine with the brake.

ANSWER: One doesn't have eyes at the back of one's head.

Very true. We have no eyes at the back of our heads, and where we have no eyes, something is bound to happen. What do we find there?

ANSWER: Darkness.

Yes, darkness and the unconscious. The unconcious is always at the rear, the rear is simply the unconscious. We have no eyes at the rear, no field of vision, and nothing is seen there. That is the unconscious. What comes to him from behind, what jumps him from behind, the black cat and all those stories, that's the unconscious, of course. One is attacked from behind. The Greek word for shadow is *synopados*,[591] an indicative name. This is the invisible brother or friend, the mortal genius that lives inside everyone, who follows behind, darkly, the cold shiver that runs down our spine. Where we are unconscious, these hens can escape most easily. Now we have a further indication: namely, that this also happens at the bottom, since the bottom edge of the fly-net must be weighted down because it is open. What do we find at the bottom?

ANSWER: The soil.

Well, you'll need to put that in somewhat more psychological terms.

ANSWER: The bottom is also closer to the instinctual than the top.

Yes, it's a fairly traditional conception that below is closer to the instinctual, because it is in fact the instinctual zone of the body. In the Indian Tantra yoga, the lowermost centers are the centers of the instinct, which correspond to the solar plexus.[592] Where instincts prevail, there is

---

[591] Gk. συνοπαδός: literally, "with-follower." In Jung's terms, "he who follows behind." They [the Greeks] expressed in this way the feeling of an intangible, living presence—the same feeling which led to the belief that the souls of the departed were 'shades'" (*Basic Postulates of Analytical Psychology* [1931], CW 8, § 665). "Everybody," Jung said in DAS, "has his *synopados*, the one who follows with and after, the shadow, understood as the individual daimon" (508). He identified it with the genius of Horace (Horace, *Epistles* 2.2.187ff.) and recalled the ancient Greeks' conception of daimon as "something tremendous, intensive, powerful, neither good nor bad necessarily; it simply did not enter the category of good and evil, it was a power. At a more primitive level, the term *mana* would be used, with more the connotation of the animus or the anima—a soul" (508). See the compelling analyses of daemons in E. R. Dodds's outdated but still insightful classic *The Greeks and the Irrational* (Berkeley, CA: University of California Press, 1951). Dodds outlined the psychic structure of the archaic Greek individual as fatally split between an "indestructible daemon" of divine origin "resident in the human head" and "a set of irrational impulses housed in the chest or 'tethered like a beast untamed' in the belly" (214); so substantiating the image of human being as a kind of "puppet"—according to Plato's *Laws* 1.644de—perhaps created for the mere fun of the gods.

[592] The *plexus solaris* (solar plexus), popular term for *plexus coeliacus*, is located near the stomach, and corresponds to the *manipura* chakra, whose element is fire. According to Jung, "the psychology of the lower centers is analogous to the one of primitives—unconscious, instinctive, and involved in participation mystique. Life appears here as an occurrence, so to speak, without ego" (*Psychology of Kundalini Yoga*, cit., 85). In DAS he described the *plexus solaris* as "the brain of the sympathetic system" and "the centre of all vegetative functioning.

no consciousness. Because where instinct resides, consciousness cannot arise. Therefore, the primitive fears sheer instinct because it involves unconsciousness, and because for him falling into instinct and into the unconscious is extremely dangerous because, as mentioned, they elude his control. Consciousness is still too weak, and can as it were be maintained only with difficulty. You can see this, for instance, if you observe the day-to-day life of primitives when nothing is going on. They either sit or lie around. One assumes that they're thinking about something, but they're not. Nothing whatsoever happens. They're simply in an unconscious state. It may happen that a white man asks a primitive, "What are you thinking? Are you thinking?" The primitive will get angry and retort, "What, do you think I'm sick?" Because if a person is quiet and thinking, then he's sick, because the normal condition is that one doesn't think at all. One thinks only if one is worked up in some way.

Thus, the below expresses itself in that fashion, so that, for instance, the unconscious is often called the subconscious. A certain class of people enjoys referring to the unconscious as the subconscious, because it is so inferior to consciousness, even though one senses that in the lower regions, further down in the depths of the body, this unconscious life of the soul takes place. It is also expressed that way in cults, of course. Therefore, many cults operate in in caves.[593] In subterranean chambers, for instance, the night journey is represented by a *descensus ad inferos*, a descent to those below, a catabasis—that is, a descent into the cave where initiation, baptism, and rebirth occur, where one thus borders on the unconscious, where the instinctual arises.[594] There is a danger that the hens will escape;

It is the main accumulation of ganglia, and it is of prehistoric origin, having lived vastly longer than the cerebrospinal system, which is a sort of parasite of the *plexus solaris*" (333–34). He also affirmed that "the contents of all the early manifestations of religion came, not from the mind, but from the sympathetic system" (334). One is reminded, finally, of Jung's famous, poignant statement in "Commentary on *The Secret of the Golden Flower*": "We are still as much possessed today by autonomous psychic contents as if they were Olympians. Today they are called phobias, obsessions, and so forth; in a word, neurotic symptoms. The gods have become diseases; Zeus no longer rules Olympus but rather the solar plexus, and produces curious specimens for the doctor's consulting room, or disorders the brains of politicians and journalists who unwittingly let loose psychic epidemics on the world" (*CW* 13, § 54).

[593] Many early Christian chapels and churches were built over caves, grottoes, and springs (it suffices to think of the several extraordinary examples cut out of the lava rock in Cappadocia). This was common too, as Jung contended, for the Mithraeums, or the Mithraic places of worship, as well as for other pagan cults (see *Symbols of Transformation* [1912/1952], *CW* 5, § 577 and n. 141; and *The Tavistock Lectures* [1935], *CW* 18, § 254ff.).

[594] The term "catabasis/katabasis" (from Gk. κατὰ [*kata*], "down, under" and βαίνω [*baino*] "go") commonly appears in Jung as a *descensus ad inferos*, a journey downwards

that is, that one will lose control over the four parts of living consciousness, and enter a condition for which the primitives have a specific name.
What do the primitives call this illness?

ANSWER: The loss of the soul.

Yes, indeed, this is a loss of the soul. When one or several of these hens
escape, then so and so many souls have vanished. In other words, as a
rule primitives have more souls, at least two; in Polynesia there is mention of at least six or seven, and if a soul is lost owing to some excitement
or a magical event, then the devil is on the loose, so to speak. For instance,
one missionary reports that a terrible scandal erupted all of a sudden in
the village market square. The missionary thought that a large brawl had
broken out. Someone was shouting at the top of his lungs and a crowd
was gathering. He went to see what was happening. On the ground lay a
native who was thumping himself with his fists. The missionary calmed
the man down and asked what was amiss. In reply, the man said that he
had just lost a soul, that it had taken leave of him. That is a frightening
experience.[595]

Now you probably won't be able to find evidence for such an experience with yourself. We wouldn't feel that way about it, of course, but if

---

or to the underworld and an immersion in the depths as the fundamental first step in initiation, whether in primitive cultures, Antiquity and late Antiquity, or Christianity, so as to
induce radical transformation (see, for instance, *Psychology and Alchemy* [1944], CW 12,
§ 42, 436, 441) [with thanks to Ernst Falzeder].

[595] On the loss of the soul, connected with traditional "perils of the soul," see, e.g.,
"The Archetypes and the Collective Unconscious" [1934/1954], CW 9.1, § 47; "Concerning Rebirth" [(1939) 1940/1950], CW 9.1, § 254; "The Psychology of the Child Archetype" [1940] CW 9.1, § 266; "Conscious, Unconscious, and Individuation" [1939],
CW 9.1, § 501; "The Psychology of the Transference" [1945/1946], CW 16, § 412. An
early definition by Jung reads, "Loss of soul amounts to a tearing loose of part of one's
nature; it is the disappearance and emancipation of a complex, which thereupon becomes a
tyrannical usurper of consciousness, oppressing the whole man. It throws him off course
and drives him to actions whose blind one-sidedness inevitably leads to self-destruction"
(*Psychological Types*, CW 6, § 384). Jung's famous, dramatic invocation in the *Liber
Novus* beginning, "My soul, where are you?" comes to mind. The notion of the "loss of the
soul" is documented in the religious beliefs of innumerable people on earth: consider by
way of example the sickness called *susto* in Central and South America, meaning a "hard
fight" and deemed to be caused by natural forces or black magic; see Henri F. Ellenberger,
The Discovery of the Unconscious: The History and Evolution of Dynamic Psychiatry
(New York: Basic Books, 1970), 7–9; Ellenberger justifiably asks whether the contemporary
psychotherapist engaging with severe mental issues and trying to "establish a contact with
the remaining healthy parts of the personality" should not be be considered "the modern
successor of those shamans who set out to follow the tracks of a lost soul, trace it into the
worlds of spirits, and fight against the malignant demons detaining it, and bring it back to
the world of living?" (9).

you observe yourself just a little more closely, then you'll discover something similar: namely, that you lack a certain amount of energy. You wake up in the morning, and something's missing. For instance, patients say that their libido has fallen off or something like that. It's as if something had gone forth during the night. Among primitives, these manifestations are developed to such an extent that doctors, doctors of the soul and medicine men, have very specific techniques to ease this grievance. Some, for instance, hang birdcages from trees every night. They leave the doors open and next morning go to the tree and close the cage doors. That's done because, at night, birds of the soul have gone forth somehow and visited these cages. If they settle there and the clients consult the medicine man the next day to reclaim their errant souls, asking, "Doctor, have you caught one of my soul birds?" he'll reply, "Yes, indeed, I have many birds in the cage." Then one goes to the cage, placing it on the head of the person concerned so that the soul bird comes out and enters the head through the fontanella.

QUESTION: How does one avoid confusion with alien souls?

That's out of my ken. It's the professional secret of the medicine man. I'm not certain that misunderstandings don't occur from time to time. Because if souls go forth, then you can never be sure whether you possess your own soul. Sometimes one possesses another one or several others. Naturally, these souls are simply autonomous complexes, autonomous figures, which perhaps exist in a random number, and come and go as they please. For instance, consider the creative imagination. He doesn't have that faculty at his free disposal all the time. Sometimes it simply leaves him for months, and then the man is at a complete loss. One day the great bird returns, and everything changes. This is the going forth of the psyche or the migration of the soul. Thus, possession is migration. If one receives an alien soul, that can be most unpleasant, because such possession exerts an incredible fascination, whose effect can be extremely agitating under certain circumstances. Naturally, this is treated by a doctor, who lures the false soul bird out of the patient, and thereby cures such possession. Myotieosmus[596] is the same. Here, too, the false soul bird is driven out with the help of particularly potent formulae.

---

[596] Unidentified term. There is an outside chance that this incorrectly transcribes *manitouism*, Arthur Lovejoy's coinage for the Algonquin version of his universal life energy among primitive peoples, extended by Jung in 1917, via conservation of energy and the hypothesis of primordial ideas in a phylogenetic unconscious, to his own theory (see

Therefore, as the dreamer fears, a loss of soul could occur which would naturally hinder the entire procedure. For the procedure that the uncoscious undertakes, it is necessary that all soul birds, all individual souls, be present. Now one of these hens, which is particularly adventurous, begins to get away. While he holds the others in one hand and attempts to confine them to the most secure compartment of the cage, he discovers that this particular compartment has a window that is fitted only with a fly-net. His associations indicate to him that it's not a fly-net at all, but instead a fishing net, a wide-meshed net, which would by no means keep flies at bay. Instead, it is rather like a fishing net, which could naturally keep the hens inside quite effectively. Such cages are a common sight at markets around Germany: poultry merchants use cages fitted with netting to keep the animals inside. Evidently, the dreamer has this in mind. Well, what do you think? Precisely the most secure compartment of the cage has a window from which the hens can once again escape. It's ridiculous, of course, that a cage consisting only of iron rods should also be fitted with a window. That's completely unnecessary. But that's the way it is.

ANSWER: This is his chosen function. But he mistrusts it.

What does he mistrust?

ANSWER: His thinking function. He now mistrusts his thinking function, because he has realized that he has overestimated it. This is his principal function, after all, his individual function.

Well, yes. But the animal that he loves most he takes to be his feeling.

ANSWER: But his thinking function is the safest of all.

We are not certain whether these compartments coincide with the so-called functions. There is no description of how these compartments are made. There are simply different compartments, of which one seems to be the most important and safest. But it's far from obvious to him that this happens to everyone. He does not know how this is actually constructed. We should not delve into too much detail. Instead, we must simply assume that there is a particularly secure compartment in the cage. And precisely that one contains a window. Would you perhaps have any association? None occurred to the dreamer.

ANSWER: Perhaps he means his marriage.

I wouldn't go quite that far.

Shamdasani, *Jung and the Making of Modern Psychology*, cit., 231–32). However, the application of this concept here, conveyed without explanation, remains obscure.

ANSWER: Precisely that which is most secure will have an opening, because consciousness considers it to be very safe. That explains why he has such a conscious attitude.

Possibly it is precisely that particular compartment, that his consciousness considers to be the safest, that is in fact the least safe. As I have mentioned, these hen cages have this particular kind of window; that is, the cage in actual fact consists of only four side walls, a floor, and the whole upper part consists of a window, a kind of fish net, under which the hens are kept. Presumably, this window is precisely the opening through which the hens are pushed into the cage or through which they could escape. It must be this kind of thought that occurs to him: namely, that what his consciousness considers to be safest is actually the least safe. Curiously, however—and I was wondering whether you notice—he used the word "window."

ANSWER: Windows can be opened.

Indeed, rooms have windows to allow light and air to enter. But this is wholly paradoxical. It could be that another idea, another image, enters here: of a closed room that has a window. Well, what kind of room could be compared to a cage that has a window?

ANSWER: A house, a prison.

A prison cell. These have a window that confines one by means of a grid of iron rods, from which one might perhaps escape, as happens from time to time. So, this peculiar idea of the window stems from the idea of the prison, through the well-known *contaminatio*[597] of images. But this could perhaps have an even deeper meaning.

So imagine youself in a prison cell. There is a window through which one can escape. Where does one get to, or if one climbs up there, what does one look out into?

ANSWER: Into the vast outside world.

Indeed, into freedom. And as a rule what impression does this make on the prisoner?

ANSWER: He experiences longing.

The longing for freedom. That's the alluring effect of the outside world. One looks outward, yet again, and the soul birds fly up and away, yet again. This opportunity is particularly enticing. Either one is afflicted by

---

[597] Lat.: "contamination"; used by Roman dramatists to indicate their borrowing and adaptation of scenes from Greek theater.

a drive—that is, an affect—and the soul bird flies off or is chased away without any psychological commotion whatsoever, or instead one gazes outside, and the soul bird has departed, just like that.

There is a very beautiful image of this gazing outward in a passage in Saint Augustine's *Confessions*, where he mentions the fate of his friend, Alypius, who, as a young Christian, could not resist the temptation to attend the circus, just once. Although he decides to avert his gaze when the gladiator falls, he nevertheless watches it happen. Augustine writes, "He opened his eyes and was wounded more deeply in his soul than the man whom he desired to look at was in his body."[598] Because it was precisely at that very moment that the soul bird flew away, and Alypius was once again intoxicated by the bloodthirstiness of the circus; well, I'm sure you can understand his feelings.

In analysis, one must thus also concentrate and reflect on what one actually is and what one actually wants, so it becomes clear that one actually faces a rather difficult task. The next thing that one always does is quickly to look out the window, and turn one's back on the center. One says, "What do other people do?" For instance, what do the eleven virgins do?[599] And stories like that. One compares oneself to Mr. Jones. He has a motor car, and he's doing fine, and, indeed, he's not in analysis. Why doesn't he have a neurosis? Just a moment's distraction, and away flies the soul bird, does it not? Whenever we remove ourselves from our most essential task, which expresses the totality of our personality, and look

---

[598] Augustine, *Confessionum libri XIII* 6.8.13 ("Et percussus est graviore vulnere in anima quam ille in corpore, quem cernere concupivit"), translation by John K. Ryan: *The Confessions of St. Augustine* (New York: Doubleday, 1960), 144. This story of Alypius in Rome also illustrates the depersonalizing and almost "bestial" power which, activated by collective emotion or identification, overrides individual focus and resolve: Jung's point immediately following. The same episode is referred in *DAS* (12), and, as noted by McGuire, in *Symbols of Transformation* (CW 5, § 102) Jung comments in this regard: "The moral degeneracy of the first centuries of the Christian era produced a moral reaction which then [. . .] expressed itself at its purest in the two mutually antagonistic religions, Christianity and Mithraism" (ibidem). Originally the Alypius episode was mentioned in *Wandlungen und Symbole der Libido: Beiträge zur Entwicklungsgeschichte des Denkens* (Leipzig und Wien: Deutike, 1912), 65–66 n. 3.

[599] Incomplete transcription for "the eleven thousand virgins," the legend of Saint Ursula's martyred band at Cologne, which Jung uses to tag our tendency to identify with the views of others. Cf. *DAS*: "The patient was trying to build up a philosophy of an old virgin about his sexuality. He tried to be one of the eleven thousand virgins just initiated, not taking nature into account" (141). Later in that seminar he deploys the same analogy in discussing the gregarious inclination of the animus, which would turn a discreet problem into "talk about what the 10,900 virgins would do [. . .]. Therefore I say to such a person: leave all that stuff alone, appear on the stage and say *I*, and not as *they* opinionate" (497–98).

outward, then we go to pieces. Our chicken-run flies away, and becomes nothing but a chicken-run, and we can start all over again. Gathering together all the hens. Such are the perils of the soul.

Thus, the primitive does not look at things, because doing so could have dangerous effects: namely, a loss of soul or possession. He seeks to avoid certain affects, avoid certain situations, and not offend certain gods, because that entails a loss of soul. In this respect, primitives are extraordinarily conscientious. It's remarkable with what tenacity and fidelity these principles are upheld by primitives. It's extremely impressive to observe how these simple people obey the intimation of the unconscious, how they strive to fulfill the conditions imposed upon them from within, to the best of their knowledge and belief, and what respect these people have, for instance, for those who must subordinate themselves to certain ridiculous inner commandments.

For instance, I recall a chief who, at the age of forty, experienced what we mostly overlook, and if we cannot quite disregard it, we close our eyes to it: namely, that the character of his life turned around. At that time the chief, who was a great warrior, a real man, a man quite different from certain other men, had a vision. Manitu, the great spirit,[600] appeared to him and said, "From now on, you will be changed into a woman, you will wear the clothes worn by women, you will eat the food eaten by women, and you will sit with the women and children." The next day, the man dressed in women's clothes, ate the food eaten by women, and sat with the women and children, faithfully observing these precepts. No one dared laugh, and his prestige grew twofold or indeed manifold compared to what it had been earlier. We'd say, what a crazy fellow. But he abided by what he had seen within. We marvel at this. The missionary who reported this story was amazed too, because he believed that nothing of the kind happens to us. However, if we observe the symptoms that emerge from people who haven't noticed this sort of thing, then it's actually the gods who must laugh.[601]

---

[600] *Gitche manitou* is Algonquian for "Great Spirit." The Northeastern and Midwestern American Algonquian tribes believe that all plants and other beings have their own *manitou*.

[601] No source text for this story is known thus far. Jung first cited it in a newspaper article in 1930, "Die seelischen Probleme der menschlichen Altersstufen," revised as "Die Lebenswende" in 1931, Cary Baynes's first translation of which was merged into Jung's *Modern Man in Search of a Soul* (London: Kegan Paul, Trench, Trubner, 1933). In *CW* 8 it is "The Stages of Life": "There is an interesting report in the ethnological literature about an Indian warrior chief to whom in the middle of life the Great Spirit appeared in a dream. The spirit announced to him that from then on he must sit among the women and children. He obeyed the dream without suffering a loss of prestige. This vision is a true expression of

Let us now turn to the further transformations of this dream. Ultimately, he manages to keep these animals together, and he has now discovered that it's safest if they're all placed in a high-rimmed, smooth bowl which they cannot climb out of without falling back down again. What do you make of that? Would you have any idea what placing the hens in a bowl could mean?

ANSWER: In the bowl, they would all be together. There is no more scattering as in the cage.

No, the situation is even more unpleasant for the animals. Imagine a bowl with a smooth rim. They'll keep sliding down and can't escape. Besides, the bowl has other unpleasant associations. What is its customary use? It's used for cooking food. We know that in the end one of the hens seems to have been roasted. So, here is what we already know: this bowl relates to cooking, and the hens have gone to the hangman. Something must be in the offing. Let's reflect that we've heard that these hens are psychological instincts, compulsive drives as it were, and are somehow related to animal functions. Perhaps we can say that the four functions are placed in a cooking pot, in a bowl, and either cooked or roasted. What comes to your mind?

ANSWER: They become palatable.

ANSWER: A transformation takes place.

A transformation, whereby they are changed into a palatable state. What motif enters matters here?

---

the psychic revolution of life's noon, of the beginning of life's decline. Man's virtues, and even his body, do tend to change into their opposites" (CW 8, § 781). Jung turned again to this report in the Children's Dreams Seminar, almost verbatim as here until this point (on the gods' laughter), and continuing as follows: "A chief, an Indian, had a dream at the age of forty, in which he was told: 'You have to sit among the women, eat their food, and wear their dresses.' And he accepted it and subsequently enjoyed an even higher reputation. He had been addressed by the Holy Spirit and turned into a woman. This is uncanny and full of meaning. One becomes a *mana* personality, that hermaphroditic primordial being [see above [ZL], 96 and n. 304]. Winthuis [ethnologist and religious scholar Joseph Winthuis, author of, e.g., *Das Zweigeschlechterwesen bei den Zentralaustraliern und anderen Völkern: Lösungsversuch der ethnologischen Hauptprobleme auf Grund primitiven Denkens* (Leipzig: Hirschfeld, 1928)] went to the Australian negroes and made a terrible mess out of that hermaphroditic being. Now he thinks it can be found behind everything. He let himself be taken in, because primitives feel what others wish them to feel and then produce it. But in essence Winthuis nonetheless came up with a very real thing, because it is a primordial idea, extremely primitive and thus obviously a fundamental truth" (Jung, *Dream Interpretation Ancient and Modern*, cit., 119). [T.N.]

ANSWER: The laid dinner table.

Well, arrangements for the wedding banquet are evidently underway. What kind of peculiar symbolism is this? Where does it come from?

ANSWER: In terms of the dream, the cauldron is the symbol of the phallus.

Indeed, as you know, witches always have a cauldron, in which they stew all kinds of odd things.

ANSWER: There is another communion.

Do you mean through the meal? But I have in mind the cooking process in particular.

ANSWER: He makes it palatable.

Yes, we have already said that.

ANSWER: A cleansing.

Yes, it is sterilized.

ANSWER: The alchemical process.

Indeed, please do not forget that alchemy is the immediate preliminary stage of this kind of psychology.[602] Our alchemistic German philosophy has long prepared this process. If you pick up any alchemical book and study the series of pictures, you can observe the very series of pictures that contain our thoughts. I have always observed how patients who had no idea about these matters produced alchemical symbols, about which we always laughed. The alchemists simply sought to interpret this mysterious process of the soul with the language at their disposal. They did not yet have a psychological language, so instead they took chemical language.[603] They even worked with the term "analysis." Chemistry dispenses with association. The old alchemists simply regarded this as the diversity of substances, whereby they quite consciously understood the diversity of substances as psychic diversities. Some later alchemists were perfectly aware that they were conducting a peculiar kind of philosophy, and that the royal art consisted in turning worthless matter, specifically lead, into gold or into the shining *lapis philosophorum*, the epitome of utmost consciousness, the *pharmacon*,[604] the elixir of immortality, the

---

[602] See above [I], 39–40.

[603] The mild anachronism here—it was not then chemistry—of course is to be understood as "proto-chemical."

[604] Though Gk. φάρμακον [*pharmakon*], meaning "drug," can refer to both a remedy and a poison, here evidently the former meaning only is intended.

elixir vitae.[605] All these attributes recur in a different shape and form in India and China. In actual fact, it is the designating and formulating of what the unconscious forever tries to bring forth; for instance, what these dreams seek to bring forth. Now you might object that a bowl containing four hens and an allusion to one roasted hen is rather too far-fetched. As you noticed at the beginning, I am not in a position to present perfections, but instead only poverties. These are meager allusions, which occurred partly in the head, or I know not where in the nervous system. In any event, they took place within a human being who had no idea whatsoever of these matters. Nowadays, we appreciate that these are serious events. We are tempted to laugh about these people [alchemists], but from the outset we know what these people took such great pains over. The same applies to association. We have no idea of the fact that association is the oldest form of psychology, dating back at least five thousand years; the ancients, I can assure you, were just as intelligent as us. They were no simpletons, and they knew what they were doing. Their intent was serious. Equally so, the alchemists and their language, which is admittedly confused, were also serious. But it depends upon us whether or not we understand it. If you have before you a cuneiform text and you have no understanding whatsoever, that doesn't mean that the Babylonians were idiots. The same holds true for alchemy. Alchemical philosophy concerned itself with this problem and discovered formulae and typical series of pictures that expressed these peculiar connections.

Naturally, I'm in no position to demonstrate what effect such an alchemical series has on a person. All I can say is that it does have an effect. For instance, if I tell you that these events brought about an essential inner transformation in this man, then I'm afraid that you'll just have to believe me. As I've already said, I have no need to carry out some kind of mission, and to persuade you of what certain truths are. I find these truths most amusing. And if they amuse you, too, that is fine. If, however, you don't want to believe these things, there is no need for you to do so. This dream, which sounds so strangely ridiculous, has a very profound meaning. I don't know whether I'll be able to offer you a complete explanation of this meaning. But I'll try to clarify matters. If I seem to be confusing you, then I apologize, but perhaps clarification will occur.

Imagine, for one moment, how a primitive person, or shall we say an unconscious person, lives and thinks and feels. The more unconscious he

---

[605] Lat.: "elixir of life": alchemical formula for the legendary serum able to prolong life indefinitely and thus referring to the end-state of the *opus alchemicum*.

is, the more he consists of compartments. Imagine someone playing. If you observe how a primitive plays, he plays as if he had always played, and as if he were always playing, like a child with total concentration and devotion. He forgets everything else. We can also gather this from their manner of living. A young man is nothing but a young man. Equally so, an old man is nothing but an old man. He's always been old, he will always be old, he is timeless. It is the perfect idea of age. Why? Because the man is completely captivated by either respective condition and has no consciousness of anything else that exists or might exist. Everything else is excluded. If he is desperate, then he is completely desperate. If he is sad, then he is utterly sad. There is no standpoint above or beyond.

This leads to an extremely contradictory psychology, which is characteristic of all human beings who, as it were, live unconsciously. They all have what English refers to as a compartment psychology,[606] that is, a split-off part, where the right hand has no knowledge of what the left one does, where there are no double or tenfold moral standards, and where a fully respectable gentleman is guilty of the most abominable dishonesty in his business life. I need say no more. Here is a state of affairs in which one must be prepared to encounter one's alter ego around the next corner, a condition in which one never knows what will happen from moment to moment. One is not safe from oneself. And if one then becomes acquainted with a moral religion like Christianity, one jumps from from the frying pan into the fire. Besides, one should be consequential. One realizes that one is unable to be so, and falls from one hole into another. For this reason, one finally must deceive oneself to avoid realizing that one has, yet again, taken a wrong step. This is an inevitable state, one in which the hens have flown away, where no totality exists any longer, but instead only fragments, sequels without beginnings, beginnings without sequels. This is an extremely unsatisfactory, fragmentary, dissociated condition.

It inspired the ancient philosophers in China and India, and those in Antiquity and the Middle Ages, to find a way to achieve such a totality: for instance, how the heterogeneity of the elements could be unified in a valuable whole. It became evident that nothing was as valuable for fusion as the force of fire in a chemical process. This was regarded as a highly appropriate symbol. The most disparate things can be stewed in a cooking pot and finally one has a very tasty soup. If they are cooked, things become palatable. Equally so, various minerals, which earlier had looked

---

[606] "Compartment [in English in the text] Psychologie": internal splitting, the divisions being watertight.

quite different and were useless in their original condition, can be fused. This produces a precious and beautiful metal.

This allegory was applied to the human soul. Naturally, the unconscious soul accommodated this procedure. Something similar also occurs within us: for instance, situations that are symbolized by fire, such as the kindling of great passion or the flaring of great pain, of great love; nothing else has the effect of such an experience: to be so synthetic, that is, so unifying and so clarifying, that one often wishes that many people, this particular person, were for heaven's sake capable of feeling such passion just once, just this once. That would melt this person together, burn all the rubbish inside, and perhaps a pure gold nugget would appear. Sometimes, however, there is not even enough wood, not to mention a gold nugget.

* * *

LADIES AND GENTLEMEN,
During the interval, a member of our seminar has called my attention to a dream that confirms what I have been saying about the hens escaping. I would like to ask Professor Schmalz to narrate this dream.

PROFESSOR SCHMALZ: The dream is exceptionally simple. The dreamer attends a lecture given by Doctor Jung, and the lecture hall is not yet very full. There is a barrier at the front where the more distinguished people are seated. He does not really dare sit down there and instead takes his seat further back; he sees copies of the printed program of Doctor Jung's lecture lying on the seats. Its subject is exceptionally simple: Doctor Jung will be speaking about "Desire as the Obstacle to Maturation." Thus the dream ended.

DR. JUNG: You see, that's how it is, and it has the effect of a brief and absolutely precise remark made by the unconscious, an *avis au lecteur*.[607] Hopefully, I have been able to make clear beforehand what I mean with this dissociated condition, this inevitable state, in which we begin, and to what extent primitive fantasy considered it necessary from time immemorial to light a fire that melts together what resists, consuming the inept and the unfit with fire, turning it into cinders or smoke, so that the fire remains constantly alight. Those of you who have *read* the second part of *Faust* will know that similar things are said there: for instance, that "all that is worthless is purged away and love's star remains forever."[608] I realize that I almost should feel ashamed for presenting something quite as meager as this dream, when I could have resorted to Goethe's words, which are so much finer. Unfortunately, however, ordinary dreamers are not poets, and we're dealing with ordinary folk, for whom it is to no avail if we recite splendid poetry, since they can't establish a connection between *Faust*'s immortal beauty and their lousy dreams.

You see, it is incredibly important that we can also analyze this meager, sometimes pestilential shabbiness, which sometimes contains gold dust, and that within the human being matters are not quite as bleak as they sometimes seem. After all, we human beings are weak and unfit things, and from time to time we are in dire need of someone who encourages us to embrace life and gives us the feeling that ultimately things also have

---

[607] Fr.: "Note to the reader," thereby with a hint of warning.

[608] Pronounced by the Pater Ecstaticus at the end of *Faust II*, Act 5, scene "Bergschluchten" (vv. 11,854ff.): "Ja das Nichtige / ganz sich verflüchtige, / glänze der Dauer Stern, / ewiger Liebe Kern" ([Lightnings, come cleave me] / Till all the nothingness / Vapours and vanishes! / Let the bright seed-stone / Of lasting love shine). Translation by David Constantine: Goethe, *Faust, Part II* (London: Penguin, 2009).

meaning, and that it also makes a certain kind of sense that we live. For this, people are grateful. I must say that I have experienced this gratitude to a great extent, which has encouraged me to persist with this unusual method and never to waver in pursuing this aim, because in this respect I know that I am in agreement with the secret workings of nature, and I find myself in the best company with this endeavor.

(Thunderous applause.)

LADIES AND GENTLEMEN,

These hens are helpful animals. When I asked him, you heard Professor Zimmer talk about this place in the forest where one loses one's way, where the *Divine Comedy* begins, and what happens afterwards. He refreshed your memory and said, "Then the animals appear." As I mentioned, in Francesco Colonna's *Hypnerotomachia* that is where Madame Polia's hero loses his soul; he, too, seeks to find his way, and he encounters a wolf, who leads him to the submerged city, where he discovers Polia. This is a matter of the *[des]census ad inferos*, that is, a catabasis; it is the old motif, yet again, where the blindness of consciousness commences, where we no longer see with consciousness, where we seem to lose all hope, and precisely at the point a helpful animal appears, and blessed is he who is capable of allowing this animal to live instead of wringing its neck on the spot, simply because it is merely so instinctive, and because one fails to notice that. Surely you know how it goes in fairytales when something happens to the hero. He either loses his way or is taken captive, at which point helpful animals show him the way out of the difficult situation; birds will lead him forth, a white snake reveals the secret to him or brings him the salutary herb, and so on. These are simply the eternal statements of the unconscious, which tirelessly assures us that everything is already envisaged if only we would let it prevail, and not only let it prevail, but also listen to the helpful animals coming to our aid.

The first reaction amid this obscure, dangerous situation is an instinctive activity. In fact, it is difficult to establish such a reaction as a fact, because it often strikes us as stupid beyond all measure. Under no circumstances do we want to look stupid, nor do we want to be blind like an animal. It offends our human dignity to be simply instinctive, or to acknowledge that something subhuman happens to us. And now, by way of explanation, following the animal there comes a sublime figure, the figure that in the *Divine Comedy* Dante assigns to Virgil, the great poet, who possessed enough dignity to lead Dante through a Christian underworld as a Christian prophet up until the last stage of the *Purgatorio*,[609] where the pure flame burns, which also strips love of the last vestiges of unconscious passion.

In the final part of our seminar, please allow me to bring our proceedings to a conciliatory conclusion, in as much as, after this animal dream, there follows a prospect that delivers the dreamer from the overcrowded confinement of the prison, and reveals that if he can sacrifice his animalistic, unconscious childish nature, he will find deliverance, so that a broad vista will dawn upon him if he is exposed to the force of the fire. The next dream says,[610]

> *I arrive at a place where a saint is revered, who, when his name is invoked, is supposed to cure sicknesses. I have come to this place because I suffer from sciatica, but I believe that healing this illness would require more than invoking the name of a saint. Then I move on with the gathering, and on our way I am told that one of the sick has already recovered. To get better, I decide to bathe in the ocean.[611] I come to a beach that is closed in by huge mounds of stony rubble. In a valley between these rocky dunes, the ocean enters in calm, powerful surges, before slowly petering out as the valley rises sharply at the opposite end. For a while I watch the majestic rise and fall of the waves. Afterward I find myself standing before the bouldery dunes once more, with my youngest boy. As we are about to clamber up the hill,[612] I see how its ridge is splashed by water and foam, and it occurs to me that if the ocean on the other side of the mound is so high and powerful, then at any moment it could cause the shaky mound[613] to collapse. And so we decide[614] to retreat from that dangerous proximity.*

Regarding this dream, he reports the following associations:

---

[609] Italian in the protocol.

[610] The dream comes two days after the previous one. Cf. *DAS*, 125.

[611] The dreamer comes to this assessment in *DAS* only after hearing the group he joins report one patient's recent cure. "I think that I need to do something more than call on the saint, that I should take a bath in the sea" (125). Jung does add this qualification here in a subsequent comment.

[612] According to *DAS*'s report, the dreamer had already started to climb the rocky hill part way before his child appeared.

[613] "Which was not made of bed-rock but heaped-up gravel and boulders" (*DAS*, 125).

[614] "On account of this I take my boy away" (ibid.).

*I can't remember the name of the saint, but I believe his name was
either Papadelanu or Papasdelianu. But I cannot really account for
the Greek and Romanian.*[615]

About the strange healing of the sick, he says that he is familiar with
such healings from Muslim culture.[616] It could be a matter of autosuggestion, which was probably very much heightened by the reciprocal influence of the friends upon one another. In the dream, he also doubted the
possibility of healing happening through miracles; nevertheless, after observing that a person had been cured, he decided to bathe in the sea. About
the sea, he says,

*The sea is the primordial foundation, the primordial mother of life.
In its innards the development of the first seed of life took place,
and one could call it the womb of nature.*

His fantasy, as you can see, is already beginning to oscillate. He believes that the unconscious, just like this hill, can prevent the waves from
entering consciousness through an open valley. He is also calmed, and
finds it interesting to observe this up-and-down movement.[617] Going down
to the ocean can be dangerous, because the frail shores could easily be
torn apart by the force of the sea. But perhaps then he would no longer
have to work his way out of the flood.[618] He mentions that his youngest

---

[615] In *DAS* "Papatheanon or Papastheanon," according to the memory of the analysand, who "knew Greek and Latin, and spoke Romanian" (133). Jung comments, "There is
much Greek in Romania, because of the mixture of Greek in the lingua rustica, the language of the peasants throughout the Roman Empire. It still remains as the Romansh in
Switzerland" (125–26). Later he says that "Papatheanon suggests the father," acknowledging also *en passant* one seminar participant's suggestion of the character Papageno from
The Magic Flute. We may note that while *theanon* carries the Greek root for goddess (*thea*),
the vagaries of transcription preclude speculation. "We have enough material in connection
with the saint [. . .]. The patient is put into an archetypal situation" (133), which is to the
point: "The ordinary technique of the dream is that it gets the patient into the archetypal
situation in order to cure him" (130; cf. also McGuire's comment, ibid., n. 5).

[616] In *DAS* the dreamer also thinks of cures at Lourdes, which he attributes to mass
autosuggestion (126).

[617] In *DAS* the patient remarks, "It is tranquilizing and at the same time most interesting to watch those waves. Speaking in that way our conscious is moved by the up and down
movement of the unconscious" (126).

[618] The patient's fear of the waves was tied helpfully in *DAS* to his fear of his sexuality
during a discussion about the meaning of the waves image (which a participant also linked
to the dreamer's "enormous transference on his doctor": 138–39). Jung pointed out the
patient's "old-fashioned, rather romantic, archetypal philosophy" in comparing the sea to
"'the primordial cause of life, the eternal mother, the womb of nature'" (139; likewise in
Psychology of Kundalini Yoga, cit., 22). Then, rather than commenting on a participant's

son wants to know what lies up there. He [the boy] is jealous of his siblings and is very eager to have everything they have. Then he [the dreamer] remarks that the sea has evidently surged very high on the other side. He thinks that a strong pressure must exist on the other side. Therefore one would need to beware of being inundated if the dune collapsed.

The first idea in this dream is that he was on a peregrination. It's the sequel to the idea we encountered earlier, namely, that now he's descending into the underworld. He has reached a parting of the ways, where pathlessness begins, where no one has ever set foot and where no one should venture. During this wandering, he comes to a place inhabited by a saint, whose name or mana heals the sick. The dreamer too is represented as a sick person. If we leave the name aside for the moment, we notice that the saint is an archetypal figure, a real archetype, that is, a medicine man filled with mana who gives off healing lightning like a charged accumulator, who as it were emits curative electricity. He is wrapped in Oriental ideas familiar to the dreamer from visiting the Orient. These Marabouts, these sanctuaries in which all the pious sheiks are buried, are charged with mana. If you've ever visited the city of Padua, you'll recall seeing a shrine next to the grave where the saint [Anthony] lies buried, where many stand with bowed heads, and you'll spot a sufferer nursing toothache.[619] This is primitive from the ground up, and it abounds in Muslim regions.

---

suggestion that this symbolic image might (also) be interpreted as "a regressive symbol, longing for the mother," Jung summarizes the development carried out thus far by the patient: "In the first part of his analysis he thought of sexuality as an uncomfortable thing in the corner, a personal difficulty which he did not feel up to at all. The unconscious is slowly trying to open his eyes to a wider vision or conception of sex. The pitiful symbol of the sewing machine is now the systole and diastole of life, which shows itself in sex, too, so he should look at sex as though it were the rhythm of the sea; the rhythm of the primordial mother, the rhythmical contraction of the womb of nature. This gives him another aspect of sexuality. It is no longer his miserable personal affair, something he has to put into the corner, but a great problem of life." Jung remarks besides of such new awareness, "Now he sees the problem as the great rhythm of life, a problem of nature, which he looks upon philosophically"; moreover, "the ocean itself which has become a universal symbol [...] gives him a philosophical attitude and he has a greater chance to deal with his problem" (140).

[619] Cf. *DAS*: "The Catholic rubs the tomb of St. Anthony in order to get healing power from it [...]. Now this dream suddenly gets the patient into a pilgrim role, traveling to a shrine, as to the tomb of St. Anthony in Padua or to Lourdes. He is put into the situation of the ordinary man of all times, and though that he is brought nearer to the fundamental nature of man" (128, 130). The Pontifical Basilica of Saint Anthony of Padua (commonly called "il Santo" by the Paduan people, as is the church itself), whose construction began in 1232, one year after the popular Portuguese Franciscan friar's death, and was completed in 1310, contains the previous small church Santa Maria Mater Domini, where the saint used

So, on his unconscious wandering he comes to the place of healing: namely, where the "saint" lives. This is the "whole one." Healing is whole, entirely complete.[620] In this instance, I wouldn't say perfect, but complete.[621] The figure somehow set before him here is what he is after, which has achieved the wholeness he has not attained. He is not healed, but still sick, yet to be freed from an impairment. The saint is beyond that. He has already become sacred and has mana and radiates something. The person who has not become complete within himself emanates no charisma. He's always sucking and is somehow empty. Thus he has a kind of suckling

---

to celebrate Mass, which was incorporated into the present basilica as the Cappella della Madonna Mora (Chapel of the Dark Madonna). Here Anthony's remains were buried until 1263, when they were removed at the centre of the basilica. Jung refers to the shrine (Cappella delle reliquie o del tesoro), containing precious liturgical objects and relics (such as the saint's jawbone, and the so-called "incorrupt tongue"), which have survived various depredations (the major ones being in 1405, when Padua was conquered by Venice, and 1797, with the arrival of Napoleonic armies). The basilica features an original combination of styles (primarily Romanesque, with Gothic and Byzantine influences) and contains works by Giotto, Mantegna, Donatello, and Bernardino da Siena. With its approximately 6.5 million visitors, it is one of the world's most prominent Christian sanctuaries.

[620] Cf. the aforementioned (in the n. 589) passage in DAS: "In German *heilig* is connected with *heil*, or 'being whole'; *geheilt* means cured. Healing is making whole, and the condition into which one gets is a whole or complete condition, while before it was only fragments held together" (128). Then Jung adds, "Here the idea of medicine comes in. The Saviour is always the medicine man who gives *pharmakon athanasias*, the medicine of immortality which makes the new man. When you take the *tinctura magna* of the alchemist you are cured for ever, you never can fall ill again. These are mythological connotations of the alchemical procedure, or the melting pot of transformation, so it is no wonder that in the next dream he starts in with a saint" (ibid.). Just before this (and this seemingly liminal space between a saving [mana] personality and an approximation to the Self), Jung had testified to how successive dreams, as these two do, work the same archetypal motif, "of assembling in the pot the sacrificial food, the alchemical procedure for the reconstruction of the new man. This is the old idea of the transformation of the individual, of the man who is in need of salvation, redemption, cure. He is like an old broken-down machine" (127–28).

[621] The difference between *vollkommen*, or "perfect," and *vollständig*, or "complete," and respective substantivized forms *Vollkommenheit* and *Vollständigkeit*, is often invoked by Jung with regard to the goal of analysis: namely, the dynamic, never-ending (and imperfect) completeness of conscious and unconscious elements. See, e.g., Answer to Job [1952]: "*Perfection* is a masculine desideratum, while woman inclines by nature to *completeness*. And it is a fact that, even today, a man can stand a relative state of perfection much better and for a longer period than a woman, while as a rule it does not agree with women and may even be dangerous for them. If a woman strives for perfection she forgets the complementary role of completeness, which, though imperfect by itself, forms the necessary counterpart to perfection. For, just as completeness is always imperfect, so perfection is always incomplete, and therefore represents a final state which is hopelessly sterile. 'Ex perfecto nihil fit,' say the old Masters, whereas the *imperfectum* carries within it the seeds of its own improvement. Perfectionism always ends in a blind alley, while completeness by itself lacks selective values" (CW 11, § 620; see also § 627, and Aion [1951], CW 9.2, § 123).

psychology. One notices this if one deals with such people. But as soon as a person arrives at synthesis, it's as if he radiates the force of fire onto others. Naturally, these are intuitive terms which we cannot place on the scales and weigh, but which can cover quite specific and simple facts,[622] which one can clearly establish in social life or treating the sick. As yet he is neither a healer nor a saint, but he's getting there. If he touches the saint, healing and making whole will transfer onto him. He will be healed, and therefore become a healer too. For the healer heals. He radiates his wholeness onto others.

Now doubt overcomes him about whether it's enough to simply invoke the saint's name. Evidently, this idea occurs to him because he notices that merely naming something is primitive. Nowadays, if it's a question of using names to accomplish things, one is referring to majority practice. We still believe that naming is the name of the game. The peril of thinking— that if you think, then that does something—still bests him. If he entertains a correct thought about the matter, then he believes that it has also got to be right. And yet nothing comes of that, because mere understanding has effected nothing. It must still be experienced, as a fact, and must somehow have traveled through him. Blood must run through it. In itself, merely correct thinking is of no avail whatsoever. He also realizes that he suffers from sciatica. Why do you think he suffers from sciatica? It's a strange affliction, after all.

ANSWER: It makes walking difficult.

Please take into consideration the dreamer's language, as well as what has happened earlier. He is a motorist, and if a motorist suffers from sciatic pain, what about that? For instance, the left rear wheel is not without blemish, and so then either his left or right rear wheel breaks down. Or his front wheels are defective. It means, quite simply, that he has had a breakdown with his rear wheel, the driving wheel.[623] That's his ailment, after all, because the hens have escaped where the driving wheel is located, the seat of the unconscious and the affects. He cannot be healed if he is still limping with his left rear wheel. Now, all of a sudden he is no longer alone, but finds himself among a larger group. What do you think he wants to achieve by being together with a larger number of people?

---

[622] English "simple" in the protocol.

[623] In DAS more attention is given to the patient's sciatica in connection with this dream, which Jung sets alongside his tendency to see the sexual dysfunction vis-à-vis his wife as a glandular matter, "a physiological trouble," rather than having to do with affects rooted in the unconscious (134).

ANSWER: Because he recognizes his problem as being a general human problem.

Yes, he realizes that his problems are quite a general, collective problem. By no means would I dispute that. If I raise further issues, then I do so only by way of supplement. If he is alone with the saint, then it's merely an individual situation, as it were. It's as if this concerns only him, as an utterly personal question. If the dream introduces other people, that means it's necessary, and that the collective viewpoint enters. To become healed, it's necessary for the collective viewpoint to come in. No one gets healed alone. If you sit down on the summit of Mont Blanc or Mount Everest, you can't be healed. For that the collective is necessary; that is, the relationship with other people, as well as the realization that your sickness is not only your sickness but also a general sickness, the suffering of everyone. In actual fact, that's how we suffer. Individual suffering in principle is a general suffering. His suffering, your suffering, is suffering as such. If he realizes that one man has already recovered, then naturally that refers not only to him, but he is one among many. He has already convalesced, so he realizes at the same time that others will also recover. The knowledge that one's own difficulty is the difficulty of all humanity reconciles one, once again, with that humanity, and also with fate. And to become completely healthy, the unconscious leads him to the sea, in fact to the very place, in a bay, where the sea foams up powerfully with a tremendous groundswell. What do you make of this? What is this supposed to mean? Where does the sea enter us, where do you feel the great movement, the great breath of life?

ANSWER: In the unconscious.

Yes, for instance, where the transformations of fantasy take place. Recently, you yourselves experienced that we were dealing with a wife and a sick child, his younger sister with the sick child. Already the next day the seamstress was sick, and she was neither the sister nor the child. This is a continuation of both [situations], and marks such a peculiar transformation that it makes one nauseous. These strange transformations induce sickness. Very often I have seen that people with these fantasies exhibited obvious symptoms of seasickness. They were positively sick, because they had completely lost their bearings. That's the way it is.[624] For instance, we assume that this wall stands upright, more or less vertically, in this room,

---

[624] Cf. Jung's account in *The Red Book* (*Liber Novus*), cit: "I clutch the sides of my bed to protect myself against the terrible waves" (Liber Secundus, ch. 16, "Nox Tertia," 298) and "This is the night in which all the dams broke" (299).

and that this is the horizontal line. Now, however, let us assume that this wall is aboard a ship and that we meet a great swell, but we fail to see that and suddenly the vessel rises up high. We don't know by how many meters, and up and down it goes. That's exactly what causes seasickness. Whereas if you stand outdoors on deck, where you can observe these events and have the horizon as a reference point, you'll endure this up and down for a lot longer without becoming seasick. Inside an enclosed space, the sheer fact that you can no longer orient yourself can immediately make you seasick. If you're walking along a corridor on board a ship in rough seas, then you'll think that the wall wants to keep touching your right shoulder. You've hung up your clothes and suddenly you are amazed to see your trousers sticking out into the cabin, and your soup will insist on slopping out of your bowl. Such things cause seasickness.

The peculiar transformation of what we assume to be solid happens in the unconscious. Two people cannot be one and the same person, after all, and yet suddenly they coalesce. Such things can induce instant sickness. So, this manifestation is caused by the movement of the unconscious. Other aspects also exist, as we have already discussed. You have heard much more about this peculiar ocean swell, only in a different form. Earlier, I even used the corresponding figure of speech. Didn't you notice? What does the sea do? It heads toward shore as a great wave and . . . ?

ANSWER: It breathes.

Yes. It's called the breath of the sea. It rises and falls like a human breast. It's the great breath of the world that we hear there. Well, what does that mean? Where did we encounter this? In the mandala.

Indeed, such movement occurs in the mandala, except that there it starts from the smallest, not from the greatest. Shall we say that there it starts from the heart, and that here we witness it on a large scale. Naturally, here we encounter a peculiarity of the unconscious, which it is impossible to explain within the confines of a seminar. Such matters lead to the paradoxical formulations of Indian philosophy: that is, that "smaller" nevertheless becomes "greater" than "great." That form is so constituted that it is the smallest, "the size of a human thumb in the heart, and all the same occupies the whole world, two handwidths high."[625] That's the definition

---

[625] In this regard, see *Symbols of Transformation* [1912/1952], CW 5, which cites the Upanishads on the paradoxical *purusha*: "That Person, no bigger than a thumb, the inner Self, seated forever in the heart of man [. . .]. Thousand-headed, thousand-eyed, thousand footed is Purusha. He encompasses the earth on every side and rules over the ten-finger space" (*Shvetashvatara Upanishad* 3.12, cited at § 178). Jung links this to Tom Thumb,

of the *purusha*, of primeval man. Therefore these philosophical conceptions and formulations have arisen from eavesdropping on the language
of nature. This vivid streaming and breathing in the human breast is also
the breathing of the cosmic ocean. What is in the smallest is in the greatest, and what is in the greatest is in the smallest. Because in primeval man,
whose consciousness of the whole was dark, the whole found itself within
him, and when he breathed he found himself in the whole. Thus breathed
the world. And if the world breathed, so did he. It is the same breath.

Those among you who attended Professor Zimmer's lecture also heard
him mention such formulations as the wild swan saying, "Ham-sa, I am
who I am, the sea, and you, you are Patuan assan."[626] In reality, this idea
is naturally not an abstract philosophy. Instead, this so-called abstract philosophy is a naive expression and a naive awoval of a strange experience:
namely, of the primeval experience of *participation mystique*; namely, of
being one with the whole, so that I am this tree, this river, that this animal
lives within me, that I am all humanity, and that all humanity is within me.
This experience is simply the primeval experience of the living being,
which does not yet differ from the whole and therefore possesses a totality, which it does not experience if we become conscious, but which we
must attain beyond all consciousness. That, for instance, is the aim of yoga,
and it's also the aim of all these efforts that we have discussed along the

---

dactyls, the Cabiri, and the phallus as "personifications of creative forces" (*CW* 5, § 180).
He later quotes the *Rigveda* (10.90) where the *purusha* is "a sort of Platonic world-soul
who surrounds the earth from outside" (§ 649), "the original 'dawn state' of the psyche; he
is the encompasser and the encompassed, mother and unborn child, an undifferentiated,
unconscious state of primal being" which must be sacrificed (§ 650). *Purusha*, which in
Hinduism defines both the cosmic or primordial man and the individual monad, is generally translated (e.g., by Hauer) as "Spirit" or "Self," and in the dualistic Samkhya system
forms the fundamental counterpart of *prakrti*, or matter, both primordial substance and
unconscious dynamism. Jung would discuss the *Purusha* (and its translations) extensively
in the ETH Lectures, noting, for example, that "[t]he Eastern spirit does not engage in logic,
it is perceptual and intuitive. *Purusha* is better rendered as primal man, man of light" (*Psychology of Yoga and Meditation*, cit., 220). Cf. *Psychology of Kundalini Yoga*, cit., where he
quotes the Upanishads: "Smaller than small, and greater than great [*Katha Upanishad*
2.20–21]" (39). At this point, considering Jung's intervening treatments of philosophic, ascetic, and yogic themes, including the mandala, this reference to Hindu philosophy via the
purusha seems to follow suit from Zimmer's initial lecture (as Jung's salute to Zimmer in
the closing remarks also suggests).

[626] The first of these expressions occurs in Zimmer's lecture (see above [ZL], 105, and
n. 323), while the second seems to be an incorrect transcription of "Tat tvam asi" (meaning
"You are that," or "That you are"), the foundational motto from the Upanishads in Vedantic philosophy and one of the *Mahavakya*s (Great Pronouncements) in the *Sanatana
Dharma*. It originally appears in the *Chandogya Upanishad* (6.8.7) and refers to the identity of the Self with ultimate Reality.

way. For that reason, the dream leads him to the sea, and says, "The way you breathe, the sea breathes too. You are one with the sea." There, he realizes that he is not alone, but instead with his youngest son, a very ambitious youth who likes to identify with his father. Why do you think he suddenly finds himself by the sea with his son, who is still a small boy? Why should he be there with him?

ANSWER: Here, the father identifies with the son, and once again feels as if he were at the beginning of his life.

Why?

ANSWER: In reality, his son identifies with him. Here the father feels like a son, once again starting out in life.

That's correct, you're quite right. He realizes that he is double, that he is both a man about fifty years old and a boy.[627] This reminds me plainly of the words Newton spoke on his deathbed: "I don't know what I may seem to the world. But as to myself I seem to have been only like a boy playing on the seashore and diverting myself now and then finding a smoother pebble or a prettier shell than the ordinary, whilst the great ocean of truth lay all undiscovered before me."[628] In this moment he realizes that in actual fact he's a boy in a certain sense. It's just as with Faust, who dies a very old man and wakes as a youth in the choir of blessed boys.[629]

This dream concerns the statement of a renewal. Such a gathering of father and son, mother and child, always means a renewal, in fact a renewal with which one should not identify. He criticizes the boy's inclination toward identification. One should not identify with this, he says. If one does, then it is longing, and that hinders both. If one identifies, one prevents youth, since one is old, and youth makes age ridiculous. For one must be as old as one is, and thus one should not identify.

Here we encounter a problem which I'd like to discuss briefly, but which I haven't at all yet considered during this seminar, because there has been no opportunity to mention it: the question of the personal and impersonal

---

[627] In *DAS* Jung affirmed that in this patient's psychology the child represents "a substitute for the father," adding that "according to the primitive idea the child is truly the prolongation of the father, the replica of the father; bodily and spiritually he is the father" (141).

[628] Simile enunciated by the father of classical mechanics, the English mathematician, physicist, and alchemist Isaac Newton (1643–1727) in fact in his memoirs, not on his deathbed: as cited in David Brewster, *Memoirs of the Life, Writings and Discoveries of Sir Isaac Newton*, 2 vols (Edinburgh: Constable, 1855), 2:347 [with thanks to Ernst Falzeder].

[629] See Goethe, *Faust II*, Act 5.

life of the soul.[630] The dreamer is the person, the I, who is so-and-so many years old, which is a fact. But within him is a life that survives him, which is younger than he is, an impersonal life, which he obviously produces[631] into the children, but which is nevertheless his own life. On the one hand, it is a strangely impersonal life, as if we were double, contingent upon and living in this three-dimensional space, in this very specific time; on the other hand, it is timeless, a peculiar life beyond, a life confined neither to spaces nor to times, a life in this unconscious, in this strange world of images—something about which this fantasy is, we might say, quite ridiculous, pointless. While on the other hand we cannot dispute that it's precisely the life of these fantasies that fills us with the sacred, brings us healing, precisely because it's the idea of rebirth.[632] By no means is this fantasy pointless, for it's an experience that transforms the human being, affording him what I would call an experience of eternal youth, an experience of timelessness, the experience of a being within him that is not "I" but "he."[633]

What is indicated here is this strange doubling, as if a being had slipped away from him, one that has another future, a future that outgrows him. As you know, similar ideas occur in *Faust*. They concern the relationship

---

[630] See Jung, *Psychology of Kundalini Yoga*, cit., 29–30.

[631] Probably a transcription error for *projiziert* (projects) [with thanks to Ernst Falzeder].

[632] On Jung's understanding of the archetypal symbolism of rebirth and its five forms, see his "Concerning Rebirth" [(1939) 1940/1950], CW 9.1.

[633] "He" most likely corresponds to incorrectly transcribed of *Er* for *Es* (It). "It" would indicate here the new awareness of otherness ingenerated by the numinous experience of the encounter with the *Selbst*, one example dear to Jung being Saint Paul's words, "I am crucified with Christ: nevertheless I live; yet not I, but Christ liveth in me" (Galatians 2:20)—meaning, he affirmed elsewhere, that "his life had become an objective life, not his own life but the life of a greater one, the *purusa*" (Jung, *Psychology of Kundalini Yoga*, cit, 40), cited in both "Commentary on *The Secret of the Golden Flower*" ([1929/1957], CW 13, § 77, and § 41: "God is two in one, like man"), and *The Relations between the I and the Unconscious* ([1928], CW 7, § 365). In *On Active Imagination*, cit. (forthcoming), Jung discusses "the spontaneous activity of the psychical" and the need to "use its forces, even submit to it, [. . .] for only the psychological not-I can lead the I to fullness. Now the subordination under a leading principle begins. Such a subordination, however, can also have an uncanny character. It seems dangerous to entrust oneself to something independent, something alive." The ambivalent character of this "subordination," distinct from the optimistic allusion to "eternal youth" evoked in the text, echoed the biblical warning that "[i]t is a fearful thing to fall into the hands of the living God" (Hebrews 10:31), recalled by Jung in a passage commenting on the absolutist idea of an extramundane "Father in Heaven" and the connected risk of inflation: see *The Relations Between the I and the Unconscious* [1928], CW 7, § 394 n. 6; see also Jung, *The Red Book (Liber Novus)*, cit., Liber Secundus, ch. 7, "The Remains of Earlier Tempes," 281).

between Doctor Faust and Doctor Marianus.[634] He realizes that he reaches a point where things become eerie: that is, that a terrific tension exists on the other side, something tremendous, something overwhelming, and if he drew too near, then the wave could reach him and undo him. What do you think about this danger? What does it mean?

ANSWER: He could lose himself in the unconscious, and it could flood him. What would that mean?

ANSWER: He could become mentally ill.

Indeed, he could lose his mind, he could go mad, and such experiences do occur, in fact; a weak brain, and immediately mental illness is at hand. These people become mentally ill, because inwardly they have beheld an image that is simply much too large for this brain, which is the size of a pitiful pack of cigarettes, so that they lack the energy to overcome it. This has an explosive effect. The French call this "un trouble sensitif";[635] that is, a gunshot exploding inside the head, as if something has broken or burst. That's the experience he cannot surmount. Here the ability to guide one's own will suddenly ends, just as consciousness does. Here, all at once, one becomes a wave oneself, oneself an experience. Other people then say, "Well, he just went bonkers," because they can no longer understand him, or he has been carried away, the wave has torn him away, and one can no longer ascertain what happened to him. No one is immune to this. Therefore, just as everything becomes decidedly dangerous, this moment arrives. So, one must agree with the dream for alerting the dreamer to the fact that he is right to withdraw and first of all to experience another problem before he receives the mystery of baptism.

Still another matter I did not discuss earlier is indicated: namely the name: Papadelanu or Papasdelianu. The essence is "papa," that is, the father, as if the father healed the totality. The father, whose *officium*[636] represents the whole, is the papa, the pope, a visible manifestation of this

---

[634] The hermit Doctor (initially Pater) Marianus in Goethe's *Faust II*, who corresponds in some ways to Faust. Doctor Marianus (the Marian Doctor, devoted to the Virgin Mary) is present, at the end of the second part, during Faust's ascent to heaven, and accompanies him in his final, supreme approach to the feminine, supramundane, all-encompassing force which will conclude his journey.

[635] Fr.: "sensory disorder," referring to a loss or abnormality of sensation, commonly numbered among the "great neuroses," which included emotional instability issues, and psychogenic conversion symptoms such as hysterical deafness or blindness. Today these would roughly correspond to troubles such as Sensory Processing Disorder (SPD) and the Functional Neurological Disorder (FND).

[636] Lat.: "duty," "commitment," "assigned task."

papa, of this father, of the spiritual father, of the spirit become flesh. Here is this saint, this ample figure who forms the apex of a pyramid, and who embodies a whole Church. It's a collective figure, a collective father, the archetype of a figure that encompasses humanity.

In the subsequent dreams, this so-called father complex at first will work like an ordinary personal father complex: that is, as a longing for authority and the like. Subsequently, a series of dreams follows in which every conceivable personal difficulty is treated, all which prevent him, in fact for a longer time, from actually tending to the sicknesses that have appeared and become known in the initial dreams. Here, as so often, dreams occurring at the outset have determined the program and the whole analysis. You have probably wondered about the speed at which this patient advances. After some time comes the standstill, once he has intuitively perceived the whole extent of the problem. Then comes the detail work. He has gotten nowhere with these dreams. They are a glimpse into the Land of Promise, and he still might die before he reaches the Promised Land.

I would like to close the seminar on that note.

(Prolonged, thunderous applause)

Index

Page numbers in *italic* refer to figures

Abraham, Karl, 12
abstract philosophy, 304
Achilles, 85n274; inextinguishable fame, 84
active imagination, 46, 65, 72
"actual" and "unactual," Indian thinking, 85
adhista, sun, 262–63
Adler, Alfred, 14
Adler, Gerhard, 31, 309
Adler, Hella, 31
admixtio omnis diabolicae fraudis, 228, 228n500
Adorno, Theodor, 7
Advaita: Advaita-Vedanta, 111n339; philosophy of, 102n319; Shankara as founder of Advaita-Vedanta, 110n337
afflictions, 88, 98. *See also* impairments
Africa: bush trails, 259–60; dream book of Swahili boys, 136, 136–37n369
African tribe: dancing "N'goma," 190–91; dealing with, 192
"After the Catastrophe" (Jung), 28
alchemical process, psychology, 291–92
alchemists, 46
alchemy, 2, 291, 292; Jung's exploration, 39–40
Alm, Ivan (Pastor), 30
Amitayurdhyana-sutra, 64
analysand, Jung's, 31–33
analysis, term, 291

analytical psychology, 72–73, 217; Jung's, 96–97; Zimmer on, 59–61
Analytical Psychology Club (New York City), 50
anima: in dream, 177; figure, 181; hypothesis of, 151–52; as seamstress, 193
animals: fairytale forest, 210–11; herdsman and, 176–77; primitives and wild, 192; primitives killing, 196
anthroposophy, 146, 146n379, 174, 218
anti-Semitism, 11, 20, 27, 28
anti-socials, 7
appearance, 85
apperception, process of, 276
Apuleius Madaurensis, Lucius, 118–19; *The Golden Ass*, 117, 117n352
archbishop: ball games, 164; bells for vespers, 167–68
archetypes, 29, 34, 36, 47, 61, 241, 242
Archive for Research in Archetypal Symbolism (ARAS), 47
Aristotle, 89n282
Artistic Form and Yoga (Zimmer), 53, 62–64, 70
Aryanization, 11
association, chemistry and, 291–92
AUM, primordial sound, 104, 104n322
Auxerre, cathedral of, 165, 165n401
Avalon, Arthur [Sir John Woodroffe], 16, 63, 64

Bad Nauheim, 71, 73
Baeck, Leo (Rabbi), 23
ball game(s): Christians, 233–34; clergy playing pelota, 163–66; as communion or community, 159–60; honoring the gods, 163; playing in hall, 216, 217; as religious symbol, 169; symbol of playing, 163. See also games
baptism, 307, 199n442; *mysterium* of, 235n510; rite of, 199–200
Bardo Thodol, Tibetan-Buddhist book of dead, 115, 115n347
Basque pelota, 143, 158–59
Baynes, Cary Fink, 51
Beleth, Jean, 169, 169n408
bells, ringing for vespers, 167
benedictio fontis, 227
Berliner Psychoanalytische Institut, 12
Berlin Seminar (June 1933), 61, 70, 72; Radio Berlin interview and Society for Psychotherapy, 16–29
Bernhard, Ernst, 31
Bertine, Eleonore, 50
Bhagavadgita, 92, 92–93n292, 93n295; Krishna's teaching in, 93n293
Bildungskraft, power of imagination, 35
Binding, Karl, *Permission for the Destruction of Life Unworthy of Life*, 8
Bingen, Hildegard von, 239, 239n523
Bircher-Benner, Maximilian Oskar, 146n377
Bircher diet, 146, 146n377
birds, *cakravaka*, 111n339
black animal, as villain, 212
black magic, 99, 226
bocce: ball game, 159, 159n392; symbol of playing, 163
body, instinctual zone of, 282–83
Boehm, Felix, 11
Böhme, Jacob, 230, 239, 239–40n524, 246
Bollingen Foundation, 50
Bollingen Press, 50
Bollingen Series, 51, 52

Borobudur, Buddhist temple at, 253n549, 253–54
Bradley, H. Dennis, 147, 147n381
Brahma, swan as sign of, 109
breath of life, 105–6
breath of the sea, movement in mandala, 303–4
Bronze Age, 244n532
Bronze Age sun-wheel, 244
Buddha, 81, 82; Amitaba, 127, 127n361; path to illumination, 84
Buddhism, 39, 42, 44; Pure Land, 64; Tibetan, 43n135
Buddhist mandala, 67
buffalo hunting, 195, 196
Bügler, Käthe, 30, 31, 55–56
bullfights, game of, 169–71

cage: chicken-run flying away, 288–89; dream of hens, 272–74, 279–81, 286, 288–89; secure compartment, 286, 287
Campbell, Joseph, 51, 52
canals, system of spiral, 251
capitalism, communism and, 9–10
Carnival, 163
catabasis. See *descensus ad inferos*
Cathedral of Chartres, rose of, 221
Catholic Church, 163, 244n532; holy water of, 228; symbols of, 227–29
Catholic convent, bodiless woman educated in, 213
Catholic sacraments, Jung defining, 71
chakras, 103, 113–14, 113n345; lotus-shaped, 113–14; symbolism of, 16, 128n362
chants, Navajo Indians, 267–70
Chapple, Gerald, 50
chemistry, association with alchemy and, 291–92
Chinese philosophy, Yang principle in, 219
Christian(s): Gnostics and, 2, 37, 39, 233, 234n509, 235, 236; overcoming fear of death, 84–85; transformation, 200n444; works of Plato and, 242

Christian church, 155, 244; Helios, 263–64; rituals in, 233–36
Christian community, 155
Christianity, 37, 41, 46, 71, 293
Christian mandalas, 240n525
Christian Science, 146, 146n378, 148, 174
Christian Scientists, 200n444
Christina Alberta's Father (Wells), 132n365, 132–33
Christmas, 258, 260
Christmas tree, knowledge of, 258–59
Church, tables for dinner, 165–66
Cimbal, Walter, 25n74, 27
circumambulatio, 223, 225; mandala symbolism, 256–57; meaning, 231; ritual circling around a person, 223n486
circus, game with audience, 172
Clarke, J. J., 9
clothes: humans expressing character through, 194; seamstress assembling, 193
coalescence, 107
collective psyche, 14, 15
collective unconscious, 34; Jung's, 44; Jung's conception of, 60
Colonna, Francesco, 251, 252n243, 296
Columbia University, 50
communion, 155, 156n389; ball game as, 159–60; Christians, 233; circus and, 172; dreamer, 157–58; gathering and, 173; marriage and semblance of, 173–74; meal, 291
community, 156; ball game, 159; dream, 155; dreamer, 183; feeling of inferiority, 160; marriage, 175–76
comparative symbolism, 2
complex psychology, 3, 5, 40–47, 71
Confessions (Augustine), 288
conscious: placing best sides center stage, 162; prayer as action, 265; relationship with unconscious, 249–50
consciousness, 104–5, 190; attitude and, 195; dreams and, 154–55,

178–79; games, 261; mental illness, 307; *participation mystique*, 214; philosophical questions, 86–87; Promethean sin of, 237; secure compartment, 287; state of, 181; theory of functions, 276–78; unity of personal, 86
contaminatio, images, 287
Coomaraswamy, Ananda, 48, 52
Corbin, Henry, 34, 47
Curtius, Otto, 19, 20, 27
customs, 265

Danae 108
dance, primitives and, 195–96
Dante's *Divine Comedy*, 65–66, 207, 209, 221, 296
darkness: as other people, 213; pushing into background, 162; unconscious, 282
darkness in the eye, 156, 156n390
Daudet, Léon, 150n385
Death in the Afternoon (Hemingway), 171, 171n412
Decline of the West (Spengler), 9
de Gobineau, Arthur, 10
"demo-liberalism," 10
descensus ad inferos: catabasis, 296; as a night journey, 283
"Diagnosing the Dictators" (Jung), 28
diagnosis, condition of society, 9
Die vertauschten Köpfe-Eine indische Legende (Mann), 49
Dilthey, Wilhelm, 57
diseases of civilization, notion of, 9
Doniger, Wendy, 58
drawings, dreamer and, 229–31
dream(s): analysis of, 142; anima hypothesis, 151–52; attention of dreamer, 141, *141*; Basque pelota, 158–59; cage with hens, 272–74, 279–81, 286, 288–89; characters in, 150; connecting path for understanding, 229; connection of, 271–74; death of a child in, 144; *drame intérieur*, 150, 150n385; drawing by people who dance, 223;

dream(s) (*continued*)
evaluation of persons in, 178; fantasy, 154; father and son, 305–6; feeling, 274–75; feeling in, 151; first dream of Old Man, 138–40; Freud's theory of, 34; functions theory, 276–78; interpretation of, 184; kindred harmonic in, 75; mana as meaning of, 262; meaning of flowers in garden, 192; meaning of gardens in, 192–93; meaning of sick seamstress in, 189–90; meaning of wife in, 188–89; message and, 245–46; monotony of, 145; music in, 151; normal person, 136–37; notion of figures, 177; patient and Kundalini snake, 217–18; performance, 154; piece of news, 181, 182; place, action and time, 140; reading the message of, 178–79; road-making machine, 202–3, 202n446, 204; sacred number four, 275–76; seamstress and clothing, 200–2; sequence of, 271–74; shadow in, 151; subject of curious hall, 141, *141*, 142–43, 216; symbolism, 213–14; underworld, 209

Dream Analysis (Jung), 31, 36, 38, 62, 65, 66, 70; analysand on playing *jeu de paume*, 143–44n376; dream account in, 143n375

dream book, 136

dreamer(s): anima of, 177; benches in the hall, 162; child in dream, 144, 145, 148, 150, 176; confusion of, 242–43; connection with humanity, 182–83; correctness of Old Man, 183; determination of, 198; diagnosis of child's sickness, 184–86; drawings of, 229–30; dream of going out with brother-in-law, 143–45, 157, 173; fairytale forest, 210–11; feminine in the soul of the man, 152; on going out with his wife, 160–61; on hall, *141*, 158, 161; interpretation as unprejudiced form, 222; lack of masculinity of, 161–62; losing own

shadow, 212–13; path of Old Man, 182; personal and impersonal life of soul of, 75; placing best sides center stage, 162; representation as sick person, 299–301; resistance toward his wife, 212; self-consciousness, 147; self-evaluation, 182; shapes and forms, 215; sister and brother-in-law of, 148–50; speculations of dream, 186–87; worldview of Old Man, 183–84

Drowned and the Saved, The (Levi), 7

drum/drumming, 102, 104, 196, 198

Dulles, Allen Welsh, 29

Durand, William, 169n408–9

Easter, 166, 228, 258; bunnies, 260

Easter Monday, 167

Eastern doctrines, superiority of Christianity over, 41–42

Eastern mandalas, 45, 64, 65, 66, 68, 223, 231, 232, 233, 239, 243, 244, 250

Eastern spirituality, 2, 5, 14, 39, 40ff.; imitation of Eastern techniques, 44–46

Eastern symbols, 129, 217–18, 251

Eddy, Mary Baker, 146n378

Edelstein, Ludwig, 49

Eidgenössische Technische Hochschule (Swiss Federal Institute of Technology), Jung at, 2

Einzelmitglieder (individual members), rule of the, 26

Eitingon, Max, 11, 12

Eliade, Mircea, 38, 51

Engel, Werner, 31

enlightening light, yogi as, 111–12

Eranos conferences, 1, 49, 51, 61

Essay on the Inequality of the Human Races (de Gobineau), 10

Esslinger, Mila. *See* Rauch, Mila Esslinger

Estsánatlehi, 269, 269n582

ethereal spaces, concentration on, 110–11

eugenic movements 8, 25n74
European mandalas, 65, 69, 70, 223
Evans-Wentz, Walter, 43
evil: black magic, 226; symbols against spirits, 227
exercises: Hatha yoga, 101–6; mood or mental disposition in yoga, 106–7
eye, darkness in, 156, 156n390
eyes at back of one's head, 281

Falzeder, Ernst, 125
fantasy, 215–16, 295; correct and incorrect, 247; dreams, 154; human being, 241; transformations of, 302–3
father: complex, 308; mental illness and, 307–8; son and, 305–6
Faust (Goethe), 54, 295, 306
feeble-minded, 131n364
feeling: dream and, 274–75; personification in dream, 151
feminine, soul of the man, 152
Fenichel, Otto, 11
Fierz, Hans Eduard, Jung and, 1
fire: Buddhist idea, 232; dream, 295; healing of sick person, 301
First World War, 9, 13
flowers, meaning in dreams, 192
Fordham, Michael, 309
formula, prayer, 265
Frazer, James George, 36
freedom: longing for, 287–88; outside world, 287
Freidländer, Dora, 31
Freud, Sigmund, 14, 61; Jung and, 2
Fröbe-Kapteyn, Olga, 1, 46, 47
Froboese-Thiele, Felicia, 30
functions: 31–32; feeling, 31; theory of, 276–78; thinking, 286. See also Psychological Types (Jung)

games: circus, 172; Spanish bullfight, 169–71. See also ball games
Gandhi, 85n274
garden: labyrinth garden, 202, 205, 206, 206n454, 207, 229, 251;

meaning in dreams, 192–93; walled garden, 55, 186, 187, 190, 192, 193, 194, 221, 221n476, 233
Geitel, Charlotte, 30
Gemeinschaft (community), salvation of, 11
General Medical Society for Psychotherapy, 1, 2, 16, 24, 26, 27, 30
Geneva Conference, 197n440
German Psychiatric Society, 25
German Psychoanalytic Society, 11
German Reich, 7
German revolution, 8
German Romantics, 62
Geschehenlassen, "letting-happen," Jung, 74
ghosts, 260–61; ghost houses and natives, 259–60
goal, 129
God: breath of, 107–8; as a circle, 240; image of, 107; Jonas on "God after Auschwitz," 7
Goebbels, Joseph: Jung's encounter with, 20, 24; myth of Führer as new "Aryan" messiah, 8
Goethe's Faust, 54, 295, 306
Golden Ass, The (Apuleius), 108n332, 117, 117n352
Goodrick-Clarke, Nicholas, on Jewish plot to destroy Germany, 7–8
Göring, Matthias Heinrich, 12, 25, 26, 27
Great Depression, 6
Greco-Roman Antiquity, 37
Greek myth, 108–9n332
Griffin, Roger, 10

Haeckel, Ernst, 10
Hague Conferences, 197n440
Halleluja: burial of, 164–65; Hebraic term for, 165n400
Hamsa: breath of life, 105–6; myth of, 75
handwriting, 214
Hannah, Barbara, 18, 30, 35–36, 56
Harding, Esther, 50
Harnack-Haus, 30

Hatha yoga, 101n312, 238; concentrating on breathing, 101–2; exercises of, 101–6; "staircase" to other methods, 101, 101–2n314; Zimmer's lecture on, 56
Hattingberg, Hans von, 30
Hauer, Jakob Wilhelm, 5, 16, 42–43
Helios, 263–64
Hemingway, Ernest, on bullfighting, 171n412
hens: bowl containing, 292; bowl with smooth rim, 290; chicken-run flying away, 288–89; dinner table, 291; dream of cage with, 272–74, 279–81, 286, 288–89; escape of, 284, 295, 301; helpful animals, 296
hermeneutics, Jung's, 46, 58
Hermetics, 46
Hero with a Thousand Faces, The (Campbell), 52
Herrenrasse ("master race"), 7
Hesse, Hermann, 49
Heyer, Gustav Richard, 30, 54
hindrances, 88, 91. *See also* impairments
Hindu, myth of the Hamsa, 75
Hinduism, 39, 44; orthodox, 64
Hindu mandala, rhythm in, 75
Hindu pantheon, superhumanity of countless deities, 54
Hindu Philosophy and Religion, Columbia University, 50
Hindu symbol, psycho-spiritual function of, 70
Hindu symbolism, 39
Hindu temples, planimetry of, 69
Hirschfeld, Magnus, 14n37
Hitler, Adolf, 6, 21, 24
Hoche, Alfred, *Permission for the Destruction of Life Unworthy of Life*, 8
Hofmannsthal, Christiane von, 48
Hofmannsthal, Hugo von, 48
Holocaust, 7
holy water, 228, 228n500
homosexuals, 7
hortus conclusus, 221

Hostioken, 257
house, windows, 287
hubris, 187
human being(s): baptism rite, 199–200; fantasy of, 241. *See also* normal person
human problem, collective viewpoint, 302
Hypnerotomachia (Colonna), 251n542, 251–52, 296

Icarus, 99
ichthys, 236
Ignatius of Loyola, 46
illness, mental, 307–8
imbeciles, 131, 131n364
immortality, 85, 86
impairments, 88–89n282; diminution, 94; freed from, as way to ideal unburdened being, 94–95; *kleshas*, 88, 88n281; means of "active" yoga, 91–93; mistaken notions, 90; partiality, 89; yoga's, 88–95
impermanence, 81–85; concept of, 85; death sentence of, 83–84; everlastingness, 85; Indian notion of, 89; inextinguishable fame, 84
incubus, 180, 180n420
Indian cosmology, world-snake in, 112–14
Indian parable, two birds sitting on tree, 110–11
Indian prince, legend of, 81–85
Indian Tantra yoga, 282
individuation process, 58, 71, 105, 106; psyche channeling, 71; Zimmer on, 57–62
Indology, 48
inferior, word, 189
inferiority: community and, 160; feeling of, 190; normal person, 131–32
inheritance, Hindu concept of, 60
inner concentration, object, 95–96
instinct, bottom versus top, 282–83
Institut für psychologische Forschung und Psychotherapie, "Göring Institute," 12

"integral man," Jung portrayal, 20–21
Interpretation of Dreams (Freud), 61
Isis: return from otherworld of death, 119; secret of, 122
Italian fascism, 10
ivory tower, symbol, 220, 220n472

Jacobi, Jolande, 12; Jung to, 12
Jaffé, Aniela, 23
Jaina-Yoga, 59
Jaspers, Karl, 49
Jehovah's Witnesses, 7
Jericho, Joshua's taking of, 72
Jericho walls, siege of, 72, 256
jeu de paume, 143–44n376, 163
Jewish doctrine, psychoanalysis, 11
Jews, 7; devaluation of Jewish psychology, 23; Nazi persecution of, 23n66
"jiva," individual Self, 106, 106n326
Jobs, Hieronymous, life plan of, 205
Jobsiad (Busch), 205
Johns Hopkins Institute, 50
Jung, C. G.: analytical psychology, 96–97; collected works of, 309–16; depth psychology, 58–61; impressions from Jung-Zimmer encounter, 53–57; (meta)psychological conception, 13; Professor Schmalz and, 295–96; Weizsäcker and, 17; Zimmer and, 237–39, 274; Zimmer's influence on, 62–72. *See also* complex psychology
Jung, Emma Rauschenbach, 1

kaivalya, 95; being separate or distinct, 94; *restitutio in integrum*, 100, 100n311; singleness, 94, 99–100; state of one, 94n298
Kerényi, Károly (Karl), 37–38, 47, 51
Keyserling, Hermann Graf, 9
Kirsch, Hilde, 30
Kirsch, James, 30
kleshas, 88, 88n281, 88–89n282. *See also* impairments
Klibansky, Raymond, 49
Koenig-Fachsenfeld, Olga von, 18

Kollwitz, Assilie, 30
Körner, Otto, 30
Kranefeldt, Wolfgang Müller, 18, 19, 20
Kretschmer, Ernst, 25
Krishna, 103, 109
Kubin, Alfred, 49
Kultbilder, sacred cult images, 63
Kummerlé, Hanna, 18
Kundalini, 102, 102–3n319, 103, 112, 114–19; ascent of, 115–17; coiled snake, 233; process, 119; secret of Isis, 122; snake, 115, 250; term for spiral snake, 217–18; as world mother, 118
Kundalini energy, 113n345, 114n346, 121–22
Kundalini yoga, 43; psychology of, 4–5; seminar on, 4–5, 55
Kundalini yogi, 121
Kunkel, Fritz, 73
Kunstform und Yoga im indischen Kultbild (Zimmer), 48, 53. *See also* *Artistic Form and Yoga* (Zimmer)

labyrinth: maze, *208*; word, 205, 206n454. *See also* garden
Lamaism: Lamaist mandala, 232, 233; Lamaist Tantrism, 232, 237; symbolism 46
Lao Tse, 204
Last Supper, 155
Law for the Restoration of the Professional Civil Service, 6
L'Étape (Bourger), 236
League of Nations, 7, 197n440
Leda, 108, 108n332
legend of Indian prince, 81–85
Lent, 163
Lévy-Brühl, 241
Liber Novus (*The Red Book*) (Jung), 2–3
libido, observation as filling, 251
Liebscher, Martin, 64
Litany of Loreto, 220, 220n471, 237
Literature Prize, Jung, 1
Lockot, Regine, 11

Lord of Spawn, 90
Lord's Supper, *mysteria*, 235
Lotus (symbol of) 65, 73, 89, 105, 106, 110ff., 122, 219, 221, 243, 255. *See also* chakras

machine(s): dreamer drawing, 215; dream of, making a path, 205–7, *208*; maze of labyrinth, *208*; road-making, 202–3, 202n446, 205–7; steamroller, 217
magic, mana, and demonism, 61; mana, living force, 262
Magic Flute, The (opera), 116, 116n349; Tamino and Pamina, 116, 116n350
Magic Mountain, The (Mann), 146n377
Mahabharata, 58
Maharshi, Sri Ramana, 52
Mahayana Buddhist temple, Borobudur, 253–54, 253n549
mana jadoga, meaning of dream, 262
mana personality, Jung's notion of, 36, 54, 59, 60, 96
mandala(s), 244; Christian, 244; concentration, circumambulations, and contemplation, 281; concentric image, 120–21; drawings, 216, 223, 232; drawing *sulcus primigenius*, 223n487; dream image connection to, 68; Eastern, 250; European, 70, 239; mosques, 254–55; movement in, 303–4; mythologem of the, 38n117; Navajo Indian, 266–68; rhythm in Hindu, 75; steamroller work, 256; symbolism, 37n116, 38n118, 69, 256–57; term, 37, 231; Tibetan, 68; yantra, and "machine," 62–72; Zimmer's definition of, 67; Zimmer's influence upon Jung, 62–72. *See also* yantra
Manitu (great spirit), vision of, 289
Mann, Kristine, 50
Mann, Thomas, 49
Marchand, Susanne, 41

marriage, 286; community and, 175–76; in dream, 153; dreamer, 158; dream involving, 138–40; problem of, 138; semblance of communion, 173–74; sewing machine and, 194–95
masculinity, dreaming lacking, 161–62
materialism, 22
Maya, 85, 85–86n275, 102, 106, 119, 122
Mead, George Robert Stowe, 164n399
meaning of life: normal person, 134–36; problem of, 145
medical psychology, 5
Medusa, 84
Meier, C. A., 26
Mellon, Mary Conover, 50
Mellon, Paul, 50
mental disposition, yoga exercises, 106–7
mental illness, 307–8
message, dreams, 245–46
Middle Ages, 179–80, 180n420, 233, 293
miraculous (or supernormal) powers (*siddhis*), 95
mistaken notions, impairments, 90
Mithras (god), bullfights and, 169–71
modernization, Clarke on, 9
modern man, Jung on transformation of, 15n38
Mohammad Ali mosque, 254, 254n551
Monist League, 10
monotony, dreams, 145
mood, yoga exercises, 106–7
moon, concentration of, 96
Moritz, Eva, 18, 19
Mosse, George, 29
Mountain Chant, 257, 257n557, 266n578, 267–70, 267n579
movement of things in time, 277
muladhara: awakening out of, 248n536; center of the body, 112n342, 250; chakra, 103n320, 128n362, 217n464, 250; Lothos, 114n346; night sun, 117n353;

"root place," 113, 113n344–45; sleeping, 218n466
Müller, Max, 41
Müller-Braunschweig, Carl, 11
mundus, 224, 224n488
mungu, sun, 262–63
music, in dream, 149, 151, 154
Muslim culture, 298, mosque, 254–55; Muslim religions, 299
Myotieosmus, 285
mysteria, 235; term, 235n510
myth of Kronos, 83
Myth of the Twentieth Century, The (Rosenberg), 27
mythology, 48, 51; Indian mythology, 57–58; mythological images, 37, 242

Nachlass (Hofmannsthal), 48
Narbonensian manuscript, twelfth-century, 167–68
natives, small houses and ghosts, 259–60
nature, 246–47; beauty and, 271; dreams, 246–47; message and, 245
nature of all things, unifying symbol, 270
Nature of Fascism, The (Griffin), 10
nature of things, 278
Navajo Indian(s), 230, 266n578; Est-sánatlehi as deity, 269, 269n582; mandalas, 266–67; memoirs of, 266–67, 266n578; Mountain Chant, 267–70; Night Chant, 267–70, 267n579; number four in, 270–71; sweat lodges, 244
Nazi Germany: 6ff.; psychoanalysis in, 11–12
Nazism, 11, 13n31; Jung's alleged philo-, 28; phenomenon of, 14
Neapolitan manuscript, ninth-century, 168, 168n406
Negative Dialectics (Adorno), 7
Nette, Herbert, 52, 53
Neumann, Erich, 16, 30, 51
"N'goma," African tribe dancing, 190–91

Nietzsche, Friedrich, 8; concept of Superman or Overman, 17; *Zarathustra Seminar* (Jung), 17
Night Chant, 267–70, 267n579
Nolde, Emil, 49
normal human life, 129
normal person: case of gentleman (Old Man), 130–41; *Christina Alberta's Father* (Wells), 132–33; consulting Jung in practice, 128–29; correctness and, 130–31; dreams of case of, 136–37; exposition of the dream, 153–56; first dream of Old Man, 138–40; importance of, 131–35; impudence, 131; inferiority, 131–32; meaning of life, 134–36; path to inner self-development, 129; question about the soul, 130; working on his "dreamer's" problem, 153–54; yearning for world-view, 153. *See also* dream(s)
number four: dream with sacred, 275–76; Navajo system, 270–71
Nuremberg Laws (1935), 7

Oakes, Maud, 52
object, inner concentration of, 95–96
occult, 32, 33; interests, 54; spiritual approach, 63; studies of, 145, 153 175
Odyssey (Homer), 143n376
Oedipus, 83
Old Man: Old Woman and, 245
"On the Psychology of Yoga," Zimmer's lecture, 4, 17
Osiris, 119
"otherness," 5, 33
Otto, Rudolf, 57
outside world: freedom, 287; longing for freedom, 287–88

Padma symbols, 219
pansexuality, 11
Pantheon Books, 51
Papadelanu, 298, 307
papal decree, clergy playing pelota, 163–66

Papasdelianu, 298, 307
participation mystique, 36, 214, 242;
old native with disobedient son,
214–15; primeval experience of,
304; primitive condition, 214
Patanjali: Kriya yoga, 91n290; *Yoga
Sutras*, 87, 87n279, 95; yoga system, 95–96n302
Pauli, Wolfgang, 34
pelota, 143, 158, 159, 162, 163, 165,
166, 216, 233
Penelope, 102, 102n318
peregrination, dreamer, 299
performance, dream, 154
*Permission for the Destruction of Life
Unworthy of Life* (Binding and
Hoche), 8
Phaedo (Plato), 154n387
phallus, male symbols, 219–20
phenomenological stance, Jung, 74
philosophus teutonicus, 246
Plato, 81n264, 154n387, 242
Pleroma, ideas of, 242
pneuma, word, 262
Poles, 7
polestar, concentration of, 96
Portmann, Adolf, 47
positivism, 22
power of imagination, 35
Prajapati, personification of life-matter
and life-force, 91n289
prakrti and purusha, 75
prayer, 265
primitive(s): beliefs on dead, 231;
dance and its meaning, 199; dancing of, 195–96; extraordinarily conscientious, 289; lacking explanations for rituals, 257–59; loss of the
soul, 284–85; Manitu (great spirit),
289; manner of living, 292–93; psychology, 37, 139; *rites d'entrée*, 195,
197; rituals of tribal elders, 261–63;
secret of medicine man, 285; sewing
machine rhythmic activity, 201,
202; thoughts or drawings, 231–32;
wild animals and, 192; work and,
195–96

primitivism, of East, 41
production, Western psyche's, 45
Promethean fire, 237
Promised Land, 76, 308
pseudopodia, 222, 222–23n484
psychic consciousness, 187
psychic hygiene, 25, 44
psychic inflation: 54, 60, 75, 97, 99
Psychoanalyse und Yoga (Schmitz), 14
psychoanalysis: Jewish doctrine of,
11; male and female symbols,
218–20
Psychoanalytic Institute, 5
Psychological Club Zurich, 16;
Zimmer's debut lecture (1932) at,
55; Zimmer's lecture on Hatha
yoga, 56
Psychological Types (Jung), 70
psycho-logos, 72
"Psychology and National Problems"
(Jung), 28
Psychology of Kundalini Yoga, The
(Jung), 16, 88n282
psychology of yoga, 4–5
Psychotherapeutic Society, 22
Psychotherapy Congress, Baden-
Baden, 3
Pueblo Indians, 230
pure-impure, terms, 89–90
Pure Lane Buddhism, 64
Purgatorio (Dante), 297
Pythagoras, 243, 243n530
Pythagoreanism, 278

"racial hygiene," Nazi, 8
Radin, Paul, 51
Radio Berlin interview (Jung): Berlin
Seminar and Society for Psycho-
therapy, 16–29; Jung's, 16, 17,
20–22, 28
Rahu, daiman, 82–83, 82n267
Raja yoga, 40, 112, 112n341
Ramayana, 58
rationalism, 210, 245
Rauch, Lukas, 57
Rauch, Mila Esslinger, 48, 55, 56
Read, Sir Herbert Edward, 30, 309

"real" and "unreal," Indian thinking, 85

rear wheel: dreamer and, 281; walking difficulty, 301

recapitulation theory, Haeckel on, 10

Relations between the I and the Unconscious, The (Jung), 3, 54, 96

religion, normal person, 134

religious groups, 200n444

représentations, term, 242

restitutio in integrum, 100, 100n311

Rigveda, 58, 69

rite d'entrée: primitives, 195, 197, 247–49; rhythm-making machine, 201; seamstress, 198; yoga exercises and, 195, 247–49

rituals, primitive lacking explanation for, 257–59

road-making machine, 202n446, 202–3, 205–6, 205–7

Roaring Twenties, 6

Roelli, Ximena de Angulo, 51

Roma, 7

Roman, cities, 225

Roman *castrum*, 225, 232

Roman Empire, 13

rosa mystica, 221, 221n478

Rosenbaum, Wladimir, 26

Rosenberg, Alfred, 27

Rousselle, Erwin, 47

sacramenta, early Church, 235

Sacred Books of the East (Müller), 41, 41n128

sahasrara, "crown" chakra, 113n345

Saint Augustine, 240, 242, 263, 288

Saint Bartholomew's Day massacre, 13

Saint Stephen, 166

saldo [sal] sanctus, 228, 228n499

Schmaltz, Gustav, 18, 30; Jung and, 33, 295–96

Schmitz, Oscar A. H., 14, 44–45

Scholem, Gershom, 23, 47

Schultz, Johannes Heinrich, 25, 30

sciatica: dream on healing illness, 297–98; walking difficulty and, 301

seamstress: anima as, 193; assembling, 193; clothing and, 200–2; figure of suffering, 212; meaning of sick, in dream, 189–90; *rite d'entrée*, 198

Secret of the Golden Flower, The (Jung and Wilhelm), 3, 42, 44, 53, 66, 73

Seeress of Prevorst, The (Kerner), 147–48, 147n382

seleteni, ghosts, 261

"Self," 58; analytical psychology toward, 97–98n306; psychological concept, 97n304

self-conscious, normal person, 134

self-consciousness, 187; dreamer, 147

self-development of the individual: Jung on, 22–31; path to inner, 129

sensory disorder, 307, 307n635

Septuagesima Sunday, 164

sequence, dreams, 271–74

sewing machine(s): as facilitator of integration, 193n435; marriage and feeling, 194–95; meaning of, 193

shadow, 156, 282; consciousness, 237; dark side and, 174–75; in dream, 151, 185; dreamer losing own, 212–13; unconscious, 175

Shakti, 121, 239

Shakti, creative force, 74

Shamdasani, Sonu, 3, 34, 43

Shiva, 74, 104, 121, 122, 238, 239, 250

sickness, dream on healing, 297–98

Silber, Hilde, 30

Simmel, Ernst, 11

simplification, 107

single-mindedness, yoga exercises, 106–7

Sinology, 48

small houses, ghosts and, 259–60

snake, yoga process for moving like, 112–14

social Darwinism, 10

Society for Analytical Psychology (Berlin), 18, 55, 79

Society for Psychotherapy, 12, 18–19, 30; Berlin Seminar, Radio Berlin interview and, 16–29; Jung in presidency of, 16
Socrates, 81n264, 154; philosophy, 81
soil, 282
son, father and, 305–6
Songe de Poliphile (Colonna), 252, 252n544
soul(s): avoiding confusion of alien, 285–86; loss of the, 284–85; questions about, 130; unconscious, 294
Spanish bullfight, game, 169–71
Speer, Ernst, 30
Spengler, Oswald, 9
spiral motif, symbolism of, 250–51
spiritualized earth, 221n477
"Spiritual Problem of Modern Man, The" (Jung), 15n38
spitting into hands, rituals of tribal elders, 261–63
Stackmann, Hildegard, 30
Steiner, Rudolf, 146n379
subconscious, 283
succubus, 180, 180n420
suffering, human problem, 302
sulcus primigenius, 223, 227, 230; drawing of, 223n487
Sultan Hassan mosque, mandala form, 254n552, 254–55
Sumeru, world-mountain, 112, 112–13n343
sun, concentration on the nature of, 96
superior consciousness, 34
sushumna, 103, 103n320, 113, 121
Sussmann, Toni, 31
Suzuki, Daisetz Taitaro, 43
Swahili boys, dream books by, 136, 136–37n369
swan, myth of, 107–9
swastika, 257
sweat lodges, Native American tribes, 244, 244n531
symbolism: dream, 213–14; mandala, 256–57; spiral motif, 250–51
symbolon, 201, 202n446

symbols: emerging from innocent fantasy, 240–41; life, 227; male and female, 218–20
Symbols of Transformation (Jung), 15n38, 37
synchronicity, 34
synopados, shadow, 282

Tantra, 42, 43, 43n136, 98–99n307; chakras in system, 250; literature, 98–99; mandala(s), 239, 244; yoga, 16, 46, 49, 107, 282; Zimmer on, 63
Tantric symbolism, 17, 39
Tantric texts, 48, 64, 103n319
Tantrism, 44, 53, 86n275, 98n307, 106n325, 120n358, 121n360, 217n464, 219n469, 231n506, 238; Lamaist, 232, 237
Taoism, 39, 42, 44; philosophy, 34; Taoistic Yoga, 46
temenos, 226
Terry Lectures (Jung), 28, 28n79
Tertullian, 199n442; rite of baptism, 199–200
theater, performance, 154
theosophy, 15n38, 32n97, 33, 40, 45, 149n384, 263n572
thinking function, 286
Tibetan culture, relics, 226
Tibetan Lamas, 67
Tibetan mandala, 67, 68, 232
Towards the Stars (Bradley), 147
Travel Diary of a Philosopher (Keyserling), 9
Tschuktschukbeutel, 230, 230n504
tuberculosis, 185–86, 188, 194, 216

unconscious: attitude and, 195; awareness of language of the, 75; confrontation with, 2; darkness and, 282; Jung's notion of, 22; life of the soul, 283; relationship with conscious, 249–50; shadow as, 175; transformations inducing sickness, 302–3

underworld, 119; Christian, 296; descent into, 209; dreamer descending into, 299
unification, 107
unity, 156; body and soul, 25; consciousness, 86, 273n585; jiva and brahman, 111; Shiva and Shakti, 74, 121
University of Greifswald, 48
University of Zurich, 2
"un trouble sensitif," sensory disorder, 307, 307n635
"unworthwhile lives," planned suppression, 8

vas, superlative vessel of devotion, 221–22
vajra, 65
Vedanta, 85n275, 87n279, 101n311, 250n538; Advaita-, 110n337, 111n339; goal of, 112
Vedic scriptures, world creation in, 90–91
vegetarianism, 146, 146n377
Vergil's *Aeneid*, 61
Versailles Treaty, 6, 197n440
vesture, expression through, 194
via decumana, 224, 243
via principalis, 224, 243
villains, as other people, 212
Virgil, 296
Virgin Mary, 221n477, 221n478
Vishnu (almighty god), 102, 103, 109
Vivekananda, Swami (Hindu monk), 40
völkisch ideology, 11

Wagner, Richard, 50
walled garden, 55, 186, 187, 190, 192, 193, 194, 221, 221n476, 233. *See also* garden
Weber, Alfred, 49
Weber, Max, 9
Weimar Republic, 13
Weizsäcker, Adolf, 17, 20
Wells, H. G., 132n365

Western mandalas, 69, 70, 239, 243
Western psychology, 4, 43, 128n362
Western psychotherapy, 5, 58; Zimmer on, 57–62
Westmann, Heinz, 31
West versus East, making of complex psychology, 40–47
wheel, dreamer and rear, 281
Where the Two Came to Their Father, 52
wife, meaning of, in dream, 188–89
Wilhelm, Richard, 2, 42, 53
will: exertion of, 247–48; power, 248n536, 249
Wolff, Kurt, 51
Wolff, Toni, 19, 20, 30
working, primitives and, 195–96
World Disarmament Conference, 197n440
World Parliament of Religions (1893), 40
"Wotan" (Jung), 28–29, 28–29n80

Yang principle, Chinese philosophy, 219
yantra: *bindu* (centerpoint) of, 120–21, 121n359; center of, 36; circle, 231; definition, 66; Jung on, 70–71; mandala, and "machine," 62–72; rituals, 68; sacred devices, 45; term, 120n357; Zimmer on, 74. *See also* mandala
yoga, 42, 44; aim of, 304–5; analytical psychology toward Self, 97–98n306; asceticism, 91–92; characteristic of, 45–46; dedicating oneself to the divine, 92; devotional exercises, 120–22; impairments, 88–95; Indian technique of, 257–58; learning, 92; means of "active," 91–93; path to afterlife, 86; path to illumination, 84; path toward *kaivalya*, 97–98; philosophical questions, 86–87; psychology of, 4–5; term, 39; traditions, 46; Western attitude toward, 45; Zimmer on, 57–62

"Yoga and Meditation in East and West" (Zimmer), 47
"Yoga and the West" (Jung), 45
yoga exercises: *rites d'etrée* and, 195, 247–49; stirring of life energy, 112n342
yoga process, moving like a snake, 112–14
Yoga Sutras (Patanjali), 87, 87n279, 95, 95n299
"Yoga und Wir" (lecture), Zimmer, 55, 79
yogic meditation, 59
Yoni, female organ, 219, 219n467–88

Zeller, Max, 31
Zentralblatt für Psychotherapie und ihre Grenzgebiete (journal), 27
Zeus, 108–9n332
Zimmer, Heinrich, 16, 18, 43, 47; analytical psychology, 59–61; biographical profile, 47–52; impressions from Jung-Zimmer encounter, 53–57; influence on Jung, 62–72; Jung and, 237–39, 274; Jung on Zimmer's seminar introduction, 127–28; "On the Psychology of Yoga," 4, 17
Zurich Psychological Club, 3

The Collected Works of C. G. Jung

Editors: Sir Herbert Read, Michael Fordham, and Gerhard Adler; executive editor, William McGuire. Translated by R.F.C. Hull, except where noted.

1. PSYCHIATRIC STUDIES (1957; 2d ed., 1970)
On the Psychology and Pathology of So-Called Occult
    Phenomena (1902)
On Hysterical Misreading (1904) Cryptomnesia (1905)
On Manic Mood Disorder (1903)
A Case of Hysterical Stupor in a Prisoner in Detention (1902)
On Simulated Insanity (1903)
A Medical Opinion on a Case of Simulated Insanity (1904)
A Third and Final Opinion on Two Contradictory Psychiatric
    Diagnoses (1906)
On the Psychological Diagnosis of Facts (1905)

2. EXPERIMENTAL RESEARCHES (1973)
Translated by Leopold Stein in collaboration with Diana Riviere
STUDIES IN WORD ASSOCIATION 1904–7, 1910)
The Associations of Normal Subjects (by Jung and F. Riklin) An
    Analysis of the Associations of an Epileptic
The Reaction-Time Ratio in the Association Experiment
Experimental Observations on the Faculty of Memory
Psychoanalysis and Association Experiments
The Psychological Diagnosis of Evidence
Association, Dream, and Hysterical Symptom
The Psychopathological Significance of the Association Experiment
Disturbances in Reproduction in the Association Experiment
The Association Method

The Family Constellation
   PSYCHOPHYSICAL RESEARCHES (1907–8)
On the Psychophysical Relations of the Association Experiment
Psychophysical Investigations with the Galvanometer and Pneumo-
   graph in Normal and Insane Individuals (by F. Peterson and Jung)
Further Investigations on the Galvanic Phenomenon and Respiration
   in Normal and Insane Individuals (by C. Ricksher and Jung)
Appendix: Statistical Details of Enlistment (1906); New Aspects
   of Criminal Psychology (1908); The Psychological Methods of
   Investigation Used in the Psychiatric Clinic of the University of
   Zurich (1910); On the Doctrine Complexes ([1911] 1913); On the
   Psychological Diagnosis of Evidence (1937)

3. THE PSYCHOGENESIS OF MENTAL DISEASE (1960)
   The Psychology of Dementia Praecox (1907)
   The Content of the Psychoses (1908/1914)
   On Psychological Understanding (1914)
   A Criticism of Bleuler's Theory of Schizophrenic Negativism (1911)
   On the Importance of the Unconscious in Psychology (1914)
   On the Problem of Psychogenesis in Mental Disease (1919)
   Mental Disease and the Psyche (1928)
   On the Psychogenesis of Schizophrenia (1939)
   Recent Thoughts on Schizophrenia (1957)
   Schizophrenia (1958)

4. FREUD AND PSYCHOANALYSIS (1967)
   Freud's Theory of Hysteria: A Reply to Aschaffenburg (1906)
   The Freudian Theory of Hysteria (1908)
   The Analysis of Dreams (1909)
   A Contribution to the Psychology of Rumour (1910–11)
   On the Significance of Number Dreams (1910–11)
   Morton Prince, "The Mechanism and Interpretation of Dreams":
     A Critical Review (1911)
   On the Criticism of Psychoanalysis (1910)
   Concerning Psychoanalysis (1912)
   The Theory of Psychoanalysis (1913)
   General Aspects of Psychoanalysis (1913)
   Psychoanalysis and Neurosis (1916)
   Some Crucial Points in Psychoanalysis: A Correspondence between
     Dr. Jung and Dr. Loÿ (1914)

Prefaces to "Collected Papers on Analytical Psychology" (1916, 1917)
The Significance of the Father in the Destiny of the Individual
 (1909/1949)
Introduction to Kranefeldt's "Secret Ways of the Mind" (1930)
Freud and Jung: Contrasts (1929)

5. SYMBOLS OF TRANSFORMATION
 ([1911–12/1952] 1956; 2d ed., 1967)
  PART I
Introduction
Two Kinds of Thinking
The Miller Fantasies: Anamnesis
The Hymn of Creation
The Song of the Moth
  PART II
Introduction
The Concept of Libido
The Transformation of Libido
The Origin of the Hero
Symbols of the Mother and Rebirth
The Battle for Deliverance from the Mother
The Dual Mother
The Sacrifice
Epilogue
Appendix: The Miller Fantasies

6. PSYCHOLOGICAL TYPES ([1921] 1971)
 A revision by R.F.C. Hull of the translation by H. G. Baynes
Introduction
The Problem of Types in the History of Classical and Medieval
 Thought
Schiller's Idea on the Type Problem
The Apollonian and the Dionysian
The Type Problem in Human Character
The Type Problem in Poetry
The Type Problem in Psychopathology
The Type Problem in Aesthetics
The Type Problem in Modern Philosophy
The Type Problem in Biography
General Description of the Types

Definitions

Epilogue

Four Papers on the Psychological Typology (1913, 1925, 1931, 1936)

7. TWO ESSAYS ON ANALYTICAL PSYCHOLOGY
(1953; 2d ed., 1966)
On the Psychology of the Unconscious (1917/1926/1943)
The Relations between the Ego and the Unconscious (1928)
Appendix: New Paths in Psychology (1912); The Structure of the
Unconscious (1916) (new versions, with variants, 1966)

8. THE STRUCTURE AND DYNAMICS OF THE PSYCHE
(1960; 2d ed., 1969)
On Psychic Energy (1928)
The Transcendent Function ([1916] 1957)
A Review of the Complex Theory (1934)
The Significance of Constitution and Heredity and Psychology
(1929)
Psychological Factors Determining Human Behavior (1937)
Instinct and the Unconscious (1919)
The Structure of the Psyche (1927/1931)
On the Nature of the Psyche (1947/1954)
General Aspects of Dream Psychology (1916/1948)
On the Nature of Dreams (1945/1948)
The Psychological Foundations of Belief in Spirits (1920/1948)
Spirit and Life (1926)
Basic Postulates of Analytical Psychology (1931)
Analytical Psychology and *Weltanschauung* (1928/1931)
The Real and the Surreal (1933)
The Stages of Life (1930–31) The Soul and Death (1934)
Synchronicity: An Acausal Connecting Principle (1952)
Appendix: On Synchronicity (1951)

9. PART I. THE ARCHETYPES AND THE COLLECTIVE
UNCONSCIOUS (1959; 2d ed., 1968)
Archetypes of the Collective Unconscious (1934/1954)
The Concept of the Collective Unconscious (1936)
Concerning the Archetypes, with Special Reference to the Anima
Concept (1936/1954)

Psychological Aspects of the Mother Archetype (1938/1954)
Concerning Rebirth (1940/1950)
The Psychology of the Child Archetype (1940)
The Psychological Aspects of the Kore (1941)
The Phenomenology of the Spirit in Fairytales (1945/1948)
On the Psychology of the Trickster-Figure (1954)
Conscious, Unconscious, and Individuation (1939)
A Study in the Process of Individuation (1934/1950)
Concerning Mandala Symbolism (1950)
Appendix: Mandalas (1955)

9. PART II. AION ([1951] 1959; 2d ed., 1968)
RESEARCHES INTO THE PHENOMENOLOGY OF THE SELF
The Ego
The Shadow
The Syzygy: Anima and Animus
The Self
Christ, a Symbol of the Self
The Signs of the Fishes
The Prophecies of Nostradamus
The Historical Significance of the Fish
The Ambivalence of the Fish Symbol
The Fish in Alchemy
The Alchemical Interpretation of the Fish
Background to the Psychology of Christian Alchemical Symbolism
Gnostic Symbols of the Self
The Structure and Dynamics of the Self Conclusion

10. CIVILIZATION IN TRANSITION (1964; 2d ed., 1970)
The Role of the Unconscious (1918)
Mind and Earth (1927/1931) Archaic Man (1931)
The Spiritual Problem of Modern Man (1928/1931)
The Love Problem of a Student (1928)
Woman in Europe (1927)
The Meaning of Psychology for Modern Man (1933/1934)
The State of Psychotherapy Today (1934)
Preface and Epilogue to "Essays on Contemporary Events" (1946)
Wotan (1936)
After the Catastrophe (1945)

The Fight with the Shadow (1946)
The Undiscovered Self (Present and Future) (1957)
Flying Saucers: A Modern Myth (1958)
A Psychological View of Conscience (1958)
Good and Evil in Analytical Psychology (1959)
Introduction to Wolff's "Studies in Jungian Psychology" (1959)
The Swiss Line in the European Spectrum (1928)
Reviews of Keyserling's "America Set Free" (1930) and "La Révolution Mondiale" (1934)
The Complications of American Psychology (1930)
The Dreamlike World of India (1939)
What India Can Teach Us (1939)
Appendix: Documents (1933–38)

11. PSYCHOLOGY AND RELIGION: WEST AND EAST
    (1958; 2d ed., 1969)
      WESTERN RELIGION
Psychology and Religion (the Terry Lectures) (1938/1940)
A Psychological Approach to Dogma of the Trinity (1942/1948)
Transformation Symbolism in the Mass (1942/1954)
Forewords to White's "God and the Unconscious" and Werblowsky's "Lucifer and Prometheus" (1952)
Brother Klaus (1933)
Psychotherapists or the Clergy (1932)
Psychoanalysis and the Cure of Souls (1928)
Answer to Job (1952)
      EASTERN RELIGION
Psychological Commentaries on "The Tibetan Book of Great Liberation" (1939/1954) and "The Tibetan Book of the Dead" (1935/1953)
Yoga and the West (1936)
Foreword to Suzuki's "Introduction to Zen Buddhism" (1939)
The Psychology of Eastern Meditation (1943)
The Holy Men of India: Introduction to Zimmer's "Der Weg zum Selbst" (1944)
Foreword to the "I Ching" (1950)

12. PSYCHOLOGY AND ALCHEMY
    ([1944] 1953; 2d ed., 1968)
Prefatory Note to the English Edition ([1951?] added 1967)
Introduction to the Religious and Psychological Problems of Alchemy

Individual Dream Symbolism in Relation to Alchemy (1936)
Religious Ideas in Alchemy (1937)
Epilogue

13. ALCHEMICAL STUDIES (1968)
Commentary on "The Secret of the Golden Flower" (1929)
The Visions of Zosimos (1938/1954)
Paracelsus as a Spiritual Phenomenon (1942)
The Spirit Mercurius (1943/1948)
The Philosophical Tree (1945/1954)

14. MYSTERIUM CONIUNCTIONIS
([1955–56] 1963; 2d ed., 1970)
AN INQUIRY INTO THE SEPARATION AND SYNTHESIS OF PSYCHIC
OPPOSITES IN ALCHEMY
The Components of the Coniunctio
The Paradoxa
The Personification of the Opposites
Rex and Regina
Adam and Eve
The Conjunction

15. THE SPIRIT IN MAN, ART, AND LITERATURE (1966)
Paracelsus (1929)
Paracelsus the Physician (1941)
Sigmund Freud in His Historical Setting (1932)
In Memory of Sigmund Freud (1939)
Richard Wilhelm: In Memoriam (1930)
On the Relation of Analytical Psychology to Poetry (1922)
Psychology and Literature (1930/1950)
"Ulysses": A Monologue (1932) Picasso (1932)

16. THE PRACTICE OF PSYCHOTHERAPY
(1954; 2d ed., 1966)
GENERAL PROBLEMS OF PSYCHOTHERAPY
Principles of Practical Psychotherapy (1935)
What is Psychotherapy? (1935)
Some Aspects of Modern Psychotherapy (1930)
The Aims of Psychotherapy (1931)
Problems of Modern Psychotherapy (1929)
Psychotherapy and a Philosophy of Life (1943)

Medicine and Psychotherapy (1945)
Psychotherapy Today (1945)
Fundamental Questions of Psychotherapy (1951)
    SPECIFIC PROBLEMS OF PSYCHOTHERAPY
The Therapeutic Value of Abreaction (1921/1928)
The Practical Use of Dream-Analysis (1934)
The Psychology of the Transference (1946)
Appendix: The Realities of Practical Psychotherapy ([1937] added
    1966)

17. THE DEVELOPMENT OF PERSONALITY (1954)
Psychic Conflicts in a Child (1910/1946)
Introduction to Wickes's "Analyses der Kinderseele" (1927/1931)
Child Development and Education (1928)
Analytical Psychology and Education: Three Lectures (1926/1946)
The Gifted Child (1943)
The Significance of the Unconscious in Individual Education (1928)
The Development of Personality (1934)
Marriage as a Psychological Relationship (1925)

18. THE SYMBOLIC LIFE (1954)
Translated by R.F.C. Hull and others
Miscellaneous Writings

19. COMPLETE BIBLIOGRAPHY OF C. G. JUNG'S
    WRITINGS (1976; 2d ed., 1992)

20. GENERAL INDEX OF THE COLLECTED WORKS (1979)

THE ZOFINGIA LECTURES (1983)
Supplementary Volume A to the Collected Works.
Edited by William McGuire, translated by
Jan van Heurck, introduction by
Marie-Louise von Franz

PSYCHOLOGY OF THE UNCONSCIOUS ([1912] 1992)
A STUDY OF THE TRANSFORMATIONS AND SYMBOLISMS OF THE LIBIDO.
A CONTRIBUTION TO THE HISTORY OF THE EVOLUTION OF THOUGHT
Supplementary Volume B to the Collected Works.
Translated by Beatrice M. Hinkle,
introduction by William McGuire

Notes to C. G. Jung's Seminars

DREAM ANALYSIS ([1928–30] 1984)
Edited by William McGuire

NIETZSCHE'S *ZARATHUSTRA* ([1934–39] 1988)
Edited by James L. Jarrett (2 vols.)

ANALYTICAL PSYCHOLOGY ([1925] 1989)
Edited by William McGuire

THE PSYCHOLOGY OF KUNDALINI YOGA ([1932] 1996)
Edited by Sonu Shamdasani

INTERPRETATION OF VISIONS ([1930–34] 1997)
Edited by Claire Douglas

Philemon Series of the Philemon Foundation
General editor, Sonu Shamdasani

Children's Dreams. Edited by Lorenz Jung and Maria Meyer-Grass.
Translated by Ernst Falzeder with the collaboration of Tony
Woolfson

*Introduction to Jungian Psychology: Notes of the Seminar on
Analytical Psychology Given in 1925.* Edited by William Mc-
Guire. Translated by R.F.C. Hull. With a new introduction and
updates by Sonu Shamdasani

*Jung contra Freud: The 1912 New York Lectures on the Theory of
Psychoanalysis.* With a new introduction by Sonu Shamdasani.
Translated by R.F.C. Hull

*The Question of Psychological Types: The Correspondence of C. G.
Jung and Hans Schmid-Guisan, 1915–1916.* Edited by John
Beebe and Ernst Falzeder. Translated by Ernst Falzeder with the
collaboration of Tony Woolfson

*Dream Interpretation Ancient and Modern: Notes from the Seminar
Given in 1936–1941. C. G. Jung.* Edited by John Peck, Lorenz
Jung, and Maria Meyer-Grass. Translated by Ernst Falzeder with
the collaboration of Tony Woolfson

Analytical Psychology in Exile: The Correspondence of C. G. Jung and Erich Neumann. Edited and introduced by Martin Liebscher. Translated by Heather McCartney

On Psychological and Visionary Art: Notes from C. G. Jung's Lecture on Gérard de Nerval's "Aurélia." Edited by Craig E. Stephenson. Translated by R.F.C. Hull, Gottwalt Pankow, and Richard Sieburth

History of Modern Psychology: Lectures Delivered at the ETH Zurich. Volume 1: 1933–34. Edited by Ernst Falzeder. Translated by Mark Kyburz, John Peck, and Ernst Falzeder

Dream Symbols of the Individuation Process: Notes of the Seminars given by Jung in Bailey Island and New York, 1936–37. Edited by Suzanne Gieser

On Theology and Psychology: The Correspondence: The Correspondence of C.G. Jung and Adolf Keller. Edited by Marianne JehleWildberger. Translated by Heather McCartney with John Peck

Psychology of Yoga and Meditation: Lectures Delivered at the ETH Zurich. Volume 6: 1938–1940. Edited and introduced by Martin Liebscher. Translated by Heather McCartney and John Peck

Consciousness and the Unconscious: Lectures Delivered at ETH Zurich. Volume 2: 1934. Edited by Ernst Falzeder. Translated by Mark Kyburz, John Peck, and Ernst Falzeder

Jung on Ignatius of Loyola's Spiritual Exercises: *Lectures Delivered at the ETH Zurich.* Volume 7: 1939–1940. Edited by Martin Liebscher. Translated by Caitlin Stephens